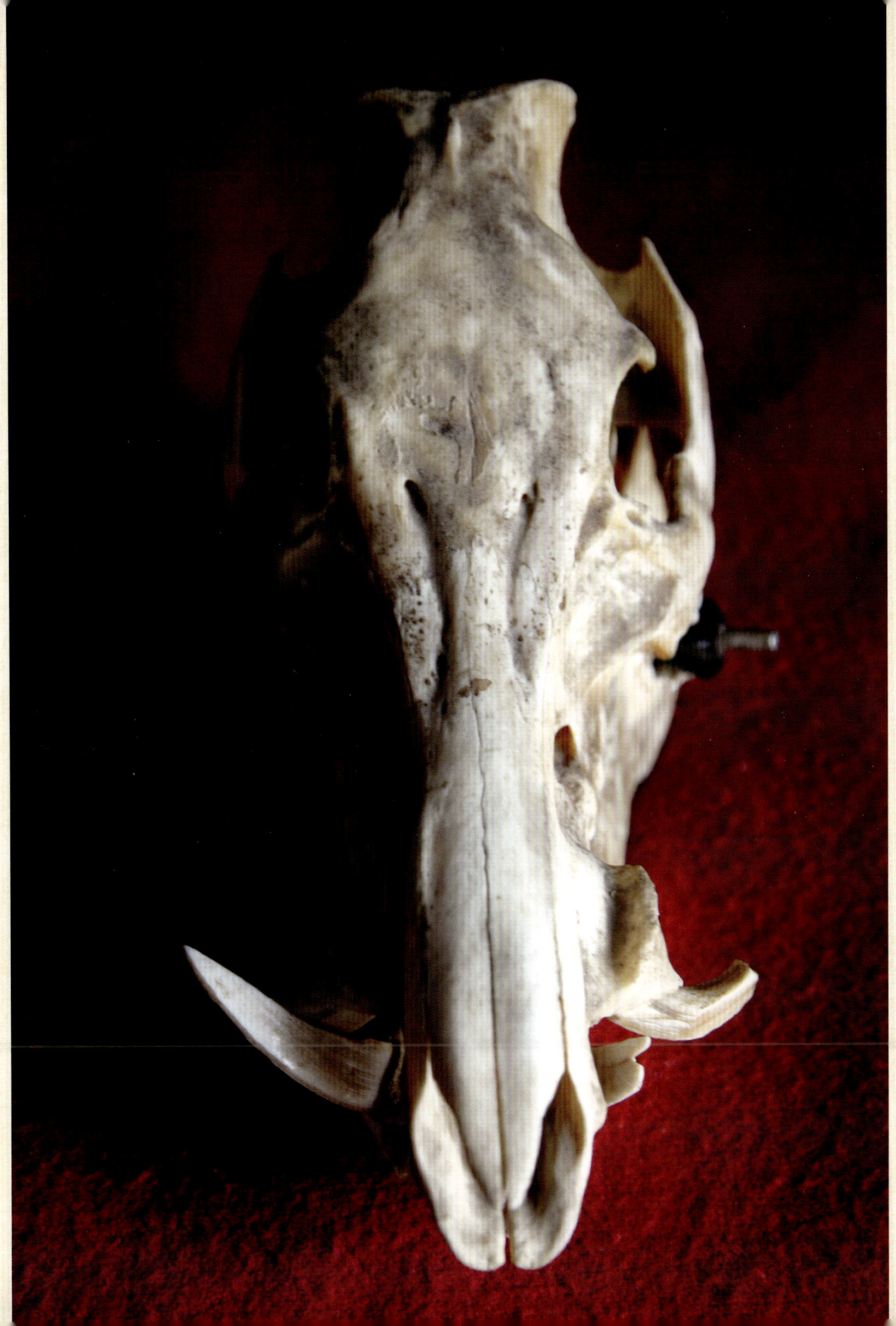

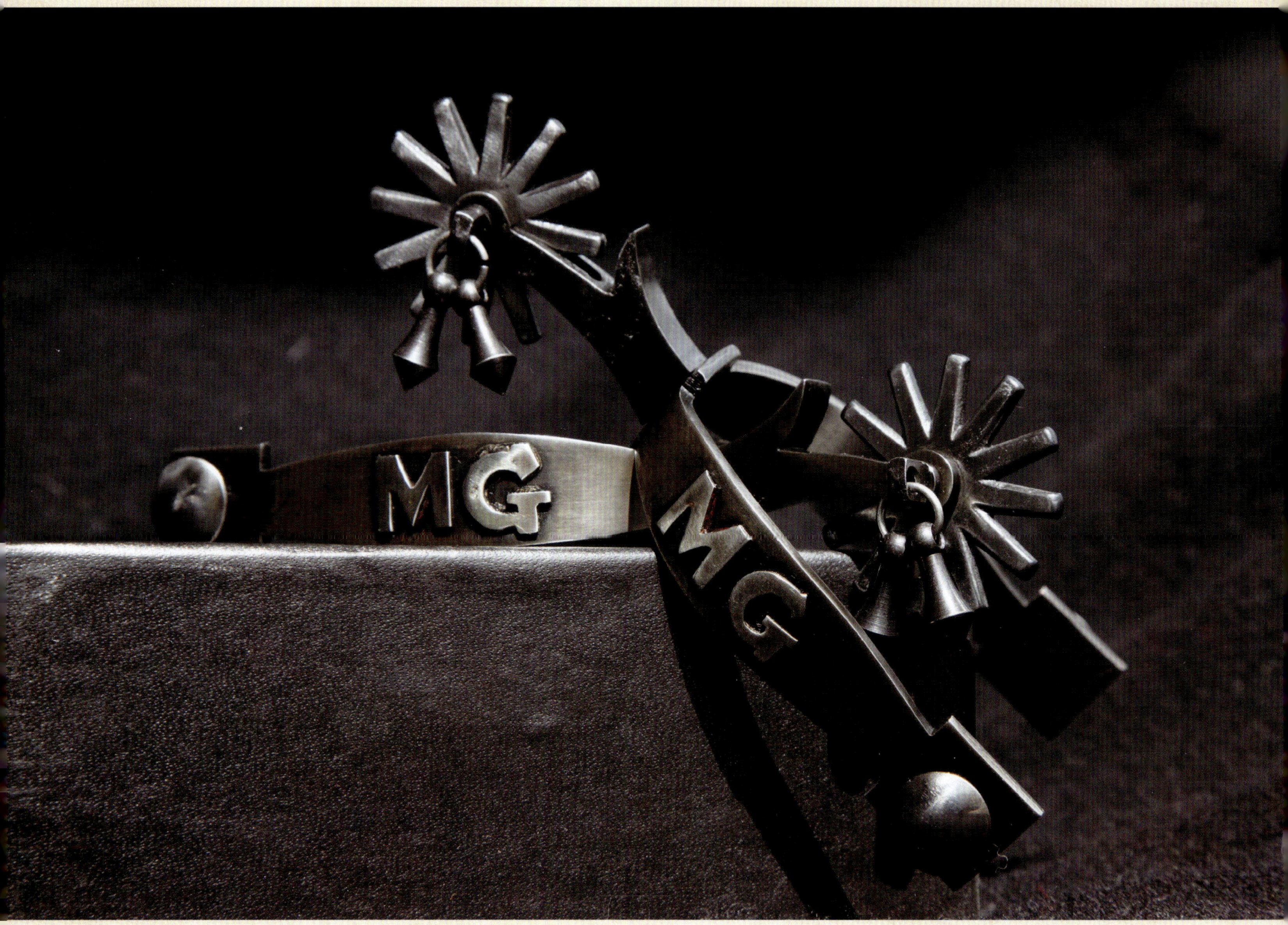
MG
MG

Panther Tract

Wild Boar Hunting in the Mississippi Delta

Melody Golding

University Press of Mississippi Jackson

www.upress.state.ms.us

The University Press of Mississippi is a member of the Association of American University Presses.

Manufactured in China

First printing 2011

Library of Congress Cataloging-in-Publication Data

Panther tract : wild boar hunting in the Mississipp Delta / [compiled by] Melody Golding.
p. cm.
Includes bibliographical references and index.
ISBN 978-1-60473-926-8 (cloth : alk. paper)—ISBN 978-1-60473-927-5 (ebook : alk. paper) 1. Wild boar hunting—Mississippi—Delta (Region)—Anecdotes. 2. Wild boar hunting—Mississippi—Delta (Region)—Pictorial works. I. Golding, Melody.
SK305.W5P36 2011
799.2'7633209762—dc22 2010030047

British Library Cataloging-in-Publication Data available

Page 1: Skull from a wild boar killed by Bill Burle. The arrow was buried within his skull and discovered after the skull had been cleaned.
Page 2: Riding spurs custom-made by Bubba Bradley for Melody Golding to use in wild boar hunting.

To Howard Brent, the Epitome of Legend

Today a legend can be a person whose achievement and fame promises to be enduring. A legend also is a person whose fame or notoriety makes him a source of exaggerated or romanticized tales and exploits. Howard Brent is truly a living legend. The stories about Howard are boundless and on most occasions true. He is worthy of inspiring fantastical stories, and his fame is certainly enduring and endearing.

Howard has been a true friend of our family for many years, and he is a true friend to anyone else who has the good fortune of knowing him. The stories Howard tells are always awe-inspiring, unbelievable, and most times very funny. He is a source of exaggerated and romanticized tales and exploits. His romanticized tales sometimes manifest themselves in the blues he sings as he strums the two chords he knows on his ole Gibson guitar or plays the inch-long baby harmonica he loves to entertain with (and is so famous for), as he carries it along in his front shirt pocket at all times. As to the exploits, well, they are mostly true too.

Never will a person meet a more kindhearted modern-day hero. Because of Howard's generosity he gave us the unique opportunity to hunt on his land and document through the photographs and escapades some of what transpired at Panther Tract during the hunts. He made us all feel welcome, at ease, and at home.

This book I fondly dedicate to Howard Brent, a friend, a mentor, a legend, and a steward of all things most people look for in a Delta gentleman. Thank you, Howard, for making this book possible and for sharing your special place under the sun and in the heart of the land found in the Mississippi Delta: Panther Tract!

Your friend,
Melody Golding

Contents

A Note to the Reader

MELODY GOLDING

This book is a photo documentary of hunting the wild boar—or hog—in the Mississippi Delta. The photographs are representative of "a day at the hunt," starting at dawn and ending well after dark, at Panther Tract, located in Yazoo County and owned by the incomparable Mr. Howard Brent. He graciously shared his land for this true one-of-a-kind documentary effort. The stories are a collection, not only of hunting experiences at Panther Tract but from different hunters, sometimes in other locations in Mississippi, from men and women who are doctors and lawyers and judges, businessmen, politicians, farmers, the son of a sharecropper, and even a Hollywood screenwriter, just to mention a few. It is a colorful and diverse assemblage of rollicking tales which will entertain, delight, and enlighten as they unfold. Each will define, in its own way, the expectations of hunting, heroism, and adventure. There are photographs and tales of the special relationships between people and their dogs and horses that will surely touch your heart.

The hunting at Panther Tract is an unparalleled opportunity not to be missed, not only for the unique social aspects which are legendary in the Mississippi Delta with a genteel hospitality that is steeped in tradition; it is also a coveted and exhilarating hunting experience. Aside from all of the "fun," hunting the wild boar is challenging for the hunter, observer, wildlife enthusiast, and landowner alike. It is dangerous and it is daring. Anyone who has ever been on a wild boar hunt will sit back and say with a smile and a twinkle in his or her eye, "You are not gonna believe what happened to me on my hunt!"

Most important, the wild boar population needs to be controlled because, for the landowner, hogs can cause extensive damage to property and threaten a farmer's livelihood. They destroy agricultural crops and native vegetation. If not controlled, the hogs are so prolific, sometimes having two to even three litters a year, that they outcompete the wildlife species that are indigenous to the area for valuable food resources. Hogs can eat just about anything, as they are omnivorous and therefore consume all kinds of plant and animal matter.

Hunting is one method of controlling wild hog populations. It is a basic necessity of land conservation and wildlife management. This book is an example of how extraordinary hunting, southern hospitality, camaraderie, and the love of dogs and horses can go hand in hand for a very exciting day in the beautiful Mississippi Delta.

Preface

If you were to ask me how I got into wild boar hunting, I would have to blame it on my love of horseback riding.

I have ridden horses since early childhood, on four continents and through countless landscapes, but never with the intention of hunting wild hogs—or anything else for that matter. Aside from presiding over the deaths of a few flies and mosquitoes, I have never intentionally killed another creature. The only joy I have ever gotten out of killing anything involves big fat mean horseflies—and for them I have no regrets. I have, I am sorry to say, been responsible for the premature passing of one six-point buck and one squirrel, both of which accidentally met with my moving car on a country road, on separate occasions, of course. Oh, and then there was the time I exterminated a four-foot-long cottonmouth snake with my shotgun. He was staring me down in my own backyard and I felt it was him or me.

I have been around hunters all my life—from my father and brothers to my husband, sons, and friends—yet the allure of the sport had managed to escape me until the winter of 2004, when I received an invitation from my dear friend Howard Brent to ride my horse on his parcel of land known in the Mississippi Delta as Panther Tract. In his invitation, Howard's emphasis was clear: *"We'll go on a wild hog hunt."* But all I heard was the chance to trot, cantor, and gallop upon forty-five hundred pristine acres—a unique opportunity no equestrian would ever turn down.

So off I went one frosty January morning with the temperature a very chilly twenty-two degrees. I arrived early. Those hunting with Howard either spring out of one of the twenty comfortable beds he supplies for his friends at Panther Tract, or arrive very early, steady and ready with steed in tow—from quarter horses to Tennessee Walking Horses, thoroughbreds, and paints—in vehicles as disparate as their personalities: top-notch rigs with living quarters, livestock trailers with dog containers and horse stalls and hay that is stored on top, four-wheelers, Rangers, camouflaged golf carts, Polarises, Suzukis, Yamahas, and even broken-down cookie trucks with all of the animals and tack thrown in together.

As to the men, many of these hunters look as though they came straight out of central casting in Hollywood. They are dressed for the part, wearing cowboy boots or green rubber muck boots (both topped off with sets of jingly spurs), cowboy hats, camouflage, orange vests, leather gloves, and long weatherproof coats. And the equipment—what sportsman doesn't love his equipment? Hog hunters come brandishing ropes, pistols large and small, sleek rifles with scabbards attached to saddles, sixteen-inch knives with blades as sharp as razors and handles made of bone, two-way radios, GPS systems, saddlebags, goat horns, and bull horns. One of these days, I trust Howard will invent the "wild hog horn," and we'll all have to be around to hear its maiden call.

Perhaps closest to the hunter's heart is his second in command, his accomplice and confidant: his dog. There are redtick hounds, bluetick hounds, bloodhounds, yellow black mouth curs, mountain curs, leopard curs, bird dogs and bird dog mixes, bulldogs and bulldog mixes, pit bulls,

Plotts, Catahoulas, and redbones, many of which have been given names as endearing as Little Joe, Homicide, Frog, Junebug, Hoover, Hammer, Hemorrhoid, Copper, Gretta, Blue, Pete, Ox, Barney, and Mabel. Some of the dogs wear heavy-buckled vests made from 1000 Denier Cordura to protect them from the wild hogs and from the forest briars and brambles, vests complete with reflector strips, transmitter and tracking collars, and antennas. I even saw some dogs wearing Kevlar, a material that often protects police officers.

The men greeted me in the near-dawn, their faces as worn and weathered as their riding tacks. All wore smiles, some with spaces like bullet holes.

"What am I doing here?" I thought to myself. "I'm not anywhere near tough enough for this."

I soon learned that hog hunters come from all walks of life. They are doctors, lawyers, judges, farmers, businessmen, mechanics, engineers, tanker men, politicians, and more. Despite the vocational spectrum, these hunters have a great deal in common. They all love the great outdoors, their dogs and horses, the thrill of the hunt, and the camaraderie found with other hunters and the special trust found among men.

I was so proud of the women who were a part of these famous hog hunts at Panther Tract. They brought their horses and tacked them up and rode right alongside the men, and had a great time doing it. Some even got to kill a hog. Their bravery and willingness to be a part of the pack was appreciated by all—silver earrings, camouflage, and all. It was a chance for them to discover something new and another path to travel.

In Mississippi, hogs can be hunted year-round on private land, and there are times when it's done less for sport than necessity. Wild hogs are prolific, sometimes producing offspring two to even three times per year, and can devastate a farmer's land and livelihood. They will eat the planted corn, mowing down row after row "like a tractor." They are ferocious, dastardly—and smart. That's where the cunning hunter enjoys his payoff: in defeating a wily adversary. There is also a payoff in the hunt's harvest. The meat from the animals is never wasted. There is always someone who can feed a family from it, usually through the winter. (One hunter told me that he once gave a fresh sow to a man begging for some of the hog meat they had harvested. He found out later that the man had traded it for sex. This opens an interesting philosophical debate, though one best left for another book.)

With the sun a wink on the eastern horizon, the dogs ran around, barking and howling. The horses pranced, danced, and snorted, raring to go. And my big gelding, Bigshot, normally one for gliding through fields in soft-focus sunlight, was hopping and jostling right along with the others. The traitor, I thought, he's one of them.

Gear packed, destination set (and my prayers said), we set off on the hunt, riding into the sweeping vistas, under the yawning, panoramic sky, the crisp air odorous with land and animals. Our pace was brisk—no posing and lollygagging here—through the slushy mud, lightly crusted with ice, the earth a Delta gumbo, so thick it could—and did—suck the shoes off of some of the horses' hooves.

It wasn't long before the dogs caught the scent of hog, and off we shot, tree branches whipping at our faces, briars, stickers, and thorns stuck in our hair and in our caps (and in one or two foreheads, noses, and cheeks). I put my head down and let my trusty Bigshot lead the way. We came upon a dark, murky, and deep slough and the dogs and horses plowed forward, undeterred, into the frigid water, trudging through the lacy sheets of ice. Hogs are excellent swimmers. Until that moment, I had no idea Bigshot could swim.

Throughout the morning, as the dogs flushed the hogs from the brush, excitement merged with chaos, brute strength, and athletic prowess—the test of man versus beast, the grunts and squeals of mortal combat. The adrenaline was as high as September cotton.

Once all was said and done, some of the hunters were able to claim victory with their kills. But not me. I didn't want to. The thrill of the hunt and being there was reward enough.

We returned to the Panther Tract camp house, where Howard graciously served up generous portions of barbequed smoked ribs dripping with his special sauce, coleslaw and potato salad to die for, hot buttered garlic French bread, and iced tea. After the meal and before embarking on the afternoon hunt, many of the hunters retired to the porch or to a seat by the fireplace to let their food settle.

And that's when the stories began. As a rule, hunters tend to be a lively and colorful bunch, but those at Howard's Panther Tract possess a unique verve and exuberance. Tales are woven in and out of time, as the great southern American folklore unfolds in the flesh. The sharing of these tales and experiences and knowledge can carry late into the afternoon and leaves the listener in awe and most times admiration. Yarns are spun of the revered and now-dwindling hunting fields; tales are told of past hunts, rambunctious dogs, willing horses, various escapades with guns and knives, nature in every imaginable form; there are stories of gripping fear and menacing, split-second action, moments of bravery on the parts of man, mount, and dog faced with a wild boss boar hog, an angry sow, or even a herd of wild hogs in stark disagreement with the hunters' intentions. There are even love stories. Once, two of the dogs—Homicide and Little Joe—were viciously slashed open by the tusks of a three-hundred-pound boss boar hog. Homicide was breathing from the side of his body, as his guts lay exposed. Little Joe was in bad shape, too. Their owner was on the ground with his hunting buddies, crying over the injured dogs. Horses, men, and the other dogs stood frozen, watching the hunters, whose hands were covered in blood as they cradled the dogs like babies. Then out came the sterilized staple guns—the kind surgeons use—and the men, with the skilled expertise of a veterinarian, stapled the dogs back together. They applied antibiotics to the wounds and covered the now-resting canines with their own coats. Even though the temperature was freezing, the care and comfort of their dogs came first. Thankfully, these brave hunting dogs survived. I have a tremendous amount of respect for those hunters who display great compassion and care for their dogs and their horses on these hunts.

These hunters hold tight to their love of the sport—their compulsion for it. Their experiences, camaraderie, and trust in their fellow hunter flow through their veins like time itself. Hunters are a part of the American character and soul. History and literature are filled with legends of hunting and hunters' lore—even in the Mississippi Delta we have our own legend of Holt Collier and President Teddy Roosevelt and his famous bear hunt.

All of the stories suggest that hunting is a lifestyle choice and a damn lucky one, as their domain is ever shrinking. It is one of the greatest privileges one can have. These hunters and participants are holding tight to their love of hunting, and, for many, this type of hunting gives them the chance to deal with perhaps some of their greatest fears—wild animals, weapons, and wilderness—and with some of their greatest loves—dogs, horses, the great outdoors, and hunting. It also gives them great adventure in what might otherwise be just a regular life.

While there is ordinarily a distinction between the hunter and the observer, I have been fortunate enough to become both. I have now been on many hunts with scores of hunters and have taken thousands of photographs of the breathtaking Mississippi Delta. Through it all, one thing has become clear to me: this part of the Delta and the tales of these hog hunters, their horses, and dogs need to be shared, need to be known to all.

As an author and documentary photographer, I know that you are in for a real treat within the pages of this book. Whether the tales that are told are true or not is left to you, the reader, to decide. But from personal experience, I can honestly say that this kind of hunting is as close to the wild west and cowboys and exhilarating, heart-pounding action as can be.

The photographs of the people and the animals are all very special moments in time captured forever. One day they will look back on these images and think how young they looked or with a tear in their eye remember and contemplate with fondness that special dog or horse depicted as in life on these pages.

All of these people were invited to share their own personal wild hog hunting experiences. The responses are amazing. The desire of southerners to articulate their "yarn" in their own way leaves the reader with plenty to think about. The stories herein are as varied as their tellers and steeped in the history, tradition, and venerable soul of the South. Willie Morris once said, "Perhaps in the end, it is the old, inherent, devil-may-care instinct of the South that remains in the most abundance." And I believe it, especially of these fearless hunters. These shared photographs were taken solely at Panther Tract, and the stories create sometimes pleasant and always dramatic memories that are significant to the life of everyone who is involved. See these hunters as fully and with as much generosity as possible, even if you are not a hunter, like me. Living to tell and telling to live is a birthright here in the South.

Contemplate the photographs as frozen memories and moments in time in the incomparable and stunning Mississippi Delta. The inherent beauty of the Delta leaves me breathless at times and to have the opportunity to share what I have seen through the lenses of my cameras on all of these hunts is an honor. The sweeping flat vistas of land, land, land! Everywhere there is land. Big land. With its thick gumbo dirt left over from the harvest a season before and tilled just so with boggy and deep ruts from the heavy farm equipment, and with the panoramic skies full of billowing clouds and the stark and leafless trees of winter that have the nerve to stand within sky and earth, it is simply breathtaking. The colors range from a dull grey to a profusion of hues screaming at you to look. Look now! With Panther Tract, and all its unique glory, there lies within a mystery, as is true about all of the Mississippi Delta. It is large and it is small. It is more and it is less. It is beautiful and its infinite landscape is a place to be studied in all its wonderment.

Enjoy these living legends from all walks of life as their stories unfold with the hearts that beat through their very own words. As you come to the end of this book you will miss these guys, these special men, women, dogs, and horses alike—I already do—but thank goodness we can open the book back up time after time for reflection and rumination. I trust you will find within these pages a reason to reflect upon the chorus of your own experiences. My experiences at Panther Tract began with my love for horses and evolved into my love of so much more. I honestly and humbly give the reader a rare look into the very heartbeat of a true Mississippi Delta hunting camp. Panther Tract!

Thank you, Howard, and thanks to everyone who shared a story.

Melody Golding

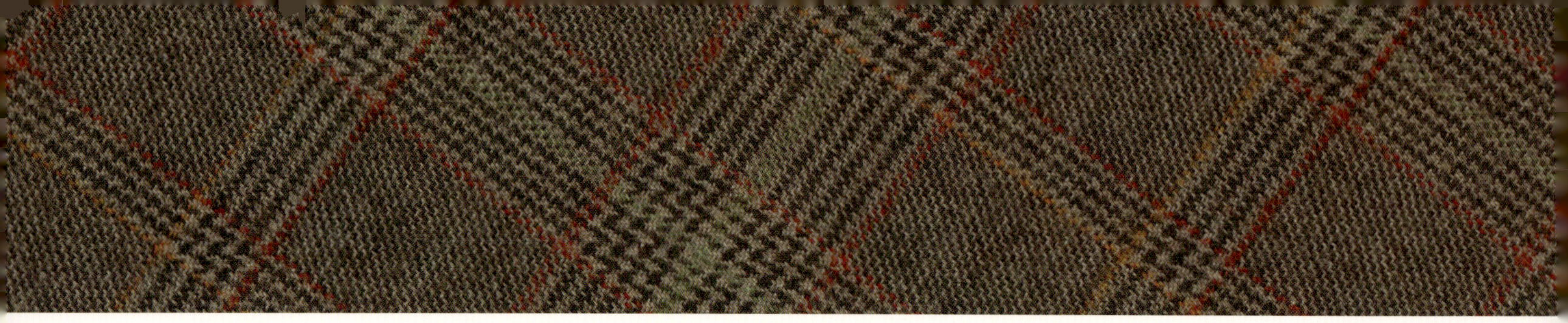

Introduction

HANK BURDINE

The Italian Christopher Columbus thought the world was round. He persuaded the queen of Spain to fork over enough doubloons to finance a trip in 1492 to prove it. But scientific research was not the reason for the *Niña*, *Pinta*, and *Santa Maria* to head west into the setting sun. Ole Columbo was trying to get to Asia in order to open up the spice trade by boat in a more direct route than going around the dangerous Cape Horn at the bottom of Africa. Spices were in high demand in those days, for without refrigeration, meats spoiled in a very short time and spices were used to hide the rancid taste of semispoiled meat.

The boats never made it to Asia; they landed in the Caribbean. What they found were angry Carib Indians and a host of chili peppers growing on the islands whose original seeds from South America had been deposited there by the excretion of migrating birds. Columbus returned to Spain and spread the news. Soon the quest to bring peppers and gold back to Europe was on full speed.

In order to eat while on the high seas and to sustain themselves while stealing and looting from the Mayas and the Incas, Spanish explorers brought along chickens, goats, and pigs. Very possibly, the first hogs came ashore in Cuba in 1493. It is thought that around 1530, Admiral Alonso Alvarez de Pinna may have run aground on some shoals outside of Mobile Bay. The pigs on his ship were offloaded and some escaped into the wild. Soon, more Spanish ships came to Florida and pigs officially arrived on the North American continent. Dutch settlers began bringing pigs into the New Netherlands, while the British began bringing walking pork chops into the northern and middle colonies.

Hernando de Soto is credited with bringing the first official swine population into the Gulf Coast region and by 1674, pigs squealed in pens and ran loose and wild in the woods in most of the settled areas of North America. Pigs were easy to care for and they multiplied more rapidly than cattle. They were also able to be herded and moved from farm to farm as settlers moved west. "Pigs soon became emblematic of the New World," stated Jessica Harris in the 2008 issue of "Louisiana Endowment for the Humanities." "However, as much as they were loved by settlers, pigs were despised by native peoples because free-roaming swine destroyed carefully tended corn plantings as they foraged in the woods." Eventually, hogs were slowly adopted by the Indians as livestock and by the time of the removal of the great Cherokee nation to the west, they were a part of the diet and economy of the Five Civilized Tribes.

From this beginning, pigs were acclimated into all lower forty-eight states and their numbers grew in pens on farms and feedlots in towns. Many pigs escaped their primitive captivity to take up residency in adjoining woods and forests, feasting on the natural acorns, roots, and grasses. From this meager beginning grew the massive pork industry we have today and from those escaped hogs we have some of the finest and most dangerous hunting in America.

A staple of the South, the swine of yesterday was as

important to rural families as Kroger or the Wal-Mart Supercenter is today. Hog killing time was revered and anxiously awaited as a festive but hard working time. But the weather had to cool down enough so the meat would not spoil once you got it all scraped, cleaned, cut up, and into the smokehouse for curing. There were no refrigerators or freezers. Time was of the essence. The farm crops were in and most of the garden vegetables were already canned. It was time to put meat in the smokehouse to cure for the following year.

My mother, Baby Jane Rule from Ruleville, Mississippi, recalled one of the worst days she ever spent as a child. It was November 1925, and there were sixteen families living on the farm. It was hog killing time and the weather had dropped below freezing the night before. At daybreak, all the working men on the place had gathered at the barn and huge pots of water were already boiling, knives were being sharpened, and gunny sacks were piled beside the steaming caldrons of water. The consistent chatter of happy faces and the occasional smack of hungry lips could be heard. In the nearby finishing pen were eighteen full-grown hogs, oblivious to their impending fate. During the previous year, the young pigs had been fed table scraps and corn, cottonseed, oats, rotten garden vegetables, and anything else that would fatten them during the year. In late October, the hogs, mostly boar pigs that had been castrated as young piglets, had been taken out of the slop pen where they wallowed around during the heat of the day. (A pig has to wallow to stay cool as he has no sweat glands.) They were then placed on a slatted wooden floor where they were fed clean, clear water and pure shelled corn to purge themselves for the impending coup de grace.

The strawboss came by early that morning with his .22 caliber pistol and ordered the first four pigs pulled out and brought under the big oak tree that had ropes and pulleys attached. As two strong men grabbed each of the three-to-four-hundred-pound pigs' hind legs, they were dragged off the floor backwards squealing and squalling to the huge oak tree that for their entire life had offered nothing but shade and solitude. Another man would press down on each pig's neck and hold him still as the boss walked by and with a single "pop" send the stout and still-squealing pig into hog heaven. Then the work began.

A sharp knife slit the two jugular veins in the pig's neck and a stick was placed between his hind legs as ropes were attached and he was pulled into the air so gravity would allow all the warm and steaming blood to drain out and be collected in waiting buckets, to be used later in liver and blood sausage. Once drained, the pig was laid back down on the ground and steaming gunny sacks that had been soaked in the boiling water were placed on the hog to loosen up the pig's hair so it could be scraped off with sharp knives. Once the hair was removed, the pig was then hoisted back up in the tree and the gutting and cleaning began, with all the innards being dropped into washtubs. No organ was thrown away.

Back then, everything was saved from the pig, from "high on the hog," the slabs of bacon, hams, ribs, and shoulders, to the entrails, organs, snout, tail, lips, and ears. It was all cured and smoked or eaten right away. The intestines were cleaned and used as sausage casings for the ground-up side meat and trimmings that had been flavored with spices and peppers from the garden. The fat was rendered down for lard and shortening. Leftover intestines were cleaned and either boiled or fried as chitlins. Bits of pig skin were saved for cracklins. Slabs of bacon were cut to be salt cured or smoked. The head was boiled down and once the bones were removed, spices were added to the remaining fat and bits of meat and then pressed into a block and cured as hogshead cheese.

The women on the place came over about noon and began preparing the casings and spices for the sausage. All day was spent working on what came from the inside of the hogs as the steaming carcasses hung high in the cold air and began to cool down. By nighttime, there were eighteen carcasses and hundreds of pounds of chitlins, sausages, tripe, and hogshead cheeses, neatly packed in the washtubs waiting to go to the smokehouse. Each family went home with enough leftovers for a feast. Everyone was eager to go back to work the next morning and butcher the hogs into hams, shoulders, slabs of bacon and pork ribs and chops. But the weather changed that night. The cold front moved out and the temperature rose into the sixties instead of staying in the thirties and below. By morning all the meat had spoiled and barn flies had begun to nest and lay eggs in the body cavities. Nothing could be done, as the meat for the oncoming year, enough to feed the sixteen families on the place, including my mother's, hung rotting in the noonday sun. The washtubs and hard work soon turned to mush. However, there were eight pigs left in the pen. There wasn't going to be much to eat until December when the weather was certainly not going to change so abruptly. A tear would always come to my mother's eye every time she told that story.

Such was the way of life back then. Not everyone in the country had the resources to pen and feed hogs out of troughs all year long. However, there were always plenty of woods around that bordered the cotton and corn fields. Often, branded and marked pigs were turned loose to roam free, breed, and grow on their own. This practice was called "free roam farming." Pigs are omnivorous and will consume both plants and animals. They have been known to eat almost any kind of food—worms, tree bark, dead carcasses, garbage, and even other pigs! Southern hardwood forests were loaded with mast-producing trees and succulent tubers that pigs love to eat. In the fall, the woods were full of acorns, pecans, and hickory nuts. During the rest of the year there was an abundance of tender grasses, turkey eggs, bugs and insects, snakes, roots, mushrooms, and wild fruits of many kinds. Country folk realized they could just turn their domesticated pigs loose in the woods and they would learn to fend for themselves. Those pigs would quite easily become feral, learning to root around and find their own food. Pigs have an outstanding sense of smell and are able to sniff out food underground.

With a strong prenasal bone and a very stout disc of cartilage within their snouts, pigs are able to root out tubers, roots, and underground goodies that are then ripped out with their teeth and eaten. A boar hog has four continually growing canine incisor teeth, two on the top and two on the bottom. These teeth can reach five inches in length and are constantly being sharpened by gnashing back and forth as he roots around under the ground, ripping and tearing apart anything that comes in contact with them. With a pig's nose sniffing and digging and his tusks ripping and tearing, a large group of hogs can easily destroy an acre of ground each day. Woe be unto any man that stumbles upon a mad hog with large tusks, as these teeth are used for defense and to establish dominance during the many breeding seasons. The tusks are razor sharp and can cut and split anything a hog rushes into as he throws his head back and forth in a sideways and uplifting motion.

A wild hog is an excellent swimmer and a very fast runner. The only thing he can't do well is climb a tree. Often, a sow will serve as babysitter and surrogate suckler for several sets of piglets. She will take care of as many as three or four families of siblings while their moms go off and forage for food. This is an extremely dangerous time, as the babysitting sow hog is most protective and if you happen to come upon a passel of piglets in the woods with one mad mama hog on guard, you better be able to climb a tree, or be very, very swift of foot.

All over Mississippi, farmers and country folks released pigs into the woods. Feral pigs begin breeding at about eight to ten months of age and a litter of piglets can number from four to twelve babies. A healthy sow can have two litters each year. Over the years, the feral hog population has exploded not only in Mississippi but all over the country, with an estimated four million hogs running wild in the woods. Years ago, a pure strain of European wild boar was introduced into the eastern mountains for sport and while pockets of that strain can still be found, the two breeds have intermingled to produce a homogenous breed that is very wild and offers excellent sport hunting. Left unchecked, wild hogs soon dominate landscapes and produce considerable damage to woodlands and crops.

Islands in the Mississippi River historically have been favorite places to turn pigs loose because of the confines and the ability to have organized hunts during the year to harvest hogs. Horses and dogs are used to track, trail, catch and/or corral wild hogs. These hunts oftentimes lead to bloodied or dead dogs, as a boar hog will fight with a viciousness that is beyond remarkable. Crossbred curs, pit bulls, purebred Catahoula hog dogs, African Rhodesian ridgebacks, and black mouth curs are used as hog dogs because of their steadfastness in the chase and bay and their own viciousness at the catch. Dogs are trained to be quiet while trailing and chasing hogs so as not to spook them and run them into deep and very dense cover. Well-trained dogs will only bark as the hog is bayed. When the handlers arrive on the scene and the command is given to catch, or, in some cases, specially trained catch dogs are turned loose, one or two dogs will rush in and bite the hog and refuse to turn loose. Once caught, the dogs will splay the hog out on the ground. A hunter will then rush in with a knife and cut the hog's throat or hog-tie him for relocation. Sometimes rifles or pistols are used, but just as the sporting bear hunters of yesteryear often used their knives to keep from spooking and making their dogs gun shy, modern-day knife-wielding hog hunters jump right into the dangerous melee as all hell breaks loose. A catch dog is trained to not let go once he has a bite on a hog. The chase dogs circle, snap, and bark, keeping the squealing hog's attention away from the oncoming blade of flashing steel.

A wild boar is one of the smartest animals in the woods and is afraid of absolutely nothing. They will run, swim, or crash through anything that comes in their way, including man or beast, with severe reckless abandon. Many good dogs have had their bellies ripped open by gnashing tusks, only to be sewed up on the spot and sent back into the chase. A wild hog hunt on horseback behind a good pack of dogs is as wild an adventure as one can imagine. The cunning and aggressiveness of wild hogs make them worthy adversaries wherever they are hunted. As was so with the Deep South bear hunts in the dense canebrakes of the Mississippi, Arkansas, and Louisiana deltas of the early part of the last century, the present-day wild hog hunt is certainly not for the faint of heart or the weak of spirit. It is for dedicated and devoted men and women who thrill in the chase, revel in the fear of uncertainty, are steadfast in their dedication to the hunt, and relish the thought of succulence at the soon-to-come wild game supper.

So why do we hunt the wild boar when we can go to Kroger and buy all the bacon and hams we want? It is because deep down in some of us, the desire to hunt and partake of the chase is inbred. It is who we are; it is from whence we have come. And as the French poet Charles Monselet once said in his "Sonnet to a Pig," "A pig is nothing but an immense dish which walks while waiting to be served." So get ready, boys and girls, it's hog hunting time in Mississippi!

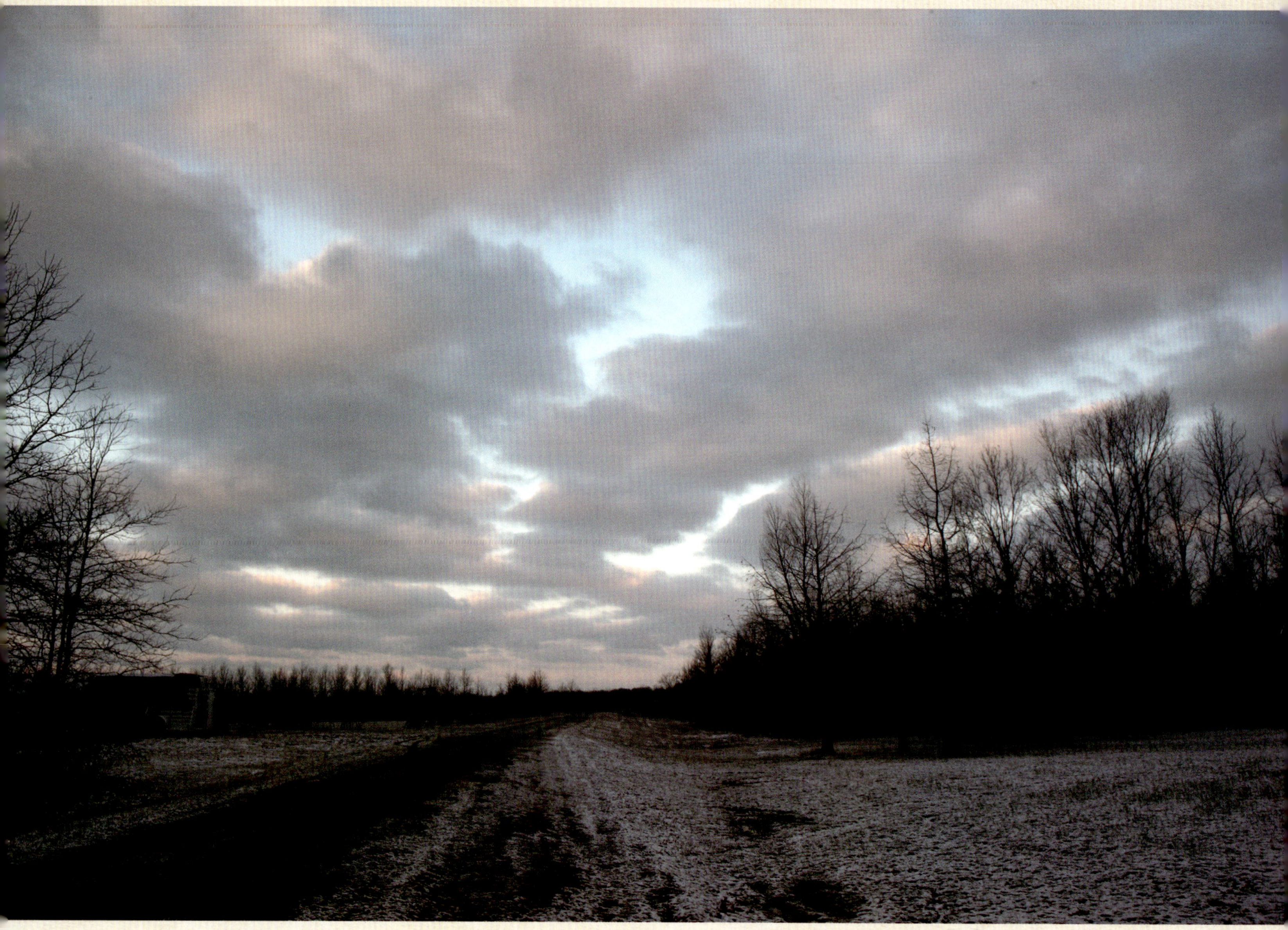

Early morning sunrise.

Freshly fallen snow. >

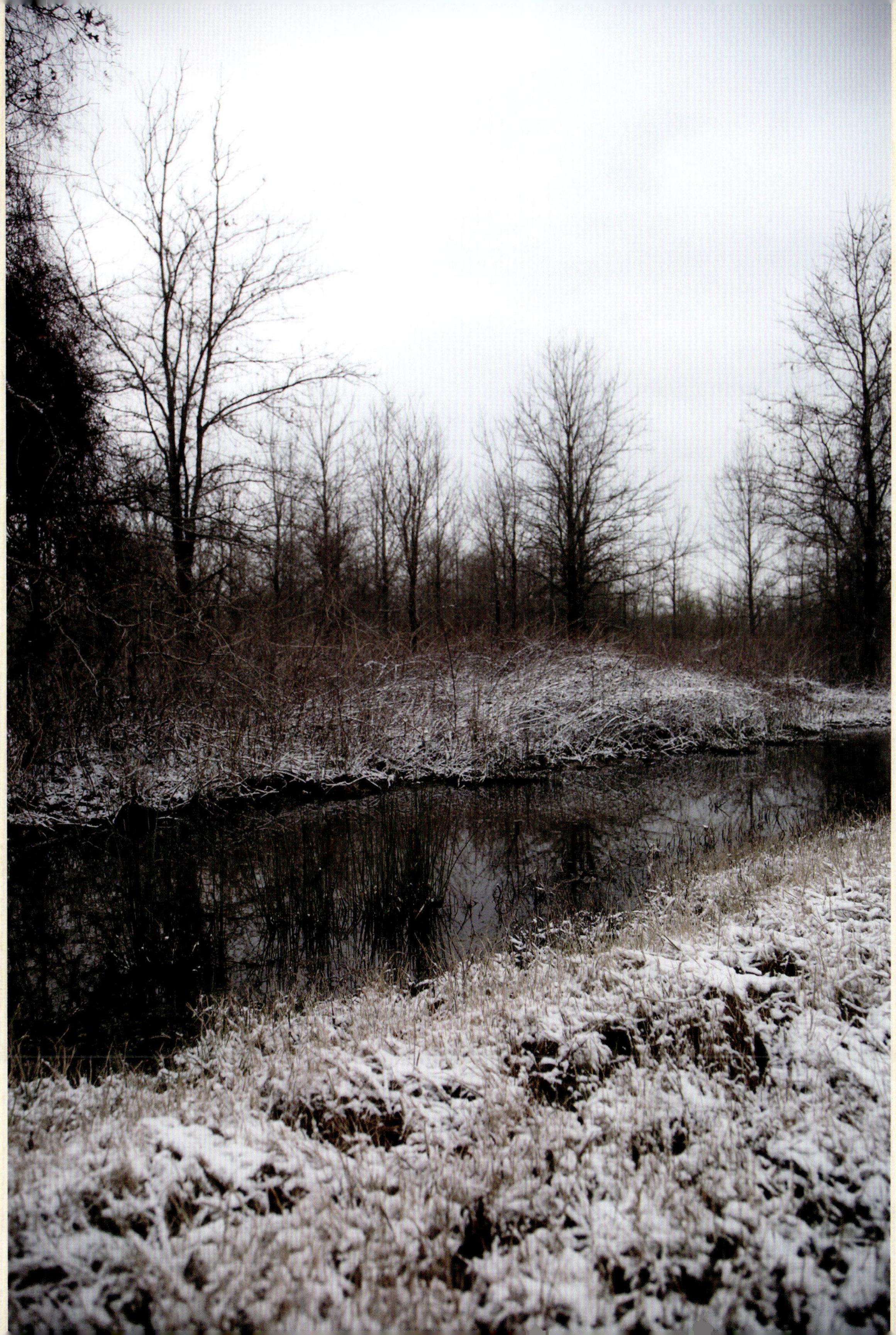

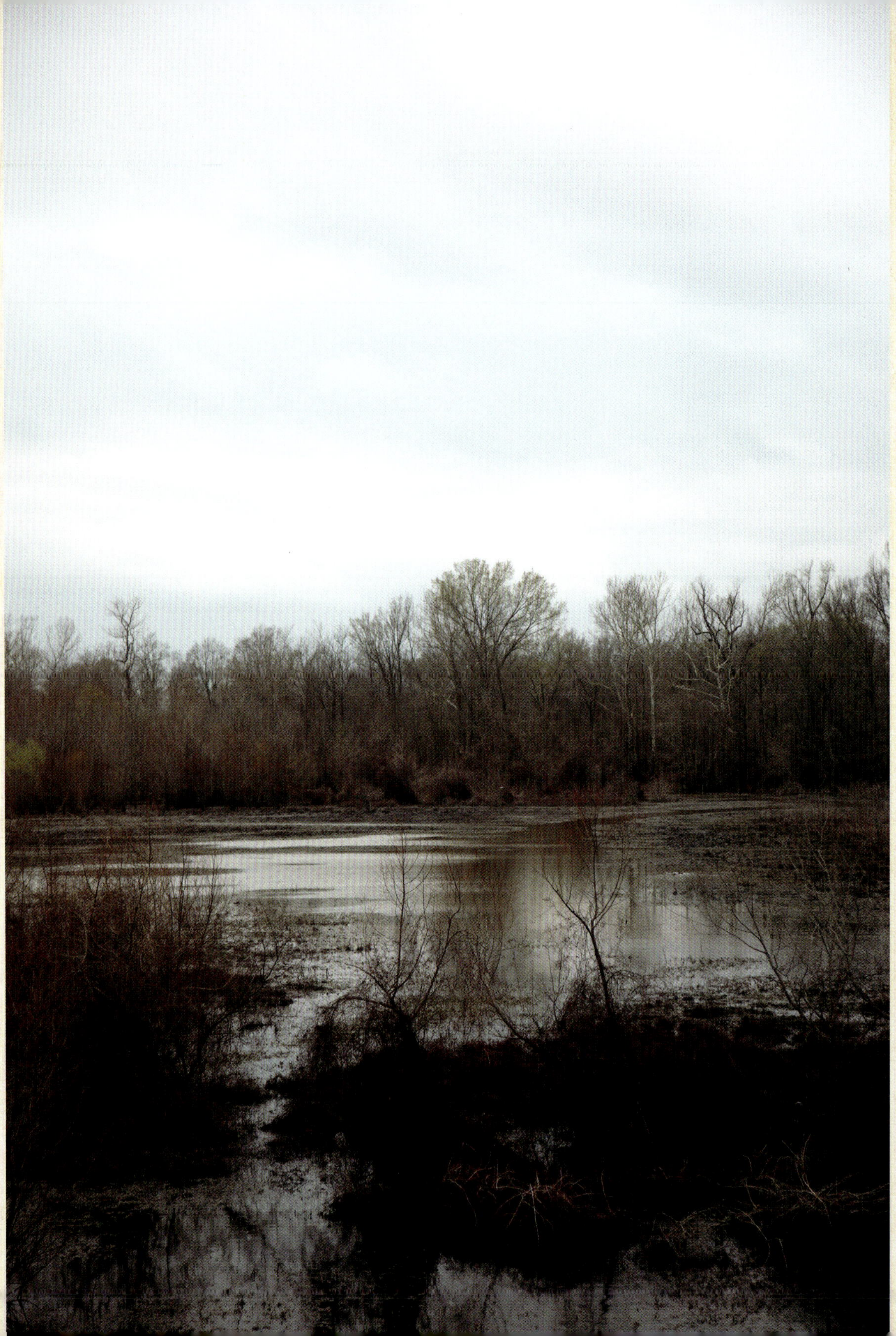

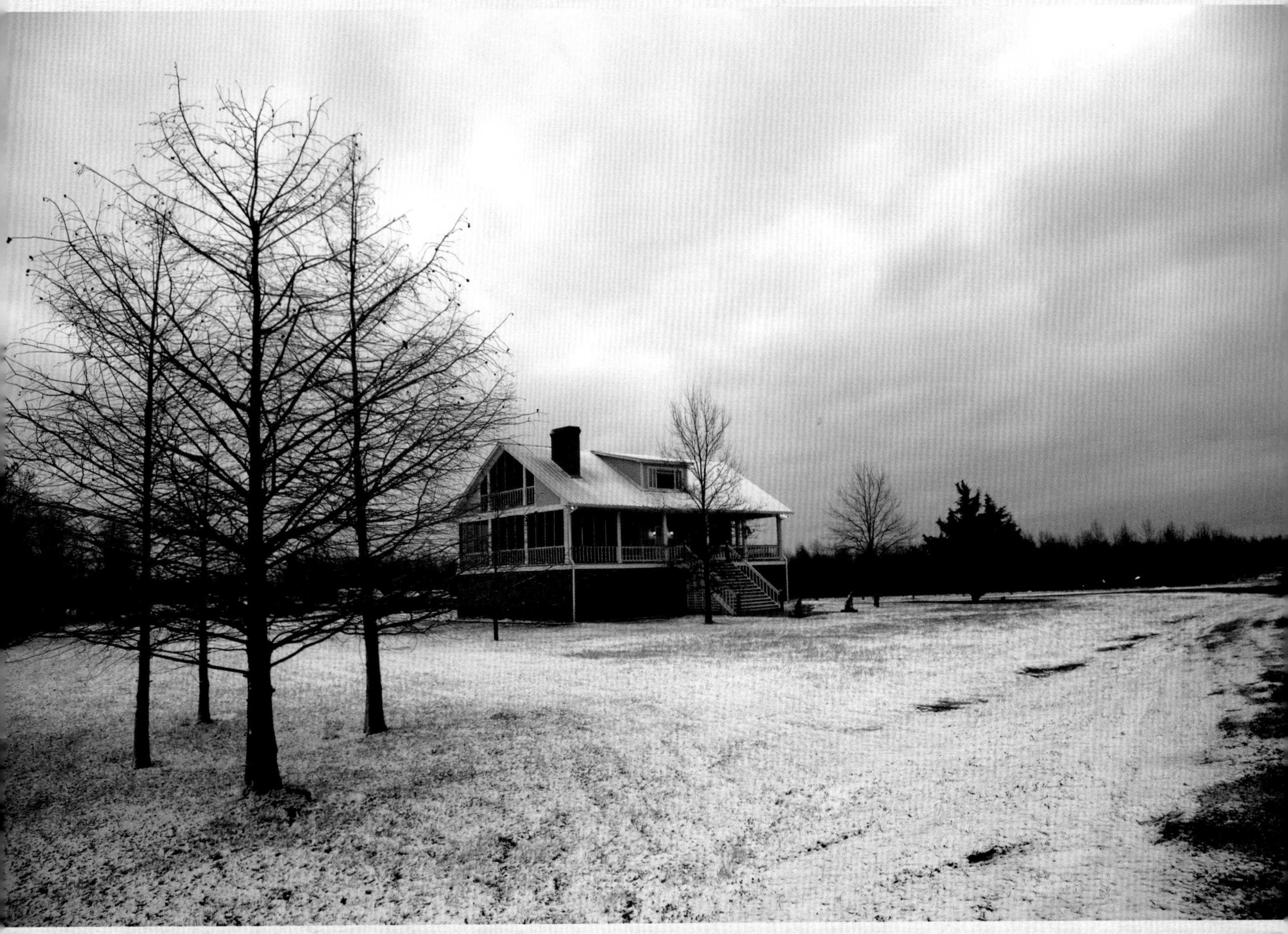

Panther Tract camp house.

< Boobie Lake.

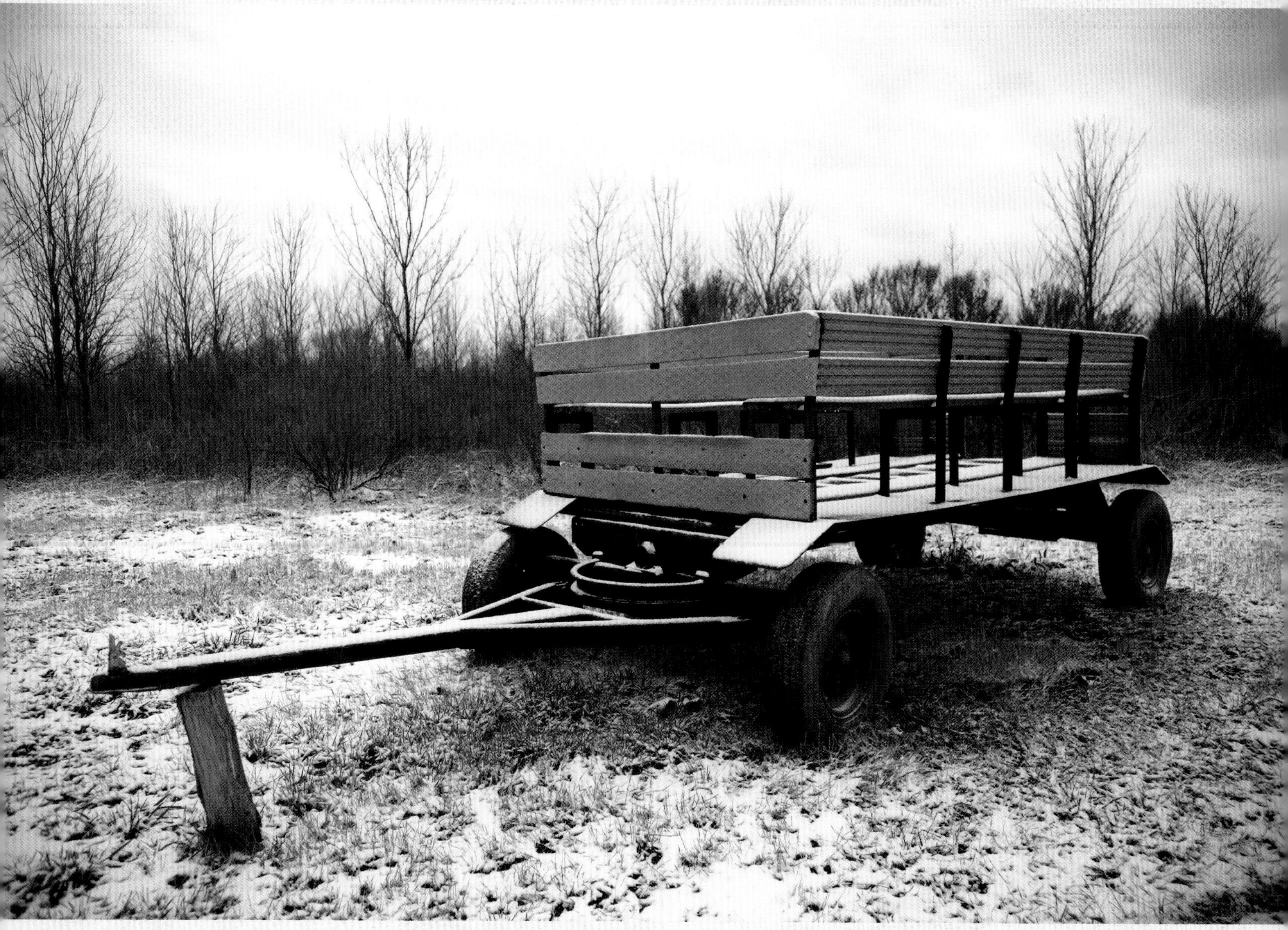

Wagon.

Horse trailer. >

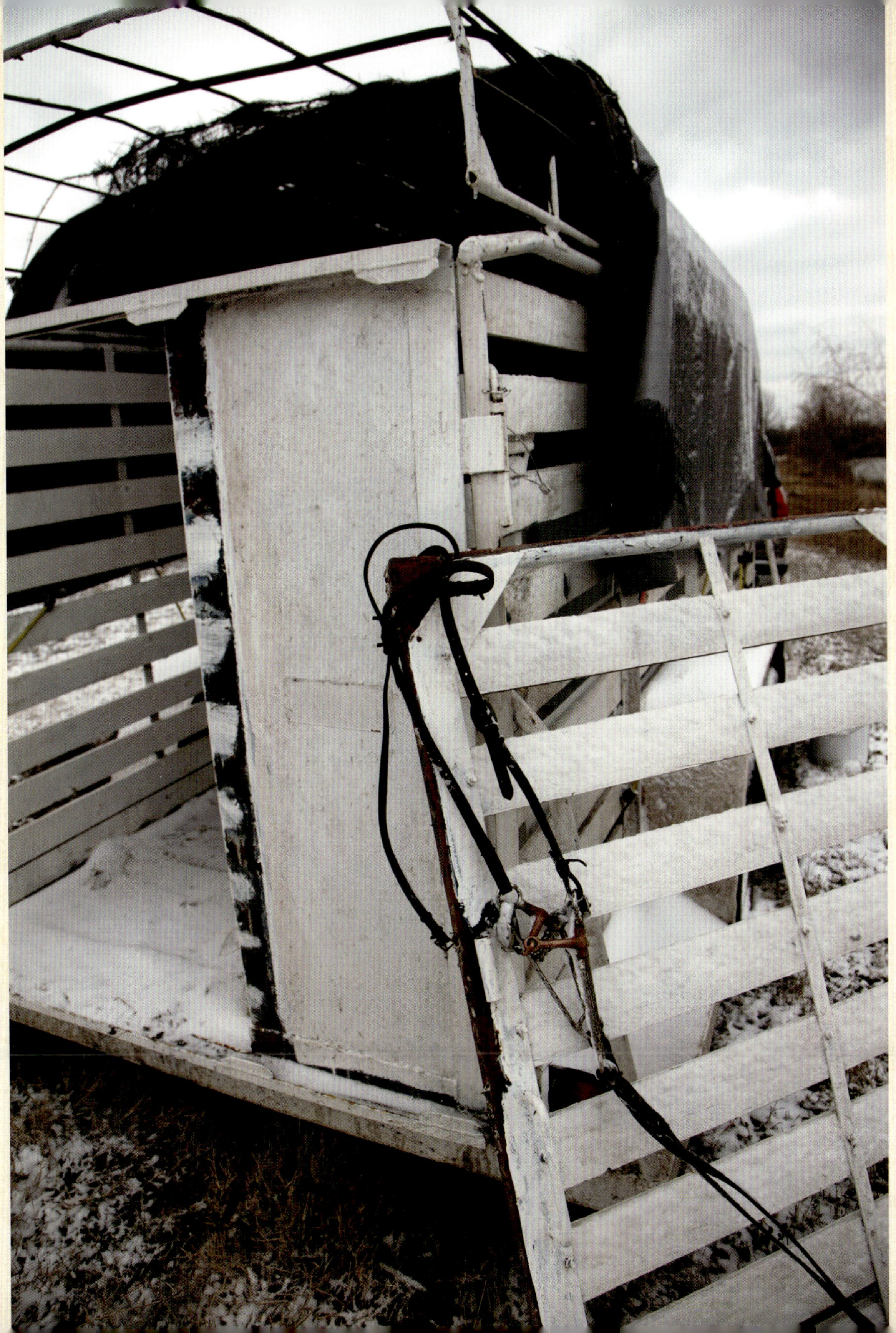

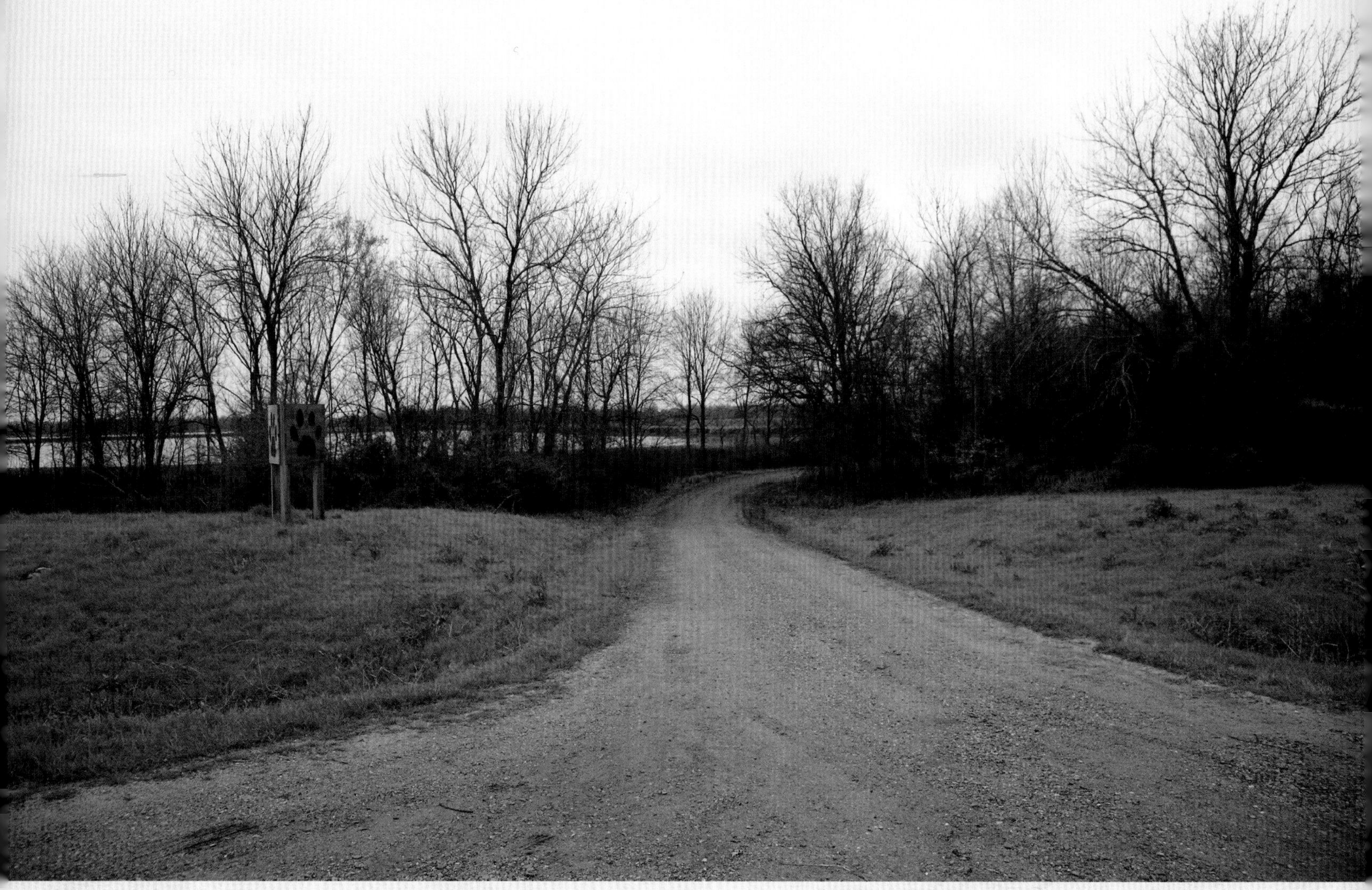

Panther Tract entrance from highway.

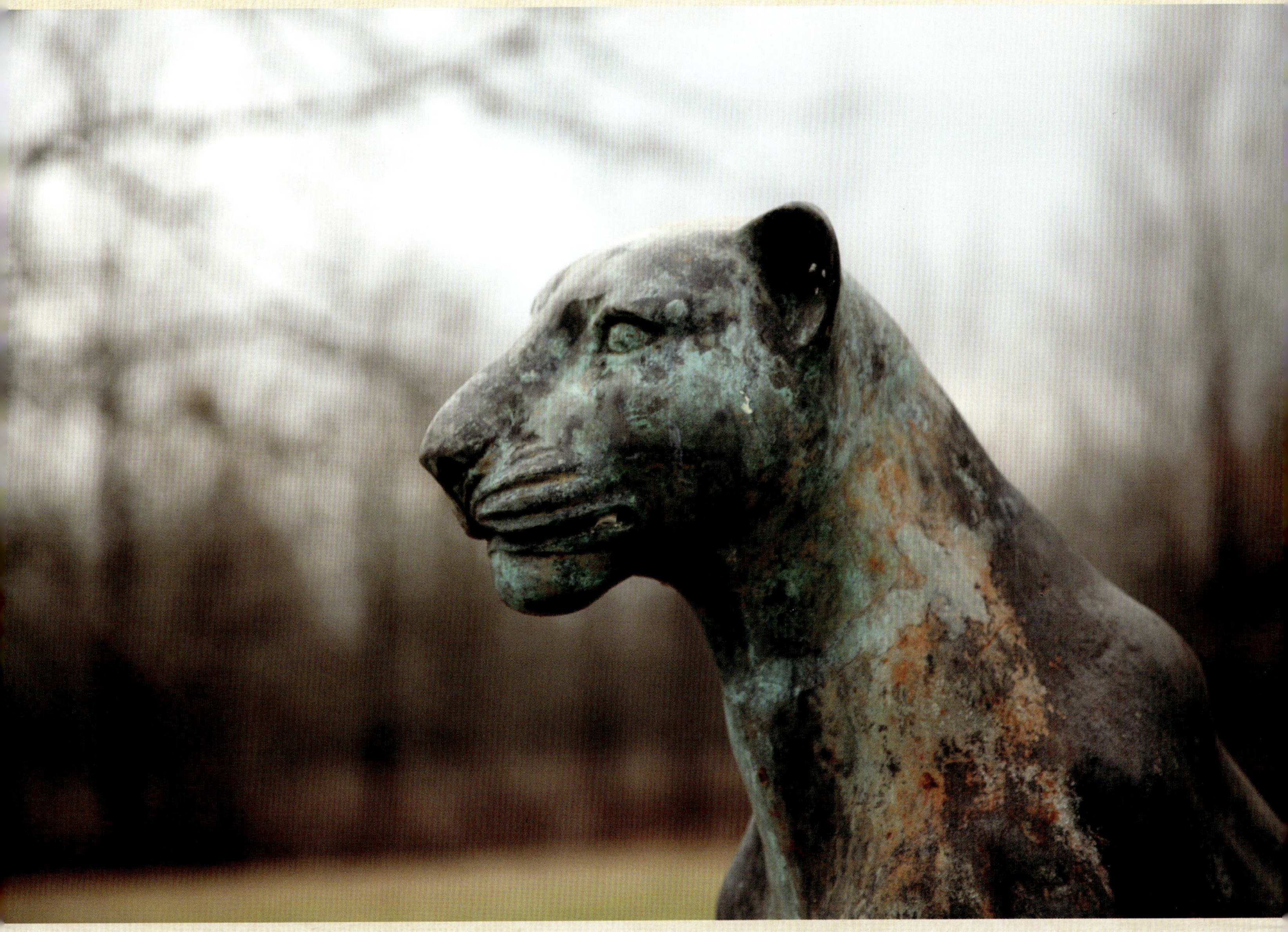

Panther statue.

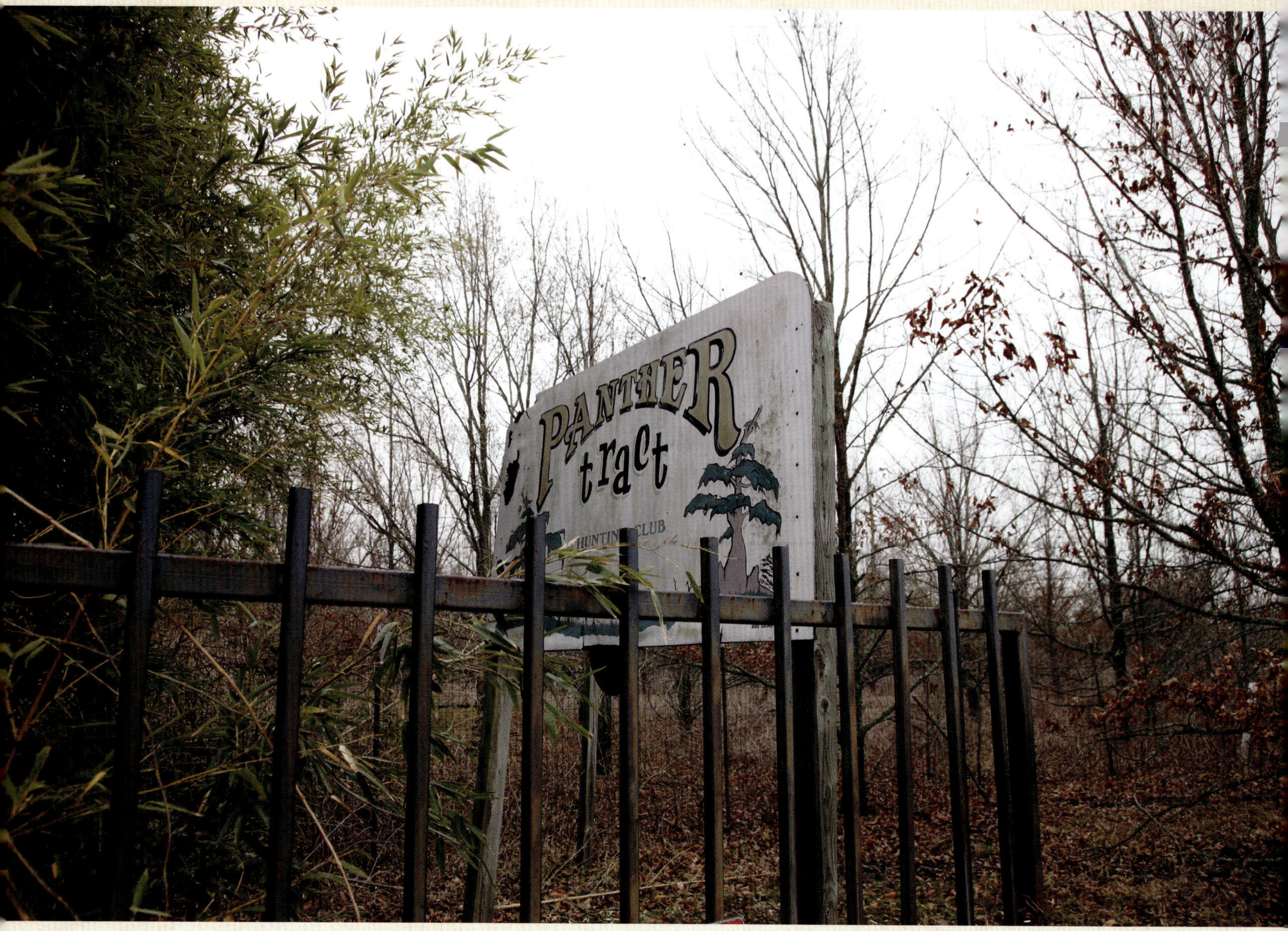

Panther Tract sign.

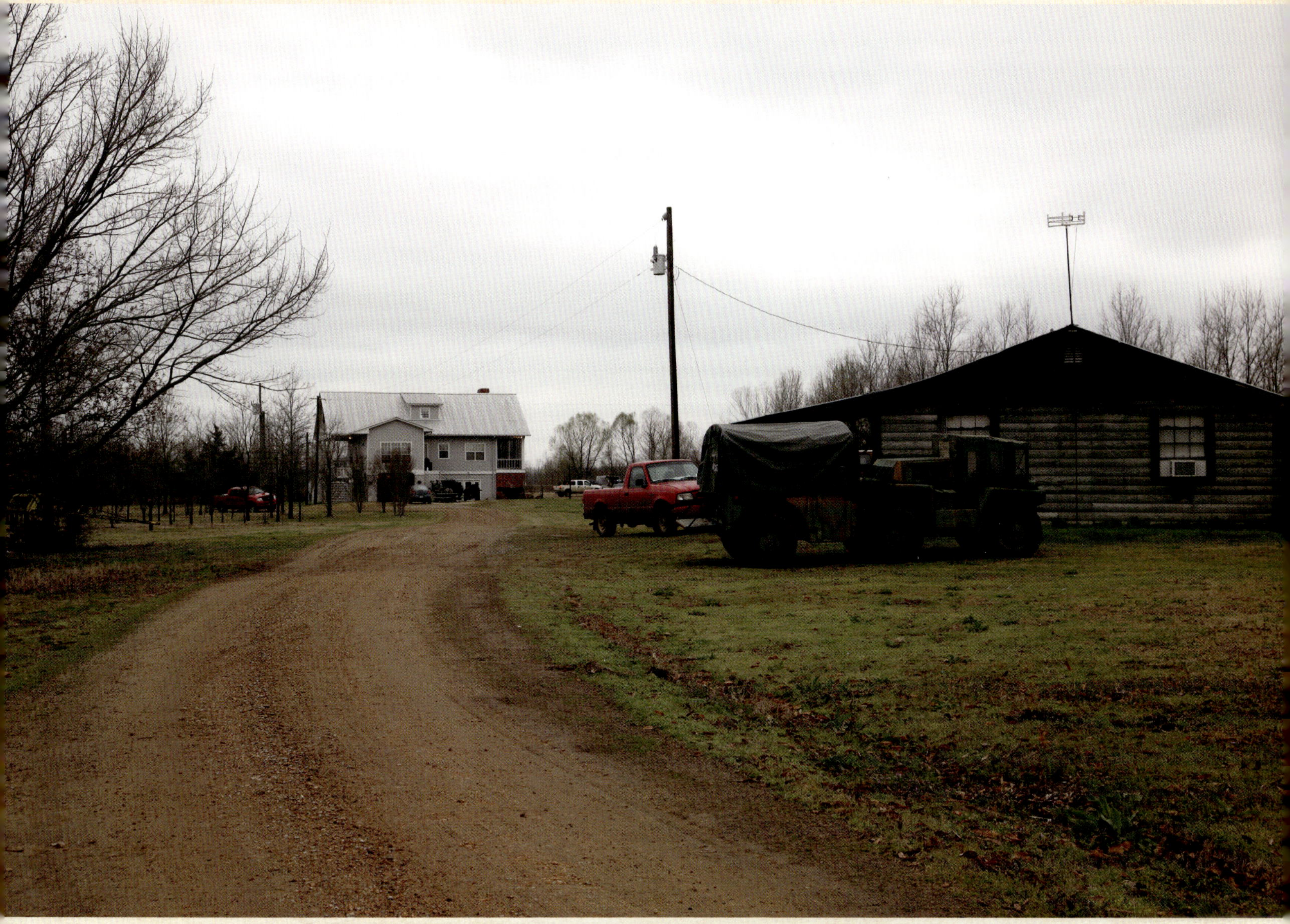

Panther Tract drive.

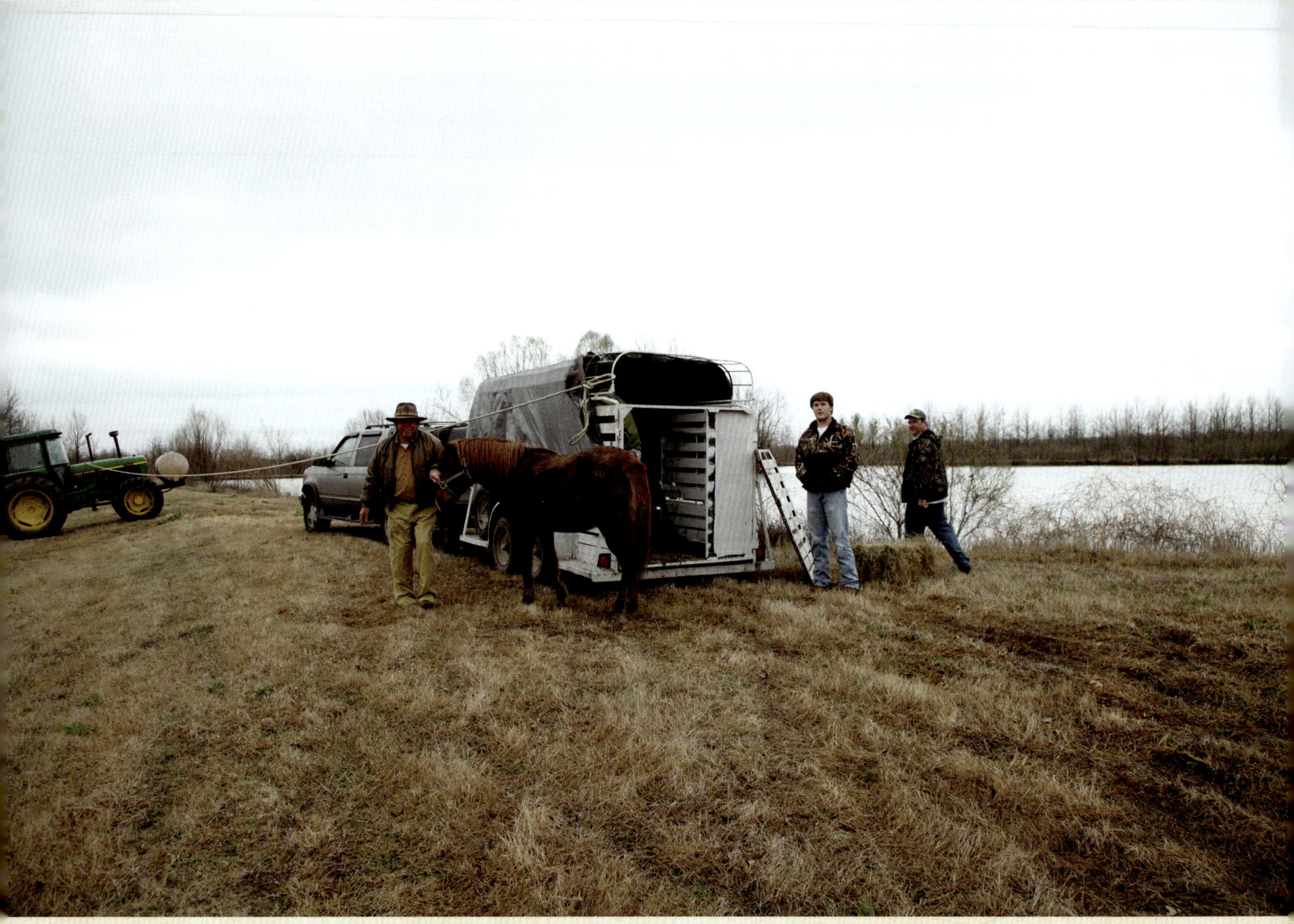

Howard Brent, Jarrad Brown, and guest with Shetland pony, Stumpy.

Howard Brent and hunting knife handmade by C. F. (Monk) Dawson. >

Overleaf left: Lea Brent, Philip Farr, and Howard Brent.
Overleaf right: John Lowery with his dogs Levi, a redbone, and Jerry, a half black mouth cur.

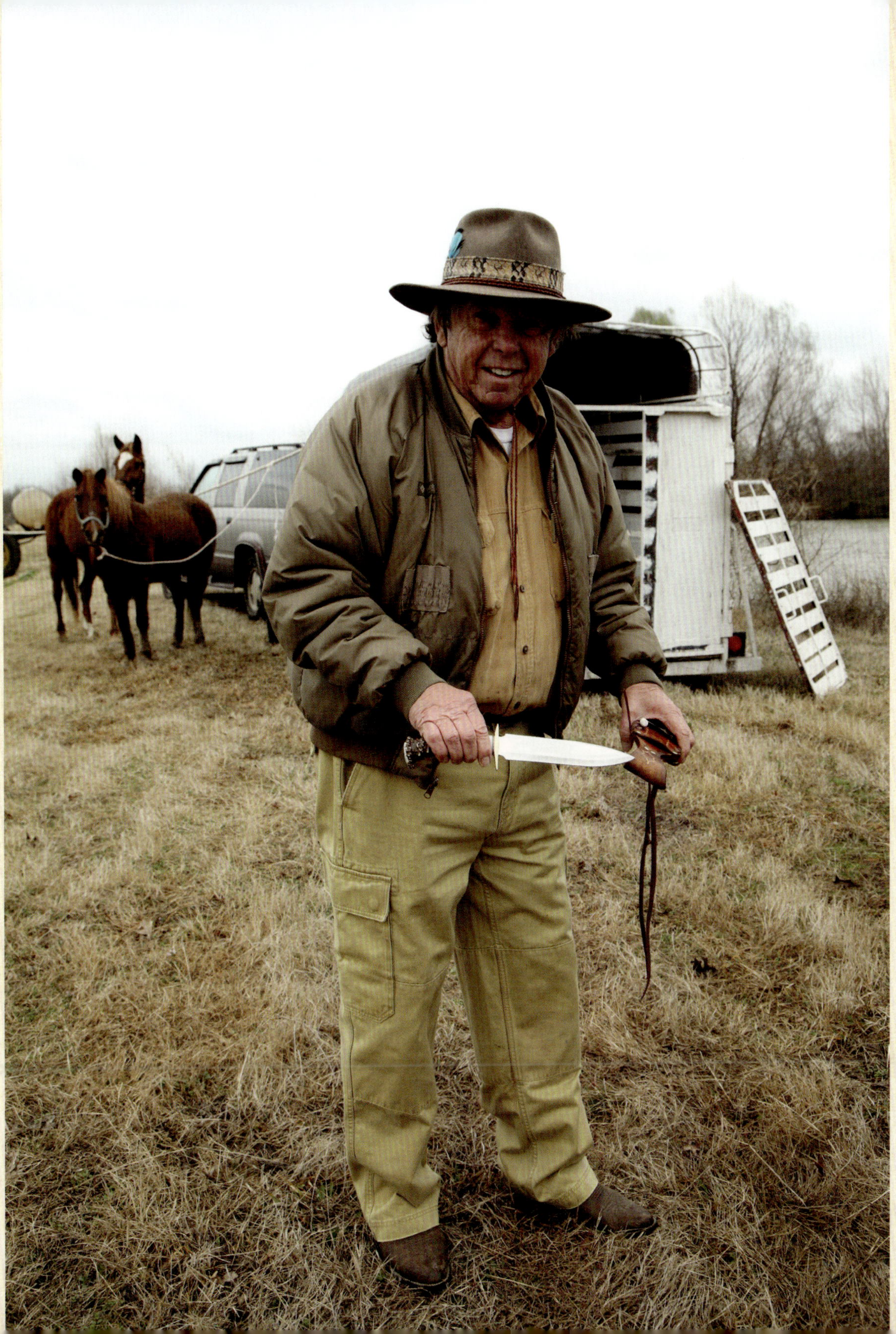

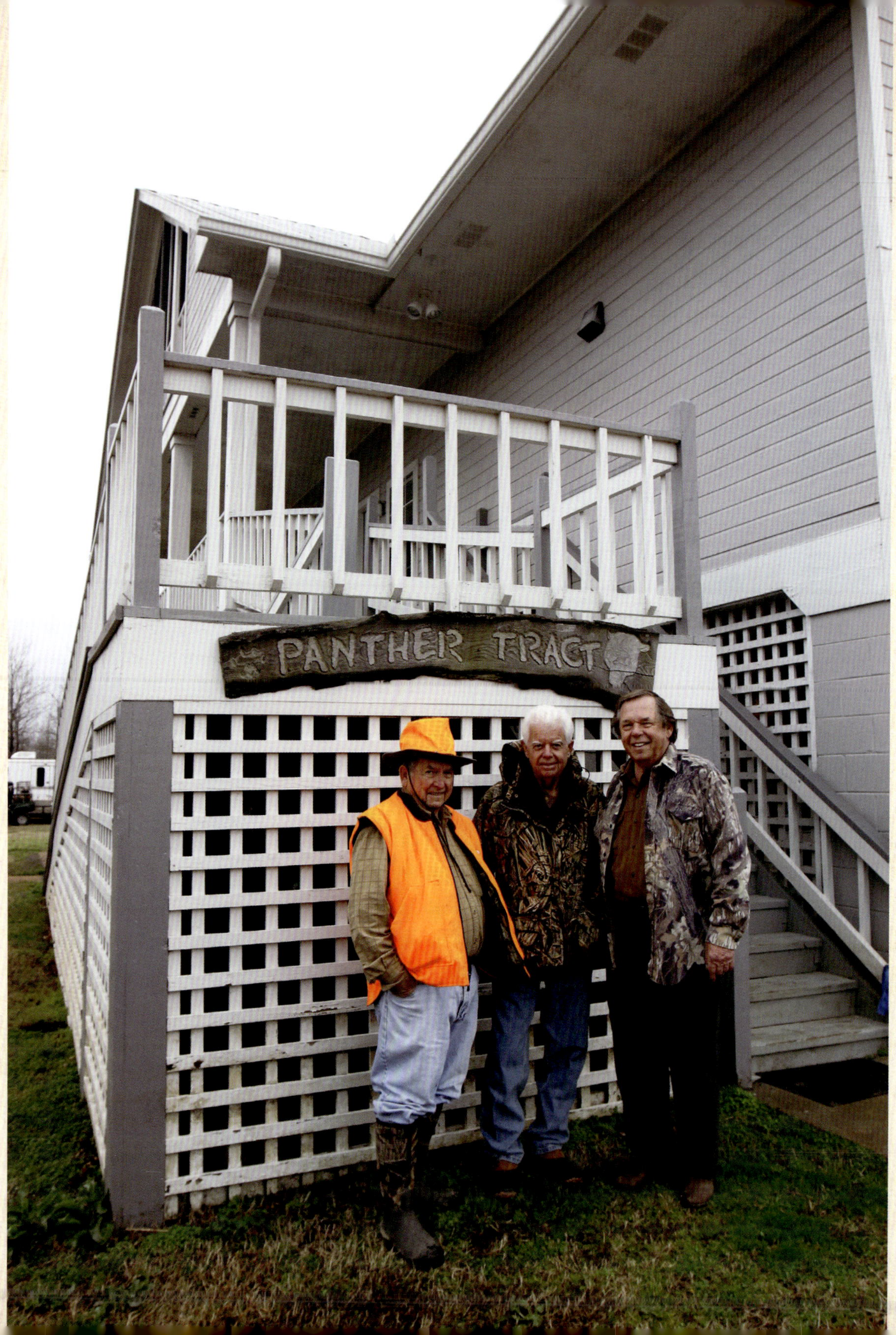
PANTHER TRACT

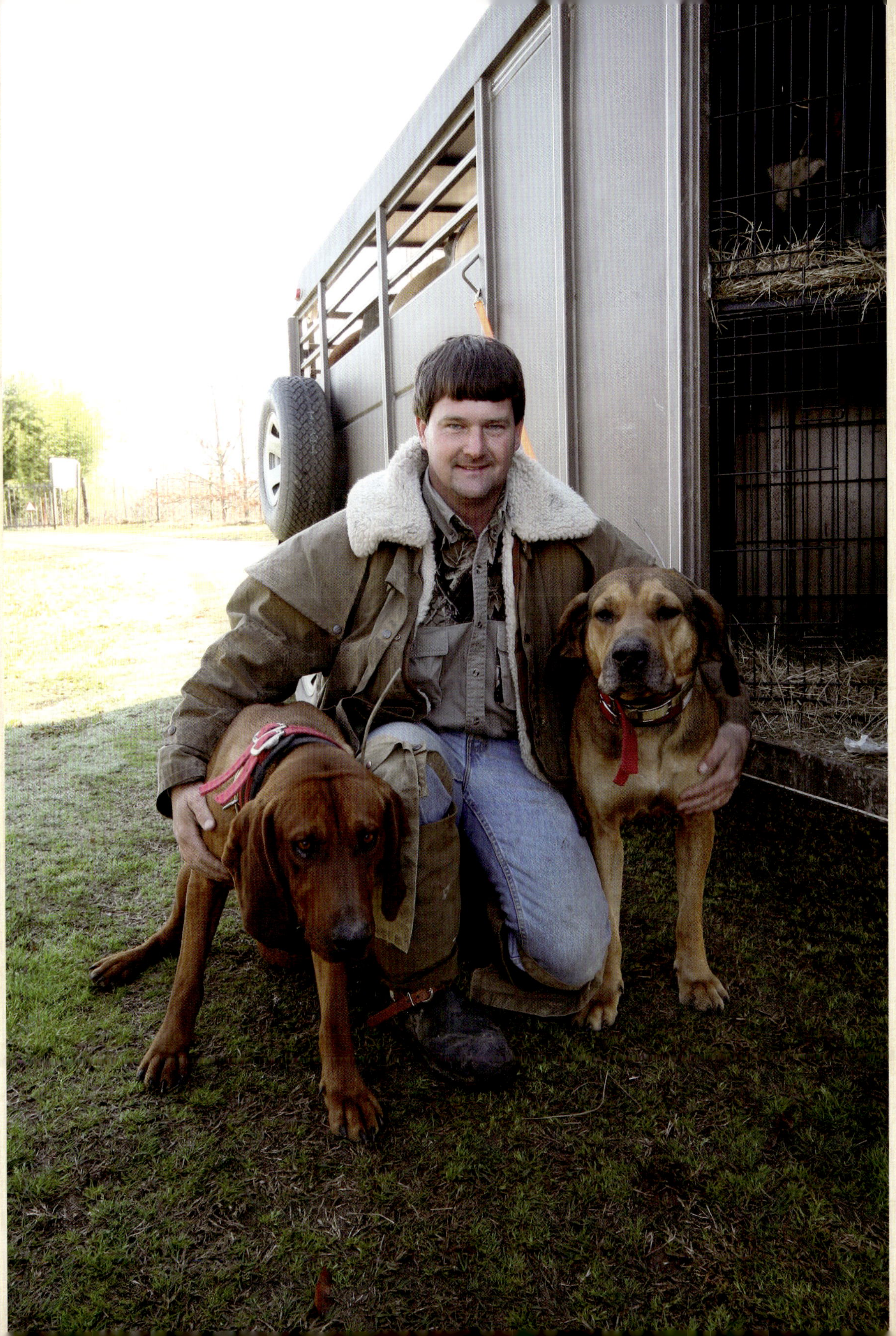

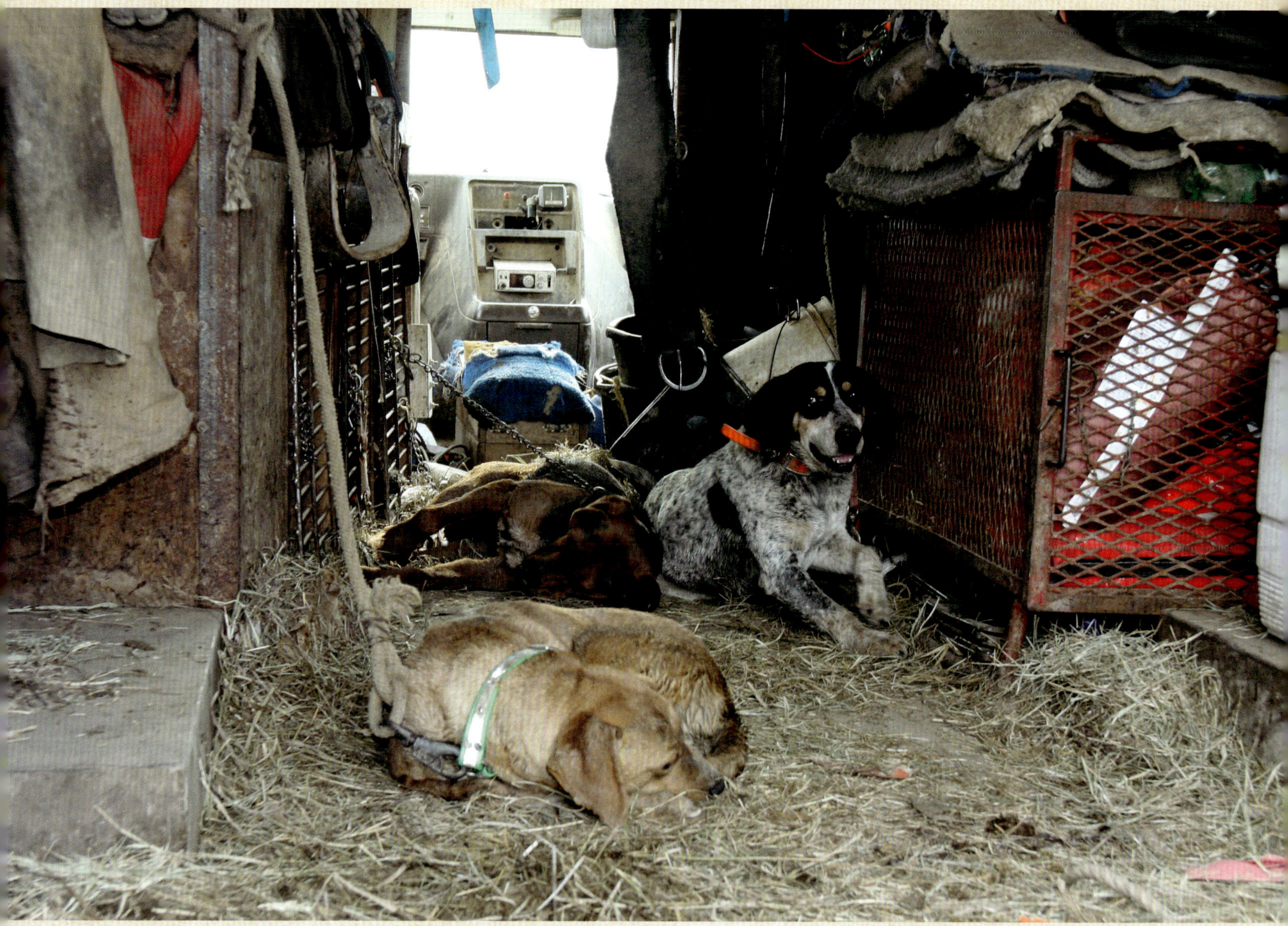

Mabel, a bluetick Plott hound, and Plott hounds in cookie truck.

Melvin (Bubba) Weeks with his dogs Mabel and Rosie. >

Overleaf left: Frank Dantone and Mabel.
Overleaf right: John Lowery and Harley, a Tennessee Walking Horse.

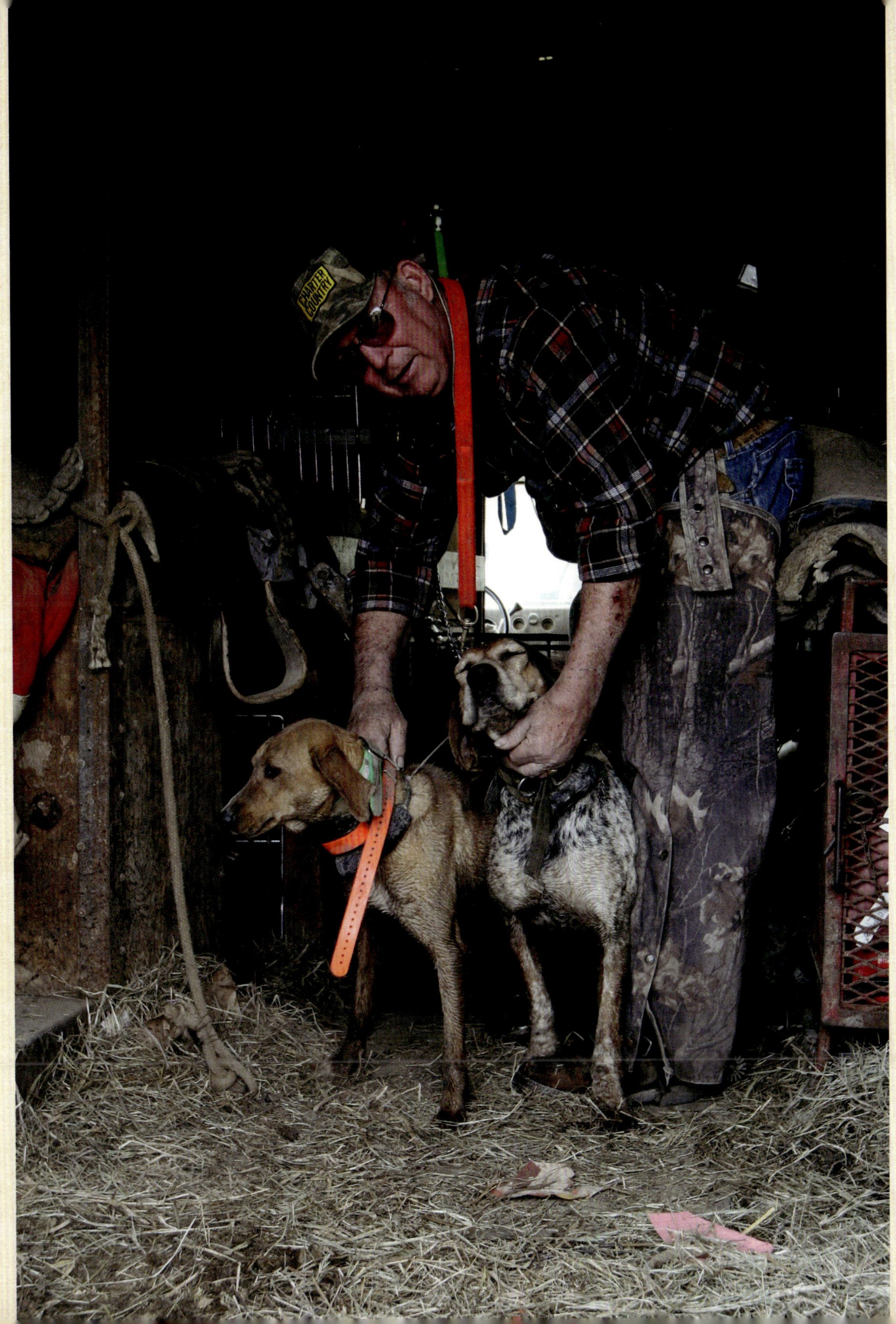
CHARTER
COUNTRY

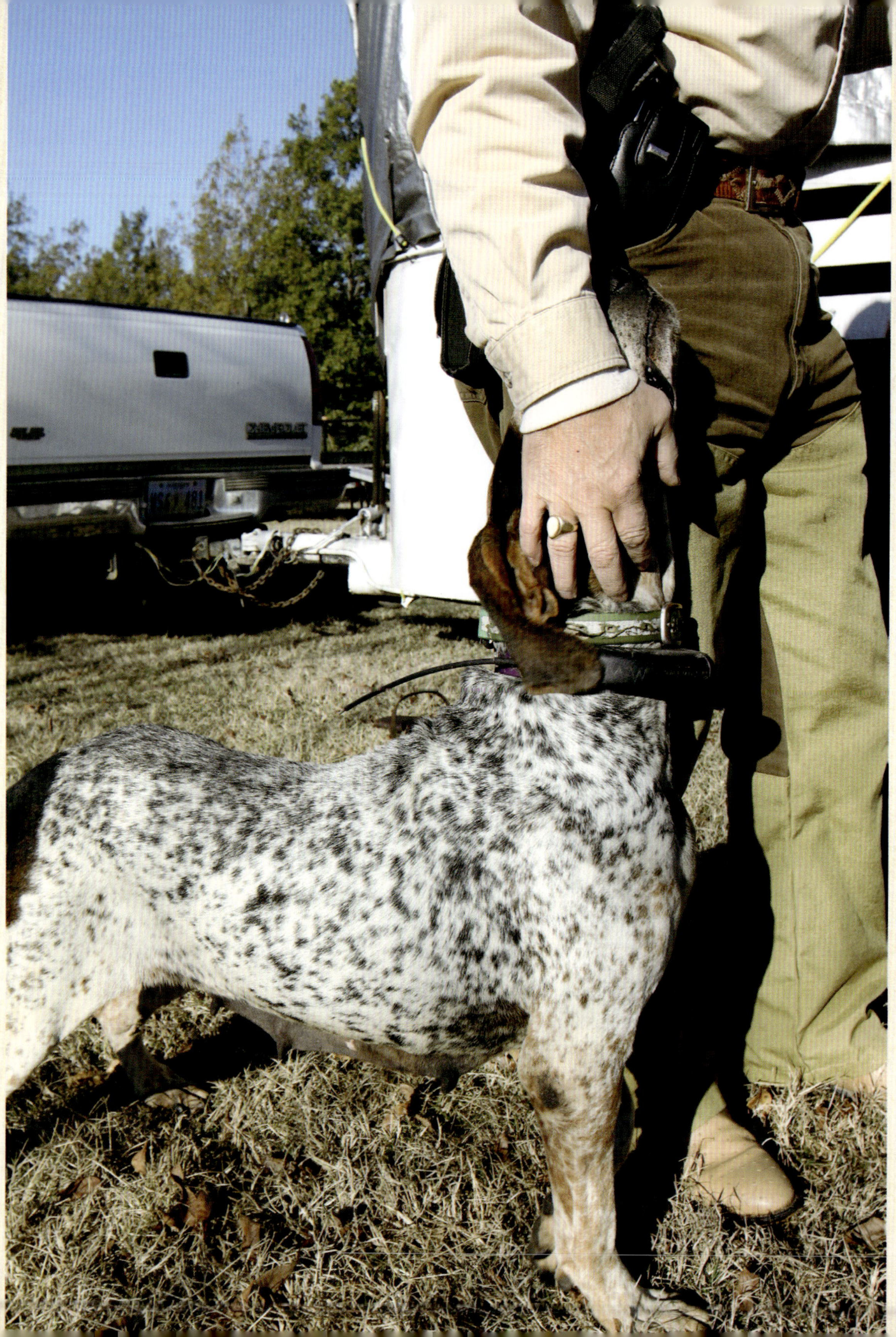

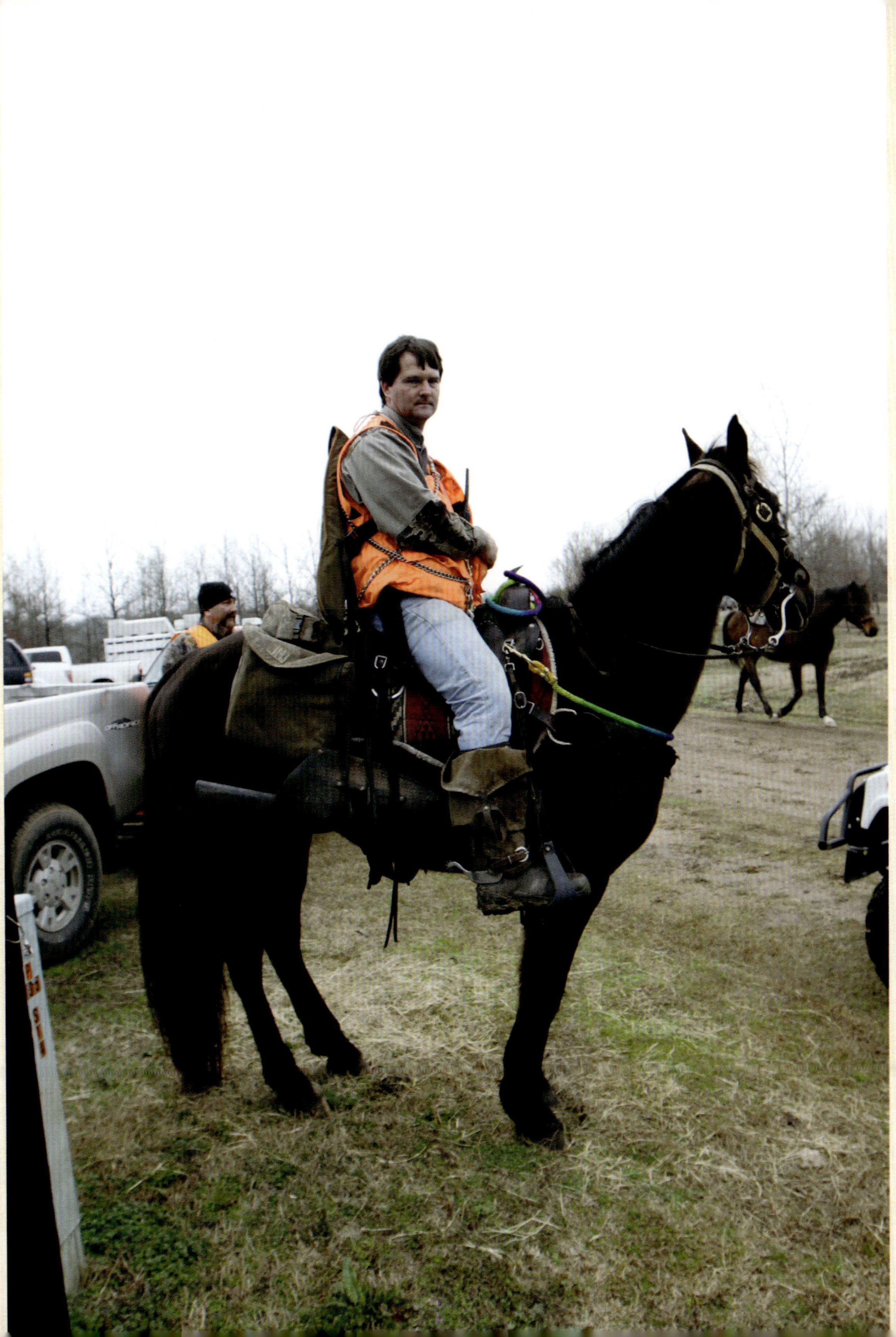

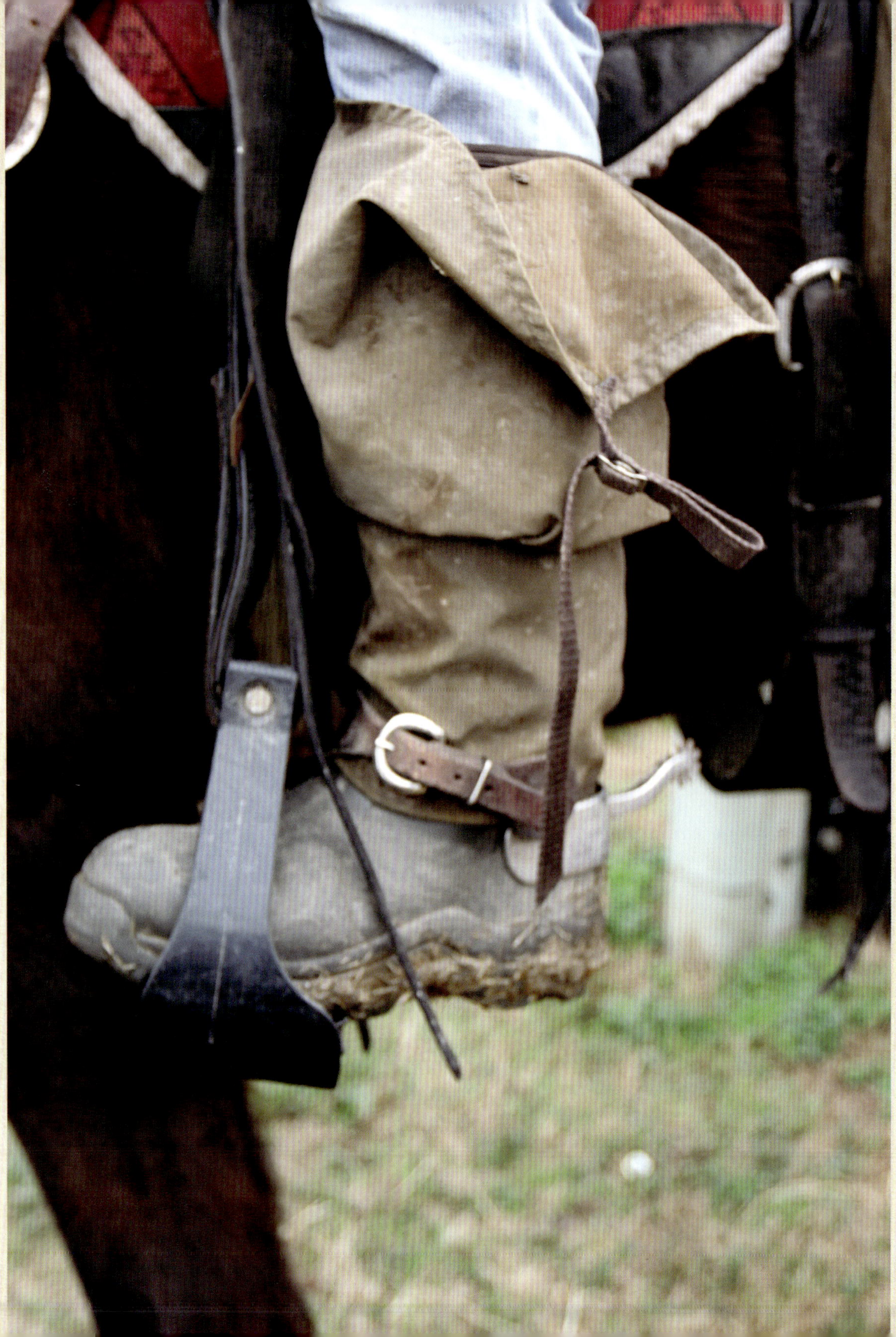

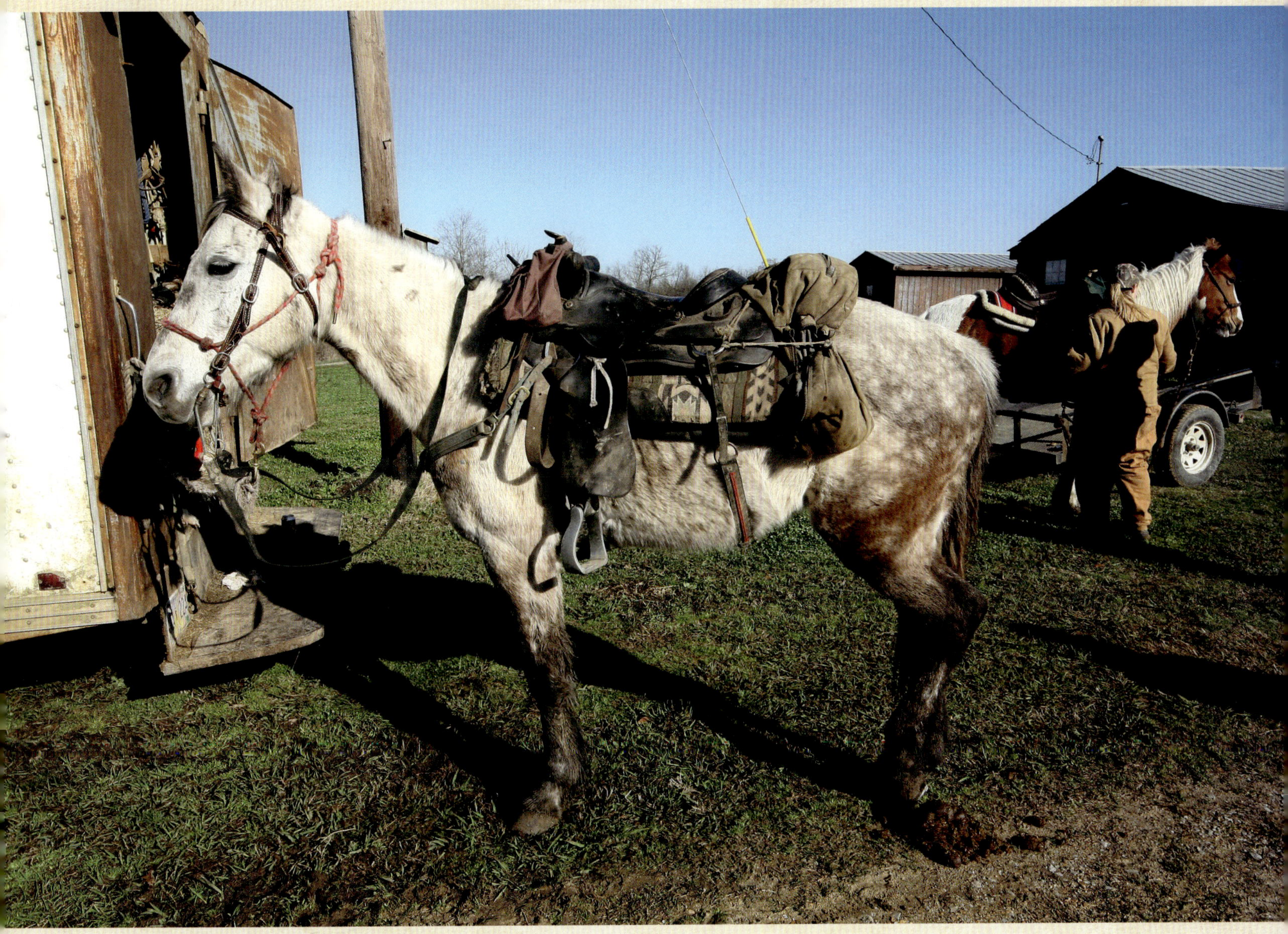

Melvin's horse King, a Missouri fox trotter.

‹ Rolled-down hip boots.

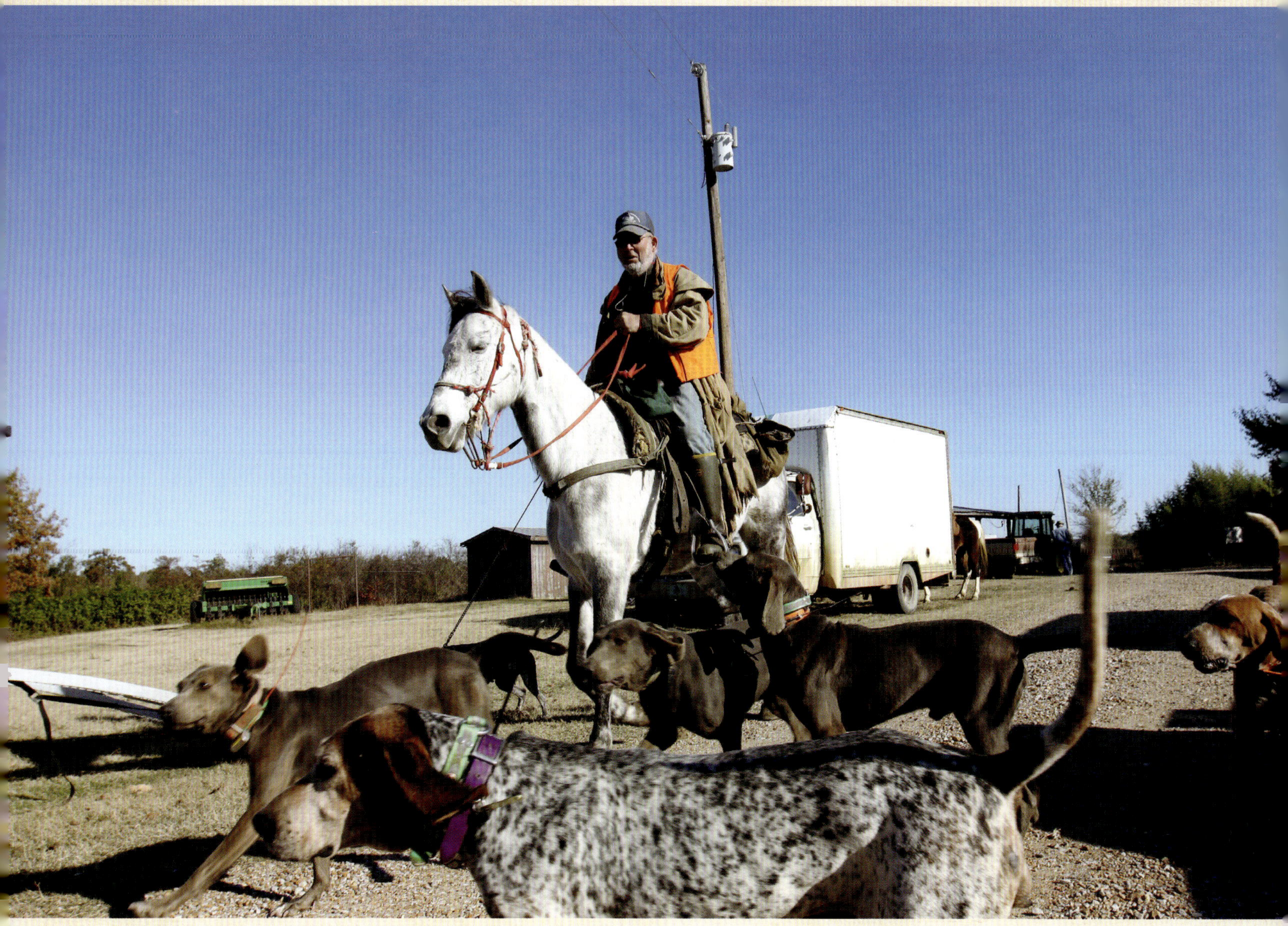

Melvin and his hunting dogs.

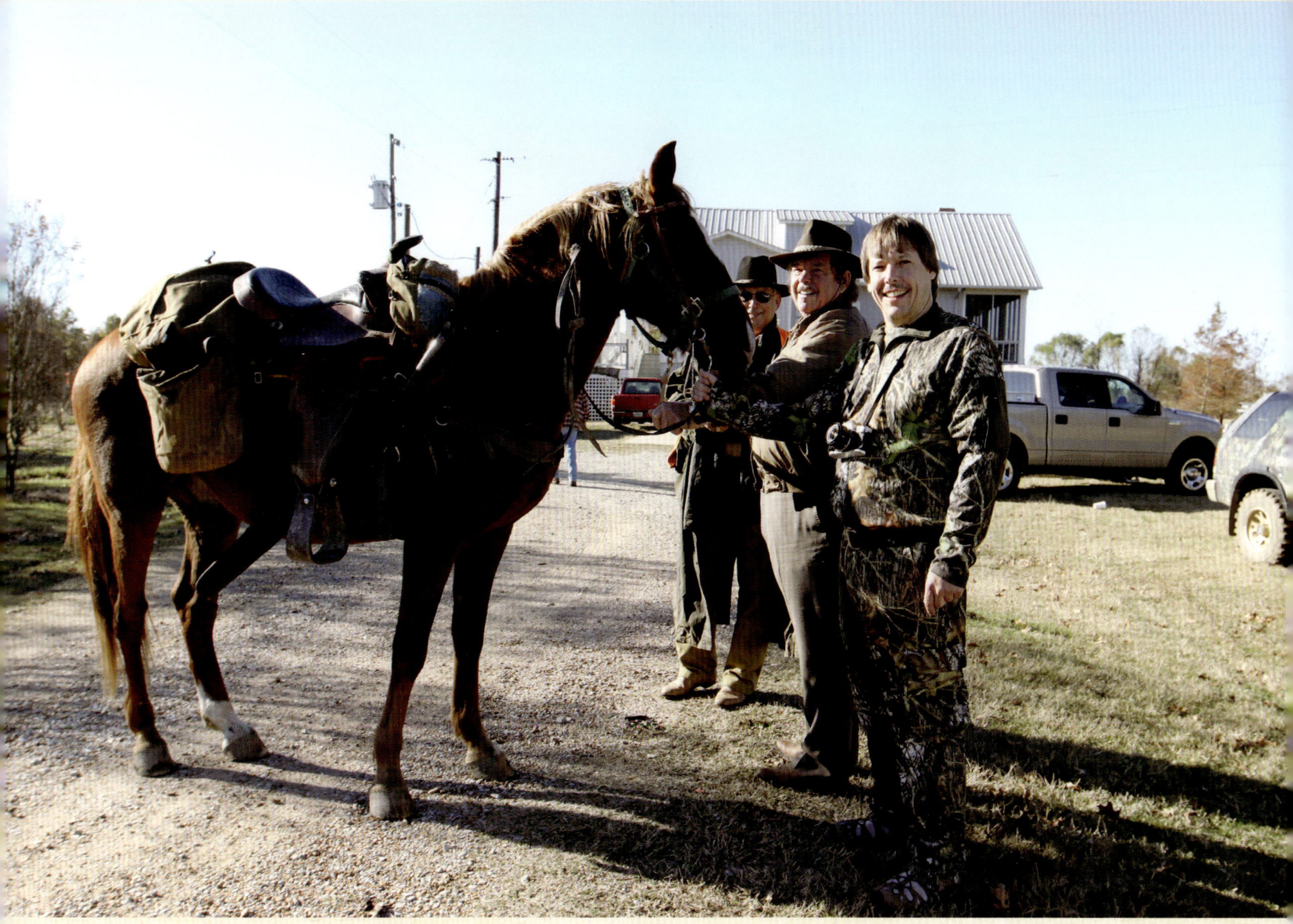

Red and Frank Dantone, Howard Brent, and Lincoln Brent.

Overleaf left: Jerry, a half black mouth cur.
Overleaf right: Collars.

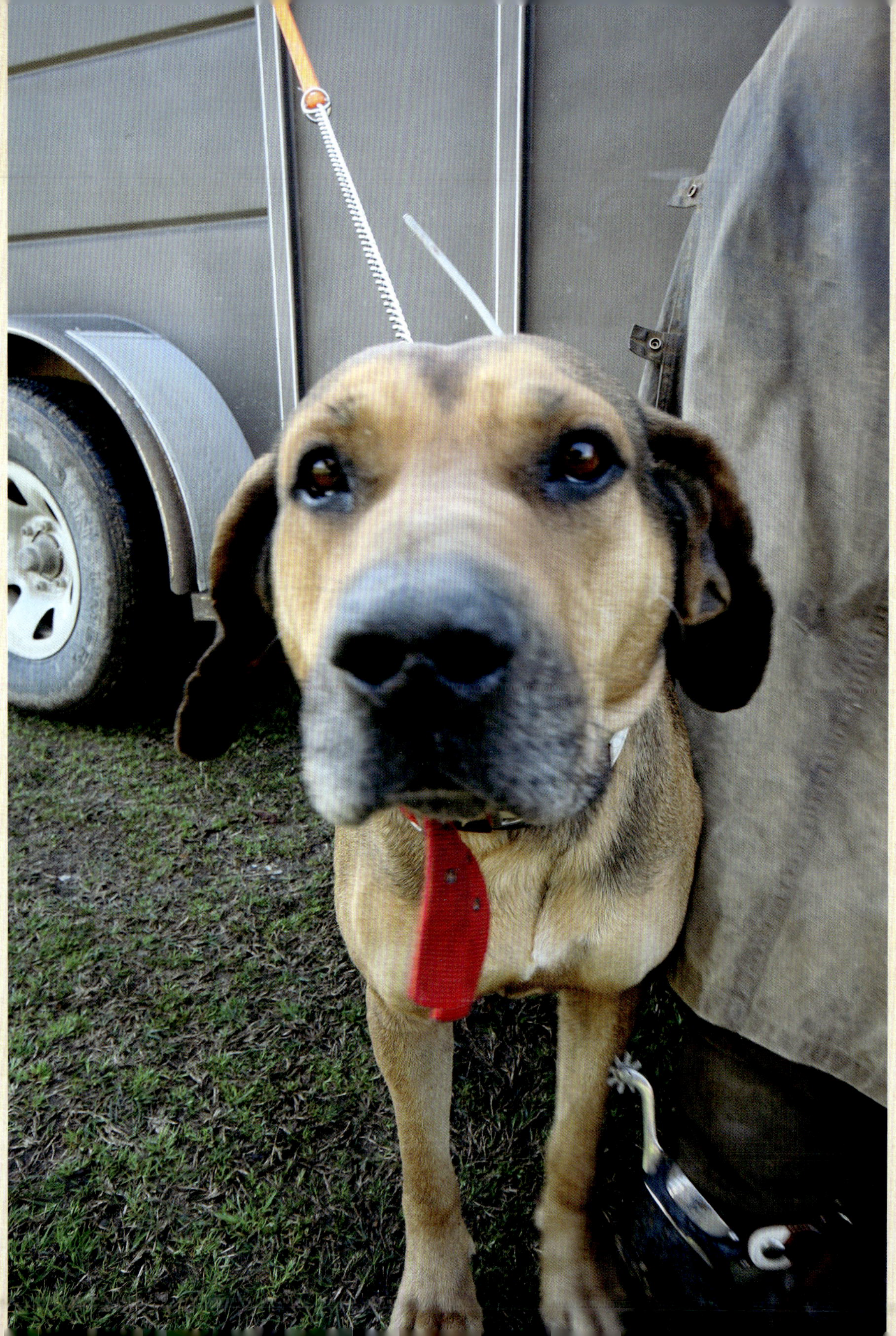

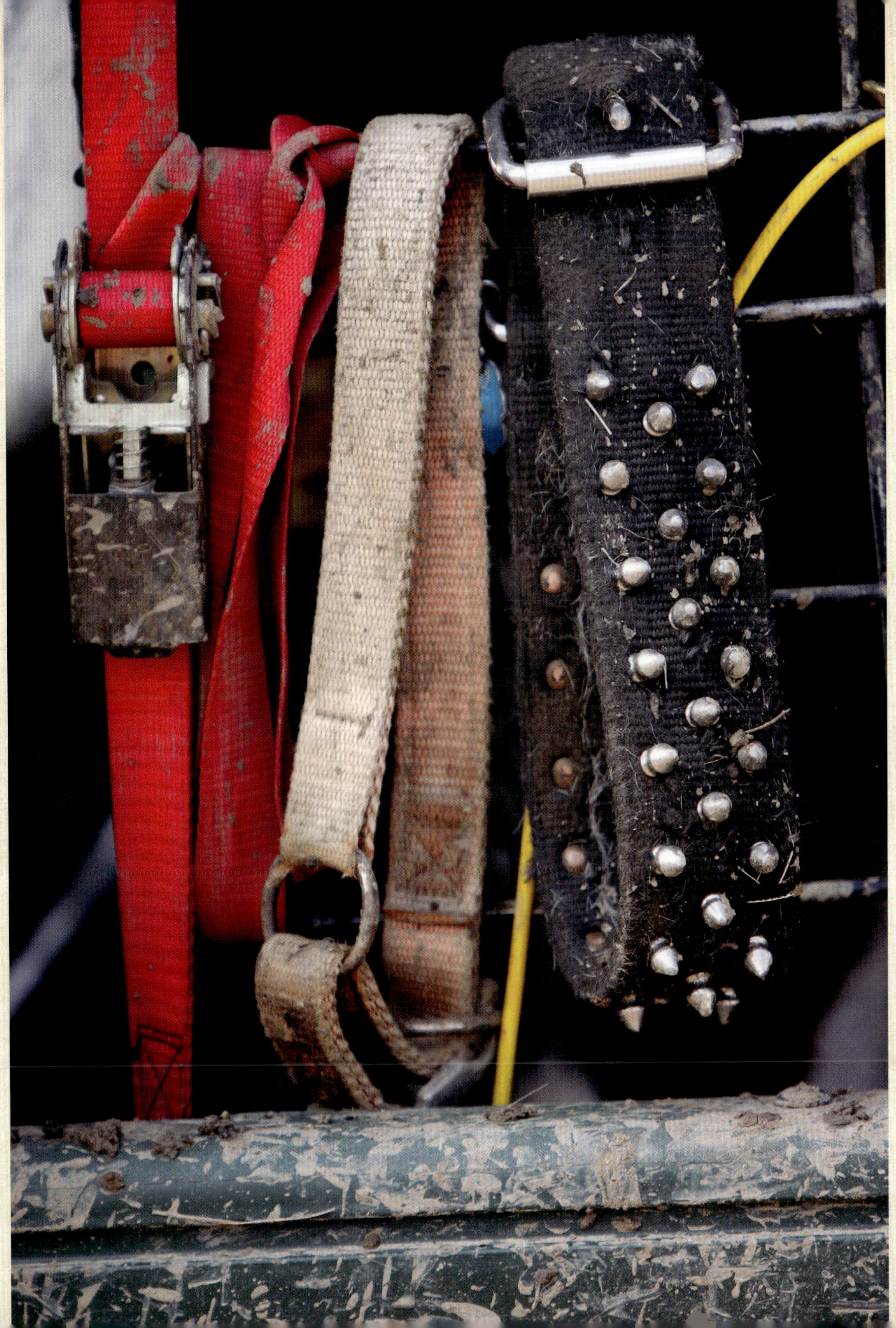

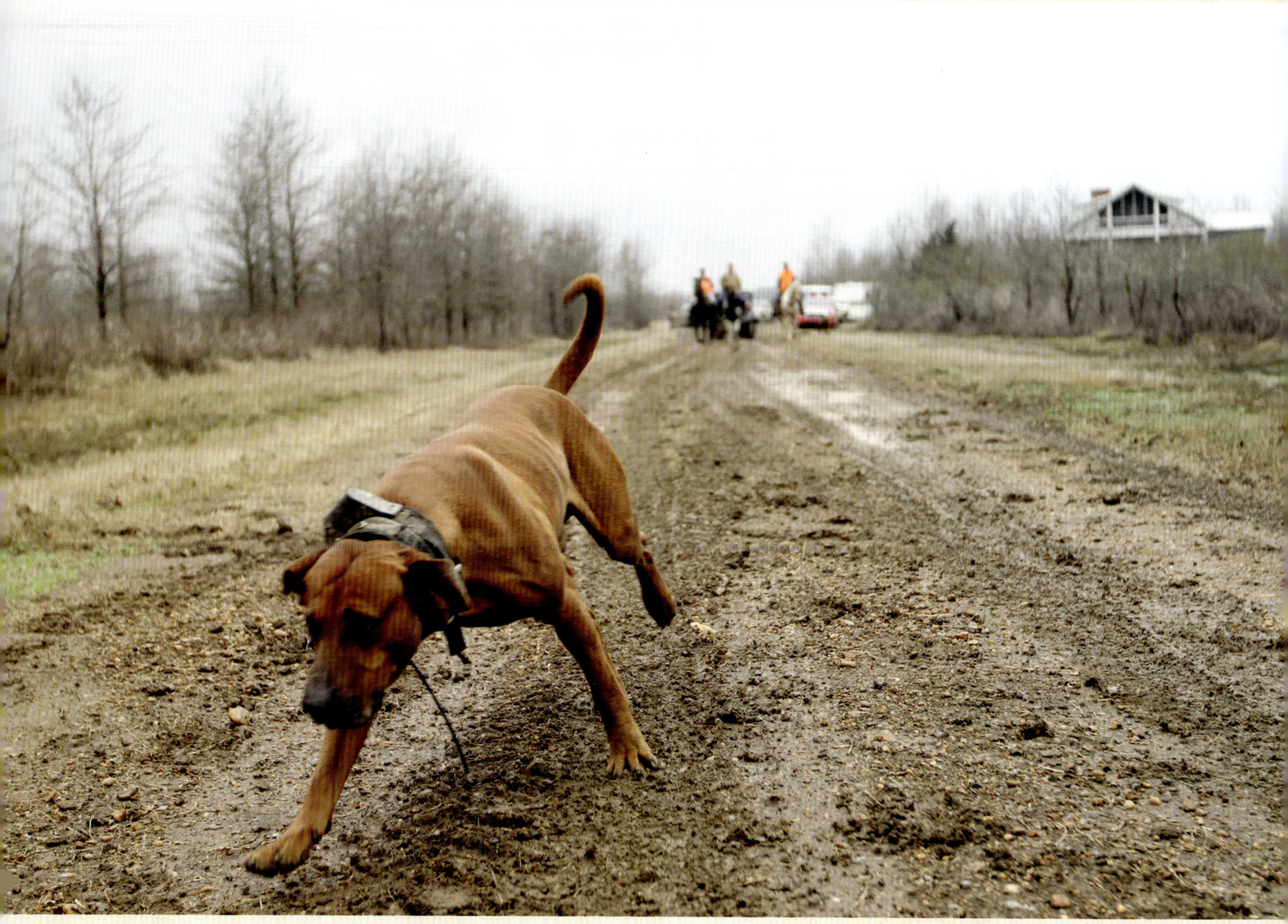

Bo, a half redbone, half black mouth cur.

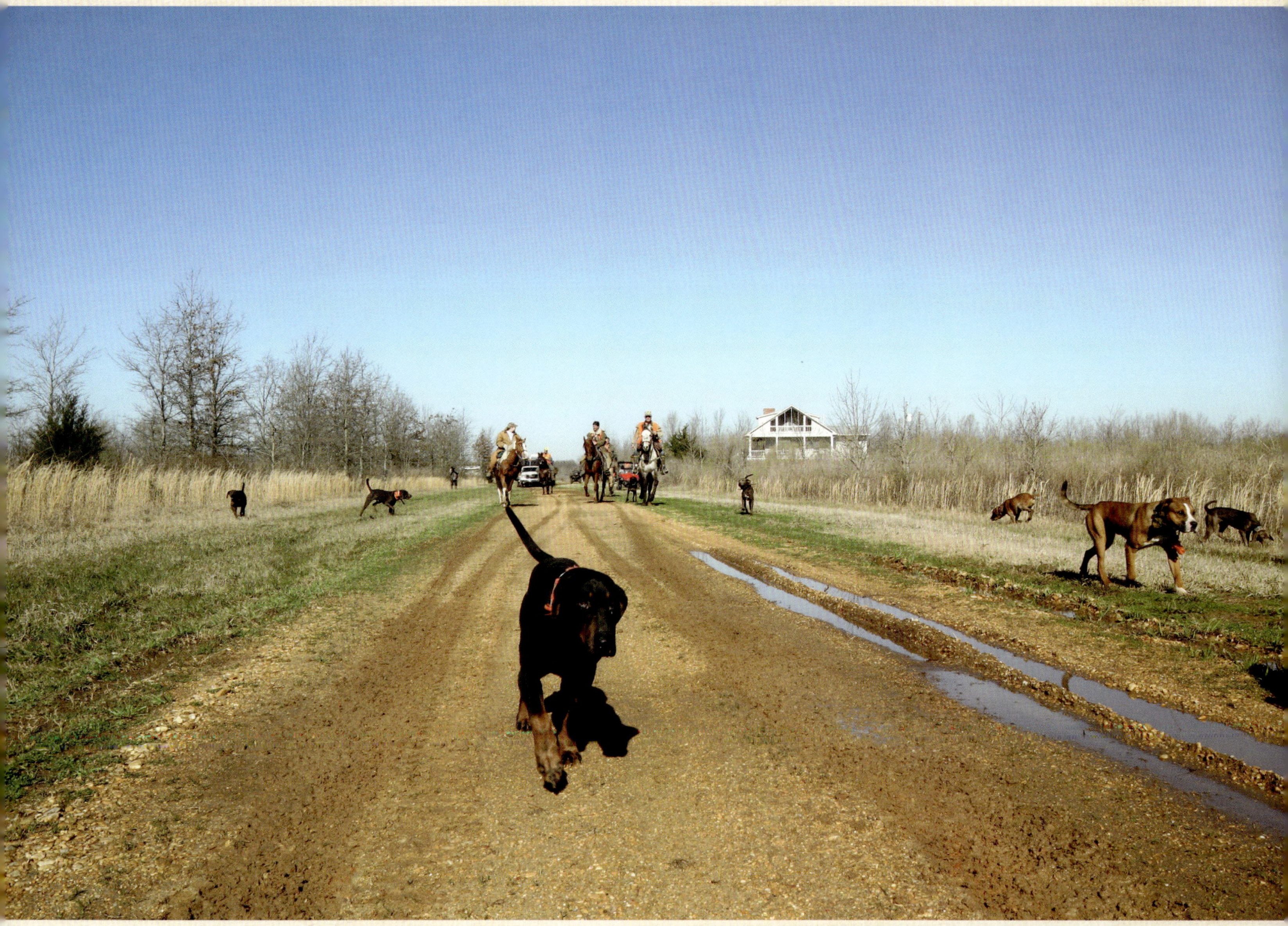

Off to hunt—Plott hound.

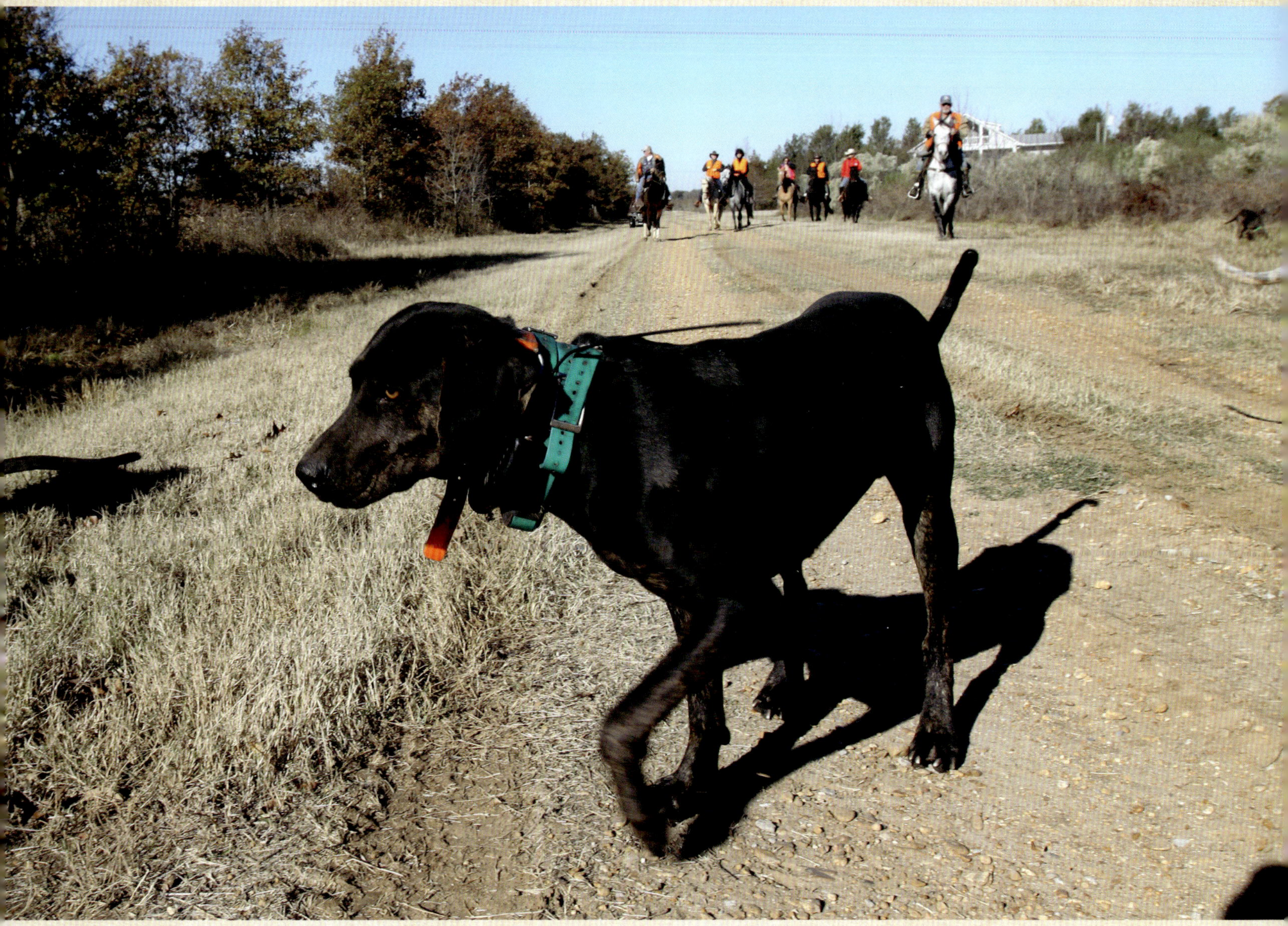

Plott hound.

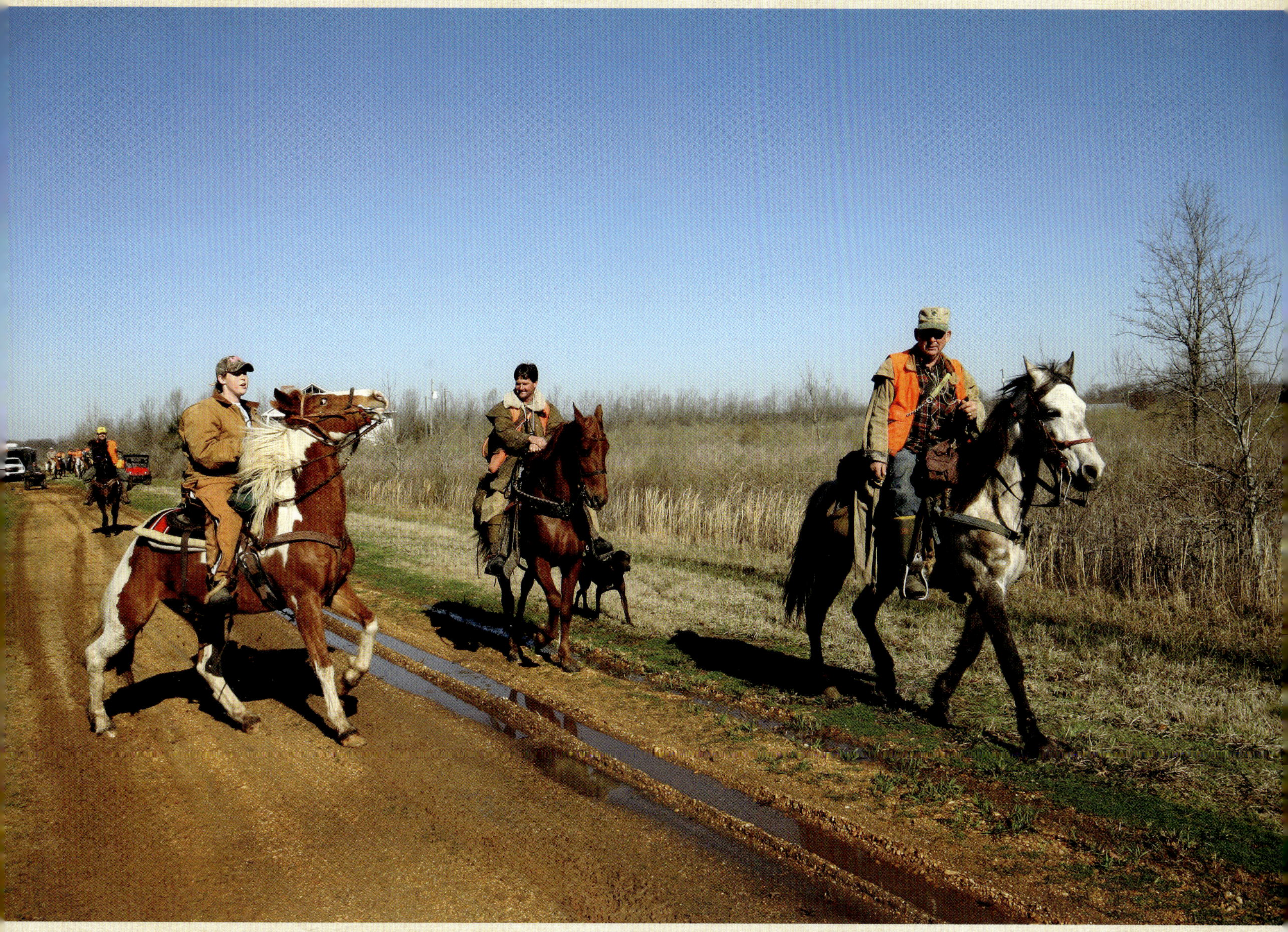

Jennifer Weeks, John Lowery, and Melvin (Bubba) Weeks.

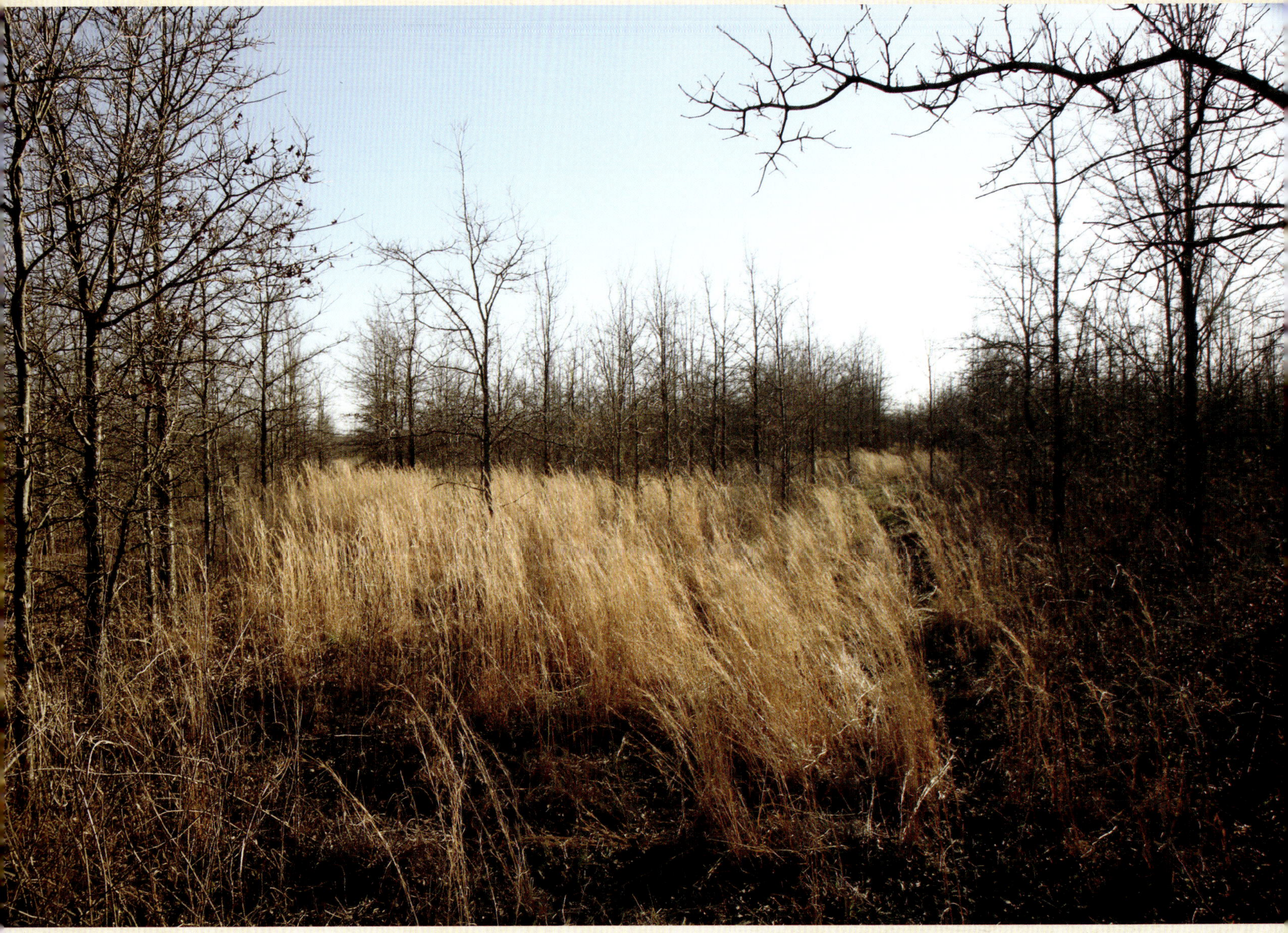

Sage grass.

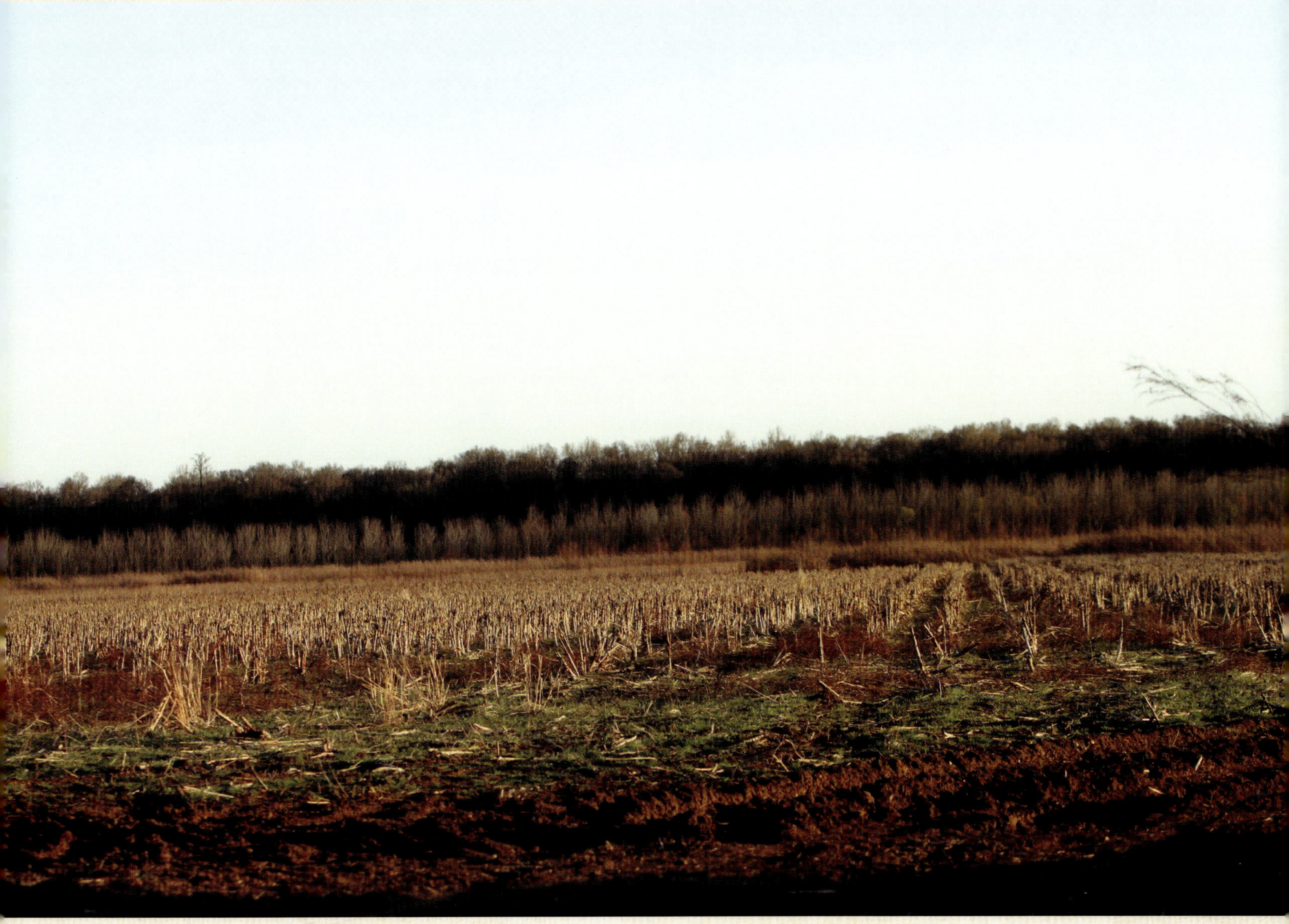

Harvested corn crop.

Panther Tract

HANK BURDINE

The Mississippi Delta was once a primeval swamp. It abounded with wild panthers, bears, deer, alligators, bobcats, raccoons, monstrous rattlesnakes, and water moccasins. Its lakes were filled with huge catfish, alligator garfish, bass, bream, and crappie. The virgin forests were home to gigantic oak, gum, elm, hackberry, cottonwood, sycamore, and cypress trees. Canebrakes were found along waterways and sloughs so dense and thick that a man could walk into them only by hacking his way with the help of a heavy bush knife.

At the lower end of the Yazoo–Mississippi River alluvial floodplain that we call the Delta lies a private hunting club known as Panther Tract. This forty-five-hundred-acre tract of land is protected from annual floods by the confining levees of the Mississippi and Yazoo rivers and the Will Whittington Diversion Channel levee system. In the heart of the lower Delta, and adjacent to Panther Tract, are the Panther Swamp National Wildlife Refuge and the Delta National Forest. These two contiguous federally protected refuges comprise thousands of acres of pristine wildlife habitat. Nowhere in the state of Mississippi is a more concentrated area dedicated to the preservation of wildlife and conservation of the natural ecosystem.

At the beginning of the last century very little of the Delta had been cleared. The lower part of the Delta was annually ravaged by backwater flooding from the Mississippi River and was the last to see the ax and crosscut saw. Those acres that were cleared for timber and were unsuitable for the growth of highland cotton were soon lost back to the counties and state for nonpayment of taxes, and the land grew back up in trees. As the levee systems became more proficient at containing floods and the price of soybeans rose, more and more virgin and second-growth timberland was bought to be cleared and turned into farmland. The south Delta was ready to be cleared again.

In the early sixties, John and Dick Twist from Twist, Arkansas, bought about fifteen thousand acres of land south of Highway 149 and west of Yazoo City. The land was due north of and adjacent to Panther Swamp. The Will Whittington Diversion Channel split the two brothers' properties. Once the land was cleared and farming operations began, the price of soybeans fell, and the farm was put on the market for sale. The larger portion of the farm, Big Twist, was eventually sold to the federal government for mitigation purposes and included in the Panther Swamp Refuge. In 1978, Howard Brent and his brother, Lea, bought Little Twist, east of the channel, and continued the farming operation. In 1990, the Brents sold their towing company in Greenville, and Howard acquired Lea's share of the farm and turned the area into a first-class hunting preserve. Historically, the area was home to the wild panther and, having seen a locally fired brick with the print of a panther's track in it, Howard decided to name his place Panther Tract in a play on words. Amidst rumors of sightings of panthers, there was a definite resurgence of the native black bear population in the area and Howard and Lea even saw one while flying over the farm one day.

Howard immediately began enrolling parts of the place in CRP and Conservation Easements. Trees were replanted that had only recently been cut down. Access roads were built to allow efficient management and travel to all parts of the farm. Bayous and sloughs were dammed up, water control structures installed, and impoundments were built to provide habitat for overwintering waterfowl. Howard's longtime best friend, Charlie King, moved down to the farm to be the caretaker and manage the crops and wildlife habitat. The old Erickson house, circa 1896, was bought and moved from nearby Wolf Lake. A large solid-brick first floor was built containing sleeping quarters and the huge Victorian two-story house was jacked up and placed on top. Howard's lovely wife, Carol, often commented, "I thought when Howard retired we would get a place in the south of France. What I got was a hunting camp in Panther Swamp!"

It wasn't long before the ducks started showing up and the deer began migrating from Panther Swamp to take up residence in the tall grasses and young trees profusely growing in the fertile soil. Sunflower fields were planted for doves and ample food plots were left for the game to enjoy. Howard spent many weekends hosting private hunts and inviting guests from all over to join him. Soon, commercial hunts were booked and the trees continued to grow. The bucks got bigger and the flocks of ducks grew denser and more numerous. People had to be careful when they hunted, as the rattlesnake population also grew. But things were soon to change.

In 1998, Charlie King noticed that one ten-acre field of corn he had planted for the deer and ducks had been destroyed. All the corn was eaten. He knew that raccoons were often seen on the place, but ten acres of corn was a little too much for the coons to eat. He then noticed the road to the back of the place had been rooted up. Hogs had moved in.

As farm pigs had escaped or been turned loose to grow in the deep woods, the hog population had exploded in the Panther Swamp and Delta National refuges. Panther Tract stood as a smorgasbord for the marauding, vicious, ravaging, and hungry feral hogs. They destroyed and ate everything they came across. One delicacy for the wild pig is the rattlesnake, often found while rooting around in the dirt looking for something to eat. So while the resident rattlesnake population was on the decline, Panther Tract was being destroyed by wild hogs. Something had to be done.

Never one to back away from a challenge, Howard called his friend Wick Eatherly and, after he had made a visit to see the damage to the property, the hog hunt at Panther Tract was on. Hog hunters are a fairly diverse group. Some enjoy the social aspect of the hunt, some relish the chase on horseback, while others look to scramble in for the kill with nothing but a heavy knife or large-caliber pistol. All true hog hunters are rabid about their chosen sport and their horses and especially their dogs. So intense are they that some refuse to hunt with other men that have a different style of hunting or types of dogs. Each hog hunter, of course, has the best dogs in the state and none can match them. They refuse to let other dogs mingle in with their packs nor do they allow anyone else to take charge of the hunt and try to tell the dogs what to do. If you are invited to go along on a hunt, do as you are told and no more. It is a very dangerous sport and explicit protocol is followed, no matter how primitive or rancorous it may seem.

It is not unusual to see a man take off his boots, strip down to his long johns, and swim across a deep slough to kill a hog with his knife that the dogs have bayed on the other side. Then he will swim back across with the hog tied to the end of a rope, redress in the shivering cold, mount up, and return to the hunt. I have seen lady hunters on horseback get lost in the swamp at Panther Tract with no one to worry. "Don't worry about them, they are big gals. They can find their way back to the house." And the hunt goes on

My Friend, Howard Brent

SAM POLLES

Comedian Jerry Clower once was asked, prior to a show, if he needed any special equipment or props for an upcoming performance. Clower replied, "All I need is a microphone." Howard Brent shares many attributes with the late Mr. Clower, but instead of needing a microphone to be able to perform, all Howard Brent needs is people.

Ever since that first group of cavemen took up their clubs to go hunt for food as a group, hunting has been a social affair. Like his primitive ancestors, Howard Brent enjoys the fruits of his hunting, but his real passion is the camaraderie and social aspects of our hunting tradition. Indeed, it seems that Howard has made it his life's mission to promote hunting as an opportunity to socialize with old friends and make new ones. Whether it is an early fall dove hunt or hunting hogs with dogs from horseback, for Howard the hunt itself is merely the kickoff. The real fun begins with his award-winning "half-time" shows and postgame celebrations.

Yet, don't think that Howard is one-dimensional, for in fact, he is a combination of many various talents. He is a conservationist, storyteller, host, and friend all rolled into a package of 100 percent pure fun. Depending on the activity or season, either the beginning or the end of a particular hunt signals the start of a really great barbecue, with music and an all-around good time. More than just the life of the party, though, Howard is a conservationist. He has spearheaded many conservation efforts such as water/flood control for waterfowl habitat, and has committed his own lands and resources to such projects. His own beloved Panther Tract is forty-four hundred acres adjoining the Panther Swamp National Wildlife Refuge. Howard has spent countless hours restoring this farmland to its original state of bottomland hardwoods and wetlands. You see, Howard Brent was "green" before "green" was cool.

Recently, hog hunting has taken off as one of Howard's new favorite pastimes. Howard can usually be found on horseback (and alongside a pretty lady), "riding to the hounds," as they pursue the porkers. Yes, hunting hogs is becoming more popular, and for Howard, what better reason can there be for yet another party.

Howard Brent will never be accused of not living life to the fullest. Neither will his passion and dedication to conserving our precious natural resources ever be questioned. To know Howard is to know that his candle always burns at both ends; however, the true enjoyment is being able to share in the warmth and light that candle provides. One of my greatest pleasures in life has been having the opportunity to know Howard Brent, and being able to call him my friend.

Panther Tract Stew

HOWARD BRENT

I have been told that the Great Flood of 1927 wiped out all of the big game in the Mississippi Delta. By the 1950s laws were passed, such as making it illegal to kill does, and game began to return. The herds of deer grew, and then by the late 1970s and 1980s the biologists recommended taking one doe for every buck harvested and also began aging the bucks. This turned out to be a good thing.

After the '27 flood, the U.S. Army Corps of Engineers made cutoffs where the river coiled into horseshoe bends. Forming almost a "U" shape, the cutoffs closed one end of the bend. The intention was to reduce the water level in high water. It is estimated that the gauge at Greenville would have been twelve feet higher during the flood of 1973 had these cutoffs not been undertaken.

My first hunting was at Worthington Island in 1957 when Elia Kazan and Karl Malden were in the Delta making the movie *Baby Doll* and staying at the Greenville Hotel. My father invited them to deer hunt. When my brother, Lea, offered his shotgun to Mr. Malden, he looked at it with hesitation. Lea reminded him that in his last movie, Malden had used an army carbine and that Capt. George Reid was willing to lend him one of those to use. Lea put a relaxed Malden on a stand where he got a big deer. This two-thousand-acre tract sold in 1959 for twenty-five dollars per acre, while today it would likely fetch three thousand dollars per acre.

After it sold, we leased Spanish Moss Point, and Lea and Daddy also joined Ashbrook Hunting Club where there were wild hogs. We decided to have a hog hunt and invited some of our industry friends down—Jack Geary with Ohio River Towing Co., Barry Hyken with Inland Oil Co., and Tony Tobin with St. Louis Shipbuilding. Barry decided he wanted to ride a horse, so we saddled one up, and Barry asked Malcolm Lowe, "Can you shoot off this one?" His reply was "hell, yeah, you can!"

When Barry took his first shot, the horse threw him so high that birds were making a nest in his back end before he hit the ground. After walking back to camp, he confronted Malcolm: "I thought you said you could shoot off of that horse."

"You can," Malcolm replied. "You just can't shoot *on* him!"

The only pain medicine we had at camp was moonshine, so I gave Barry a shot and then took a couple of sympathy shots myself since men don't like to drink alone. I took off to catch up with Lea. The year before, he and I had been hunting on horseback and Lea had a Browning automatic in the saddle scabbard when we stopped to water the horses. We had just put the reins over their necks and turned them loose when Lea's horse came out of the water, rolling on the ground and bending the barrel of that shotgun. After the bent barrel was sawed about twenty inches, it made a darn good "jump shotgun."

We were riding about thirty yards apart when Lea spotted a big boar hog. I didn't see the hog, but when I heard the shot, I looked around and his horse started bucking. Lea pointed the shotgun straight up in the air, but every time he came down out of the sky, the butt of the gun hit his leg, and with his finger still on the trigger, it was going off. I knew he was loaded with double O buckshot, but I didn't know which way the next shot was pointed, with the rodeo we had going, so I jumped off my horse, grabbed the saddle horn with one hand and the bridle just below the bit with the other, and was dancing around, keeping my horse between me and the shotgun. After the last blast, the horses finally settled down, and we started looking for blood.

We searched for about forty-five minutes and I told Lea he missed the last four shots, since I was watching, and suggested we head back to camp. I was hungry. Lea had cleaned his freezer out and trimmed off some of the freezer-burned meat and put it in a white garbage bag which he intended to feed the hunting dogs. We had other groceries we had brought over and early that morning had told the cook where all of it was before we saddled up and left. By the time we got back to camp, everybody was already eating except Jerry Brown, who was with the Greenville Police Department and always brought his hunting dogs to camp. Jerry was in the chow line and I stepped in right behind him. As he dipped into this "working man's stew," I asked the cook, "What happened to the freezer-burned meat we brought over for the dogs?" The cook looked surprised and then sheepish, pointing over to the big pot of stew. Jerry was about to pour a ladleful into his bowl when he saw the cook pointing. His face turned green and he eased the ladle back into the pot. I checked the rest of the hunters, some with empty bowls and some still digging in, so I got a bowlful. When you're real hungry, "freezer-burnt stew ain't bad."

A Family Affair

DAVID WEEKS

I was born down in Carroll County by Greenwood. I've been hunting since I was nine months old. When I was a baby, Mama had Melvin and Tommy, and Bubba was a blue baby—nowadays they call them preemies—and she couldn't take care of Tommy, Bubba, and me, so Daddy just put me in his hunting sack and carried me coon hunting. We're all one year apart—it was five of us. I remember going hunting in Boils Lake Swamp when I was four, that I can remember. We didn't have but one little ole horse and a couple of crazy mules. We sharecropped for a man named Mr. Ben Pennington at Boils Lake Swamp, and over there on Opossom Bayou for Barclay Hicks, Mr. Ben's son-in-law, and we sharecropped up in those hills in Carroll County for Mr. Earnie Meeks.

When I was young I hunted with my daddy. We'd hunt at night and ride the horse into Boils Lake Swamp and, well, the dogs they'd bay up out in the swamps and lakes and the slough and Daddy had a little ole dog named Poochie and I don't know how Daddy did it to this day, but we'd build a fire and wait till daylight. He'd send that little dog Poochie out, and he'd make that hog run right over the top of Daddy. We'd field dress the hog right there and then salt it down and put it in the box and smoke it. We had a smokehouse. They'd cure it out in the smoke. Daddy liked black walnuts and if he didn't have any he'd use hickory. Daddy didn't put the walnuts in the fire but us kids did. We had a big ole walnut tree up in our yard in Carroll County and when those walnuts would fall off that tree, we'd wait until they turned black and we'd get out there and hit that ole black hull with a rock or a hammer and peel it off and then put it in the firebox in the smokehouse. Then we killed the hogs, back when they had hog killing days for everybody, and the neighbors would all get together one day and we'd kill a hog for this family and the next day we'd kill a hog for the next family and so on. A family would kill 'em and salt 'em down for meat for the wintertime. We'd kill four or five hogs to carry our family through the winter. In those days, almost everybody raised their own food. If somebody was running low, and somebody else had a little extra you might do a little trading—hell, you might trade a dog for a hind quarter.

That was way back there in 1954–55. The wild hogs were different back then. Back then, we'd go hog hunting just to get meat to eat, down at Brazil, down by Webb and Charleston. They didn't have no sho nuff wild hogs back then. It was domestic hogs that had done gone wild because they got turned loose. People got tired of feeding them. They got this "open stock law" in Arkansas and Louisiana where your stock is registered up at the courthouse and they'd let 'em go. That way they'd get their crops in and their stock be safe. They'd cut their initials in their ear or hock their tail a certain way that showed it was theirs. The pure Russian hogs that are here now are because people brought 'em in here. Mr. Vernon Jones and Mr. Bubba Cunningham.

Hunting hogs is more exciting now than back then. You get that adrenaline rush now. I caught a hog down in the Homochitto down in a wild lemon thicket—I crawled up in there to get that ole boy.

That wild hog that you kill, you drag him off in some water and let him stay there for fifteen to twenty minutes and let him cool off to keep that musk from going through him. If you don't prepare him like that out in the woods, when you cook it, you are going to smell it.

I had a dog, one time, named Hemorrhoid because he was a pain in the butt. Well, when he was younger he knew I didn't like to run no deer, but if a deer hopped up, he was going to have to run him a couple of hundred yards and he knew I was going to get on his butt, so he'd come back and get in front of my horse and lie down. I'd get off my horse and hit him in the head and call him a son of a bitch and he was just as happy as a lark then.

I was nose to nose with a hog once. I was crawling up under some briars to the cutover and I crawled up against a little log lying there and when I raised my head up to look over that log, that hog was standing there on the other side of that log just smacking his chops at me. He had long tusks and the only thing that saved me—it was the summertime and you couldn't have no gun so I reached for my knife and I done lost the knife. He took one step backwards—they'll take one step backwards before they charge. I had a little dog named Copper. He came running by that hog. And when that dog ran by him that hog turned on him and cut him. I mean cut him bad. That's the only thing that saved me was that little dog Copper.

The dogs love to go hunting. That's what they do. That's all they do. They love to hunt. Back when I was living down by the river, before I got run off, I had seventeen dogs. Now, I have five.

I've lost dogs to alligators. Those alligators will get that dog when he crosses the water after the hog. That alligator—if he's a big one will get whatever he wants—he'll wait there by that deer crossing. Everything in those woods with four feet is going to use that deer crossing and that's where that big alligator will be. If you see a big gator he's gonna be at that crossing. I've had alligators come after me. I was wading in a slough, frog hunting up off Highway 12 near Belzoni and Thornton over there and it was my twenty-sixth frog I'd gigged and five or six of those big ole frogs would be sitting there on those water lilies and I'd come over and gig 'em. Well, I just kept getting a little further out and a little further out and I tore up this little ole open hole where the four or five big ole frogs were and I got three of 'em and put 'em in that tracker I got. One of those frogs started making a racket like rrrraaaackkkk—rrraaaaccckkkk—rrrraaaaccckkkk and those big ole bushes started to rising down. That big ole sow gator had a nest in there and she came out of there and I started moving back as far as I could. I couldn't turn my back on her—she just kept coming and I had a little .22 pump rifle on a hay string over my shoulder and I came off that rifle and she opened her mouth and went hhhhhaaaaaaaaaahhhhhh and blowed like that and turned around and went back and that's why I know it was the twenty-sixth frog 'cause that was the last one I gigged. There's more alligators now, because they protect 'em.

I used to carry my horse in the back of a pickup truck. I didn't tie him. I'd just put him in there and go where I wanted to go and let the tailgate down and tell him to get out. When I'd go around curves, he'd lean with the curve. He'd put all four feet in the four corners. Sometimes if it was real cold I'd put a croker sack around my horse's head to keep him from getting cold when I was carrying him down the road. I've hauled horses around four states in the back of a pickup truck. I had a Chevrolet Love and a Toyota and a Nissan and a Ford. I had two horse trailers and used them if I had company. If I didn't have company, I'd just put 'em up in the back of the truck.

I didn't go crazy over hog hunting. I still like to deer hunt, coon hunt, and squirrel hunt. But my brother Melvin, when he started hog hunting he quit everything but hog hunting. He don't let his dogs hunt. He does the hunting for them. He gets in a hurry and I don't. I just ease along. I got dogs to do my hunting.

I know the Delta National Forest better than any man alive. Ain't too many trees I hadn't said "good morning" or "good night" to.

My daddy he went barefooted all his life and hunted barefooted too. He was born and raised here in Mississippi up there in Winston County. He'd go hunting barefooted with the snow a foot deep and his pants legs rolled up. Kicked briars and bushes all day long and came in at night with thorns all in his feet and he'd soak em in a pan of bleach, Clorox. Then, he'd cock em up on the edge of the foot of the bed and tell Mama to get them thorns out of his feet and she'd get that ole knife and start plucking those thorns and it sounded like flicking something out of an old board, that hide was so tough and thick. Us kids would have croker sacks wrapped all around our legs up to our knees, 'cause we didn't have shoes neither, but Daddy'd wear nothing.

I squirrel hunted barefoot one time and I got up on that horse and got blood poisoning. I figured out that you don't ride a horse with bare toes showing—the soft briars will get you every time right between the toes!

I hunted barefoot most of the time, even in the wintertime. I don't like shoes. They burn my feet up—wearing these Crocs, my feet are sweating now! I just don't like shoes. My mama wore shoes. They weren't worth much, just ole work boots. She had some slippers she wore when she and Daddy went out dancing. Daddy called it "sliding the foot."

I ride a gaited horse. She's twenty-five years old now—born on the twenty-sixth day of April 1983. She can carry you up the road so smooth it'll make your butt laugh.

A boy from West Virginia came up to squirrel hunt with a boy from Michigan and they asked me to take them hog hunting. They came back four years in a row to hog hunt. The last time he came back he had a .30 caliber pistol. Hell, I've never seen one. He said, "Here, you need this more than I do," and he gave it to me. It'll hurt your ears when you shoot it. It's a whole lot louder than that .30-06. I shot it off my horse, April, and she just shook her head. It hurt her ears too. Ain't got but a nine-inch barrel on it.

I don't waste any meat. If I kill it, I eat it. I got nine rattlesnake hides in my freezer right now. It's wrong to just kill stuff to kill it. There ought to be a law against it.

Two or three weeks before Christmas, I'd load up the back of my pickup truck with hog meat and go down the road and tell people, "Come on, get what you want." We don't waste any meat.

We, my brothers and me, started picking cotton when we were three years old. When you were three they put a flour sack over your shoulder. When you were four, you got a one-hundred-pound gunny sack. When you got to be five or six you converted up to a seven-and-a-half-foot cotton sack and by the time you were nine you were pulling a nine-foot cotton sack. Me and my brothers and my mama picked two bales of cotton a week and had to finish by Saturday 'cause Mama didn't pick cotton on Saturday and Daddy gave us a choice—"Boys, if you get that bale of cotton finished by 12:00 you can go to the gin with me. If you don't, I'll beat your brains out." So, we had a good choice.

I could pick three hundred to four hundred pounds of cotton a day. So, hunting was like a vacation for me. It was a real treat for me after working all day long in the fields. We missed the first six weeks of school because we had to pick cotton in the fall. We needed the money. Back then when we got to be ten or twelve years old we got fifty cents a day for picking chunks of wood out of the fields from them clearing the fields of woods.

I miss the good ole days sometimes. But my brothers say—the hell with the good ole days!

Joe Frank Carson and I went hog hunting and he couldn't get his horse to cooperate so I rode my horse up in a briar bed and there was a pretty good-sized hog up in there. I didn't want my dogs to get cut up so I pulled my .22 out and shot him twice and he hit the ground. The dogs ran in there with the hog and about that time Joe ran up beside me. That hog wasn't dead and looked straight at Joe Frank and started to charge him! I shot him one more time in the head and killed him. Joe Frank said, "I bet I don't come up in a briar bed again without my gun!" He was just shaking! It was really scary for him, but funny as hell to me.

I had a hog to chase me around a tree. I was shooting at him by holding my gun behind my back and pointing it at him because I was running and I shot at him six times! Around and around the tree we went, with him trying to cut me in the butt. I was using a .22 Ruger. A little ole automatic. You pretty well got to hit 'em between the eyes to kill them.

I used to be a first mate on a towboat and also a tree surgeon.

My First Hog Hunt

MARY HARDWICK

My first hog hunt at Howard Brent's Panther Tract was an experience from another era! We arrived on a stormy morning, hoping for the skies to clear. The skies did cooperate and we celebrated with many Bloody Marys and hors d'oeuvres on the porch, pretending to be grand dames from a time gone by.

The first inkling of what the day was to bring came when the master of hog hunting, who was to orchestrate the hunt, drove up in a truck that resembled a plumber's truck. This leader for the day looked like someone from the movie *Deliverance*—overalls, craggy face, missing teeth, and a big scary smile. We peeked inside the back of his truck and were amazed to see eight or nine horses and about a dozen dogs all crammed in there together. We glanced over at our sleek aluminum slant-load trailer that had a spot for every horse and a hook for every bit of tack, and we knew that we were in for an adventure. We asked him how he got all those animals in the back of that truck (about three feet high with no loading ramp) and he replied, "I tell 'em to get up in there! No big deal."

The hunting began with a flurry of excitement, dogs barking and running everywhere, riders and horses following. It was amazing to see the dogs work—they would flush out a hog from the brush and encircle it, holding the hog until the riders arrived. You have never heard such a commotion—growling, snarling, barking, and squealing—as the dogs circled the hog and held it at bay. We watched incredulously as one of the men then leaped into the middle of the fray and stabbed the hog with a knife about as long as my arm! This scene was repeated several times during the day as we hunted over many acres. The climax to the day came as the dogs rooted out a huge boar. This hog was gigantic and really ferocious! The dogs snarled and growled and snapped, but they could not contain this big beast. He kept running, turning to fight, and running some more, until he came to the borrow pit. He just kept running and jumped into the river. I didn't even know that hogs could swim! The dogs never hesitated—they jumped into the river in hot pursuit of the hog. Some of the riders had shotguns and they began shooting at the hog from the top of the bluff where we watched. Our leader began screaming at them, "Cease fire, cease fire," for fear that they would shoot his prized hog dogs. I was thinking that this hog had won the battle when this hunter showed why he is considered the master of hog hunting. He suddenly jumped off his horse, pulled off his coat and boots, put his huge knife in his mouth, and jumped in the river. Holy cow, I have never seen anything like it! He swam Tarzan-style in that freezing cold water after that hog and his dogs! Well, I guess that swim tired out that old hog, because when he dragged out on the opposite bank, the dogs were able to follow him out and contain him. Then here came Tarzan dragging out of the water. And it was cold! All of a sudden he jumped into the middle of all that snarling and fighting and stabbed that hog with his huge knife. We were all watching from the high bluff across the river; our mouths were hanging open and none of us could believe our eyes. He proceeded to gut the hog, and then dragged it into the river. He swam back across the river with that big knife in his mouth, pushing the hog ahead of him through the water. His dogs all swam happily alongside him. When he got to the bank on our side of the river, he and his buddies tied a rope on the hog, hitched the rope to his saddle, and had his horse back up until the horse pulled that hog up the steep bluff. When the hog was finally at the top of the bluff, the cheers went up. One of the men pulled out a bottle of some amber-colored liquor, passed the bottle around to all who would partake, and declared the day a success. What a day to remember!

Hogs, Mules, and Horses

MELVIN (BUBBA) WEEKS

The Mule Story

It was one Thanksgiving and we were down there going hunting on Francis Domino's place and he told us to go kill us some deer off of his place. It's a small place and so I told my brothers when we got there we had to be real quiet so we didn't scare the deer off.

My brother had a little three-hundred-to-four-hundred-pound mule he took along. When we got there we were walking along and I crossed a big ditch and so did my other brother and he climbed up in a tree on the other side of the ditch. I kept walking along. Well, my brother with the mule started to cross the ditch and the mule didn't want to go. So he got off and had to lead the mule down and across. When they got almost to the top my brother got back on the mule and then the mule threw him off and back down into the ditch. It made my brother so mad that he went to shooting that mule with his .30 caliber carbine. He shot him thirteen times and killed him dead.

I heard the shots and went back to where my brothers were. I asked, "What's going on? Did you get a deer?" My brother in the tree said, "Bubba, he just killed his own damn mule! It throwed him in six inches of water down the ditch and he went plumb out of sight!"

I said, "What you gonna do now?" He said, "We gonna cut him up and eat him! You ain't killed no deer, have you? Well, we will eat the mule!"

It took two eleven-hundred-pound horses to pull that mule out of that ditch. We skinned and field dressed him right there, then put him in the trailer with the two horses and headed for home. A man saw us going down the road with what he thought was a black deer. He ran right off the road seeing it!

We took it to my mama's house and hung it up in the front of the house. Isola people thought we'd killed a big black deer too! Some people asked us, "How many horns did it have?" We said, "Thirteen points!" Then they asked us where the head was. "We took it to the dentist's office and left it in the deep freeze there so it wouldn't ruin!" (We made this up, of course.) So, right there in my mama's yard we skinned the mule up and quartered it and had to put it in six different deep freezers! My mama marked each wrapped piece "David's Ass."

Well, we had this friend who came down and his daddy-in-law was a preacher man. He asked us for some deer meat and we didn't have any but we told him we had some mule and he was welcome to that if he wanted it. He said he'd had some in the army and it was good. So, he took three or four of the mule steaks home and fed them to his wife and his in-laws and of course himself. The boy's conscience started to bother him and so he felt like he had to tell them he fed them mule meat. "The Weeks brothers gave me some of that mule meat and that's what we ate the other night." His father-in-law said he didn't care, that it was good! So, the boy then decided to tell his wife. He said "Honey, remember, when the Weeks boys from down at Isola killed their mule?" She said, "Yeah, I heard about that." He said, "Well, we ate some of that mule the other night." That woman got so mad! She said, "You son of a bitch! You fed my family mule! You fed my family *mule*!?!" She went to throwing clothes and went back home!

She was a beautiful blond-headed woman!

That boy said to my brother, "David, you did me a big favor giving me that mule meat! I'd rather her leave now than later!"

My Horse Joe

I have a cookie truck—you know, the kind they sell cookies out of for quick stops—that I take my horses and dogs in when I go hunt hogs. I put them all up in there together. Sometimes three or four horses and ten or twelve dogs, in there together. Well, my stallion Joe was a very smart horse. He would ride in the middle of the back of the truck and he would stick his head through the middle of the two front seats when I was driving along. That's how he rode. He knew where we were going and when we got close to home he'd start to whinnying. I'd say, "Shut your damn mouth, Joe!" but he just kept on whinnying! He knew where he was all of the time!

Hog Hunting

I've been hog hunting for twenty years. My daddy started us out, my brothers and me, when we were just kids. It's been sixteen years since I've been deer hunting. Ain't nothing else fascinates me but hog hunting. I don't do anything but hog hunt. I like the hogs and the dogs. I like the knife and the bulldog. That bulldog will help you out.

I like to go to these million-dollar hunting camps and hunt like the big boys do! I'd rather use a knife when I'm hog hunting. Some of the biggest hogs have been killed with the knife.

People who have the land don't want the hogs there because they are so destructive. They are a nuisance and everything, and we'll come kill them and if we kill them then we are going to eat them. We have a lot of fun. When you get a bunch of Mississippi rednecks together we are gonna get a hog. There's a lot of sportsmanship, horsemanship, and fellowship with hog hunting. Watching the guys and the dogs get the hog is a lot of fun.

Catching a hog is a lot like catching a coon. You make up your mind you're gonna get it. Doesn't matter which end, you just gotta let it roll. Get one and let it roll. The way of not getting your dog cut when hunting a hog is to get there quick, in the first couple of minutes. You get there when they bay the hogs.

We go out and look for where the hogs have been rooting the night before a hunt. The dogs smell where they have been and then when we hunt they walk them up. Early in the morning we'll track them four or five miles and usually find the whole herd, the big ones and the little ones, and then they start to split up and we usually get four or five hogs when we get a herd. We find the hog with the hound. I'd rather use the knife than the gun. It's a little more fun and a little more action. You grab them by the leg and then there you are. I hunt year-round for hogs. There's no season because the hog is not a game animal. It's livestock.

My Wildest Hog Hunting Story

My wildest hog hunting story has to be when a big ole sow bit my finger off.

I'd caught a big ole three-hundred-pound sow that had inch-and-a-half-long teeth. I went down to Holly Bluff to get something to drink after the hunt because it was hot. We had put the hog up on the dog box on the back of the truck. She was still alive because we were moving her to another place and we had tied her feet together. Well, when we got down to Holly Bluff, Joe Gambel was down there at the store and he asked me if she had any teeth and how big they were. He didn't believe she had any teeth because she was a sow. She was a big ole silver hog. Well, when he said she didn't have any teeth, I climbed up there and pulled her lip back with my finger. I guess she had gotten real mad by this time, being tied up and all, and she bit my finger plumb off!

I got down off the truck and went in the store and asked the lady behind the counter for some Band-Aids and some Dr. Tichener's. She said, "You got a cut?" I said, "Yeah!" and showed her my bit-off finger!

Hog Hunting in Silver Earrings

MARY JANE WOOTEN

Well, my motto is "Even if you don't know how to play the game, at least show up looking good." Dressed in black from head to toe—boots, chaps, shirt, saddle, rhinestone belt, down to my black bling headstall and breast collar—I was riding my gorgeous black show horse with titles out the wazoo, and yes, my silver earrings. Needless to say I have never been hunting, of any kind. So "clueless" would have been an understatement, except for the fact that I knew I could ride a horse well and enjoyed "fast and furious" from time to time. It never occurred to me I might come face-to-face with this prehistoric-looking animal and have to do something about it.

That afternoon, I decided to challenge myself to keep up with Melvin, our head tracker and point man. It wasn't long until the dogs were on the scent of the hogs and the ride got really fast across some of the toughest terrain. But I never left the right side of Melvin and before I knew it, we had left the others way behind and no one was in sight. All of a sudden we caught up with the dogs and they had a 275-pound hog bayed. Melvin jumped off and was holding the hind end of the hog when he hollered at me to "get down quick, grab my knife out of the holster and stab the hog." I wasn't quite sure that was a good idea. What if the hog started to chase me? What would I do then? Melvin hollered at me again, "Hurry, hurry." So I jumped down, grabbed his knife from his holster and tried to stab the hog. Nothing happened. It was like the hog had a coat of armor on. Melvin hollered at me again, "Under the arm, under the arm!" I tried one more time and bingo, I slayed the giant beast. Finally the rest of the hunters caught up with us just in time to load up our meal. I look back, and although I will always ride on the right side of Melvin anywhere, I hope I am not the only one.

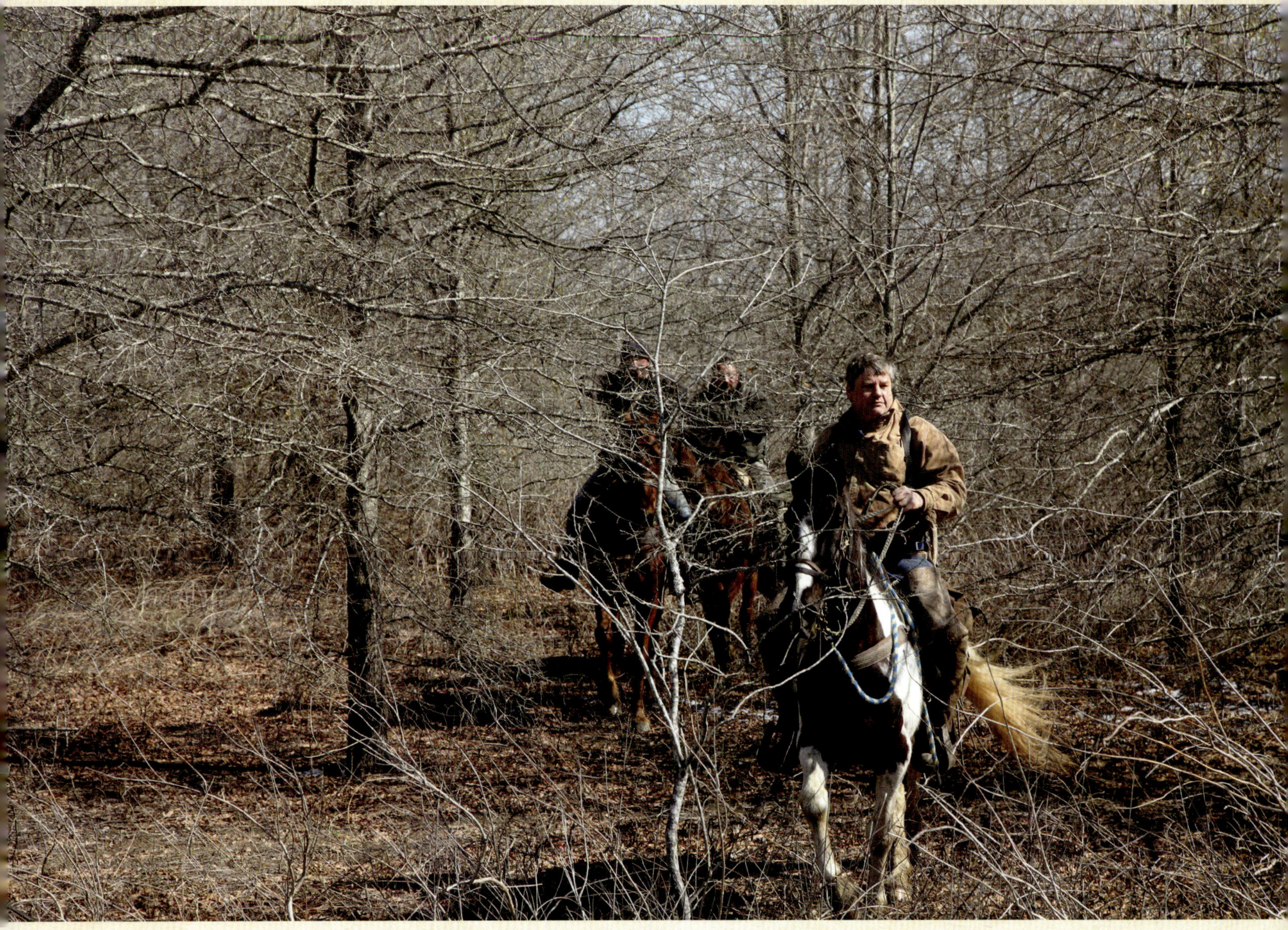

Steve Ratcliff and Chief and Jeff Cremeen.

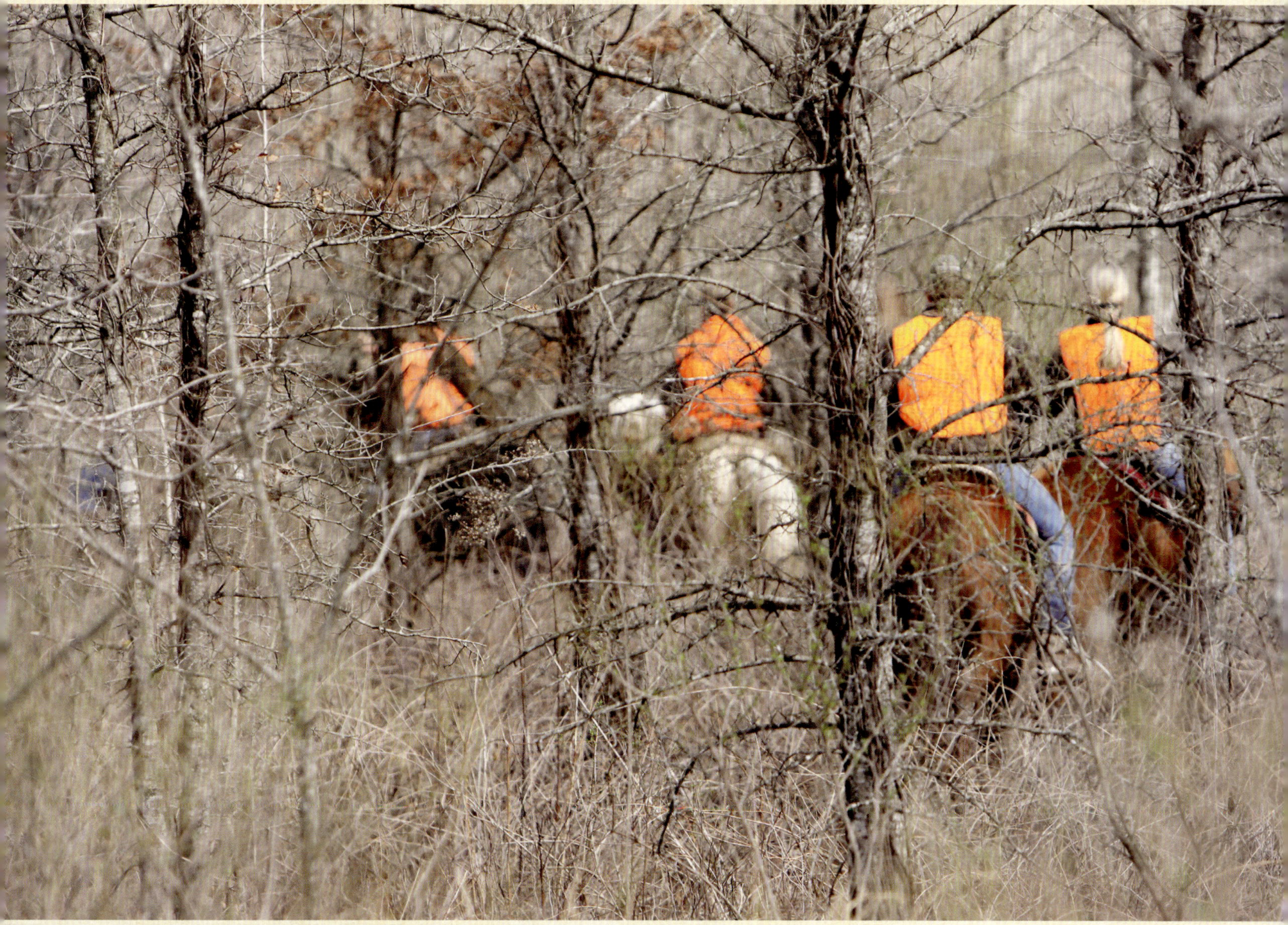

Hunting on horseback through the trees.

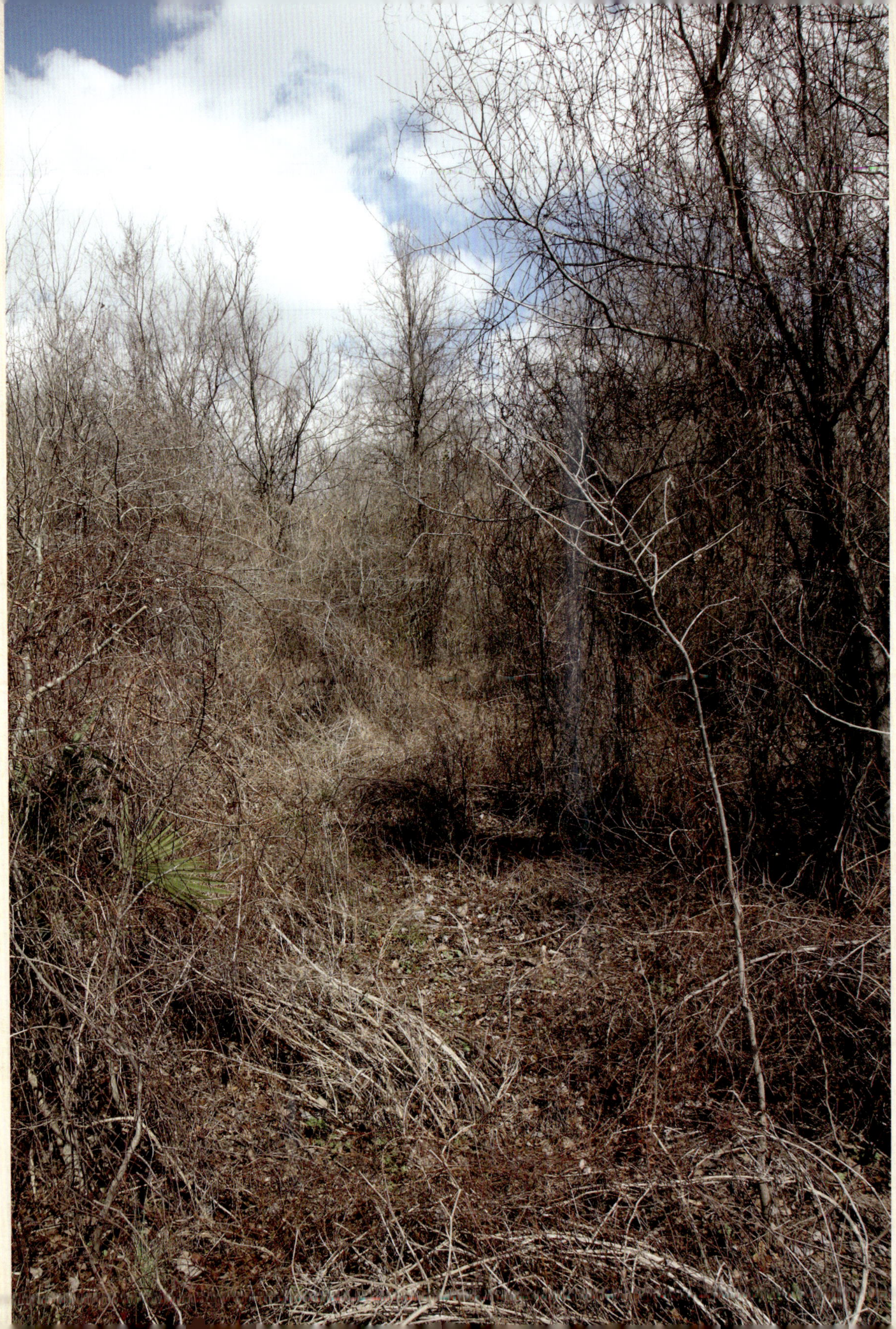

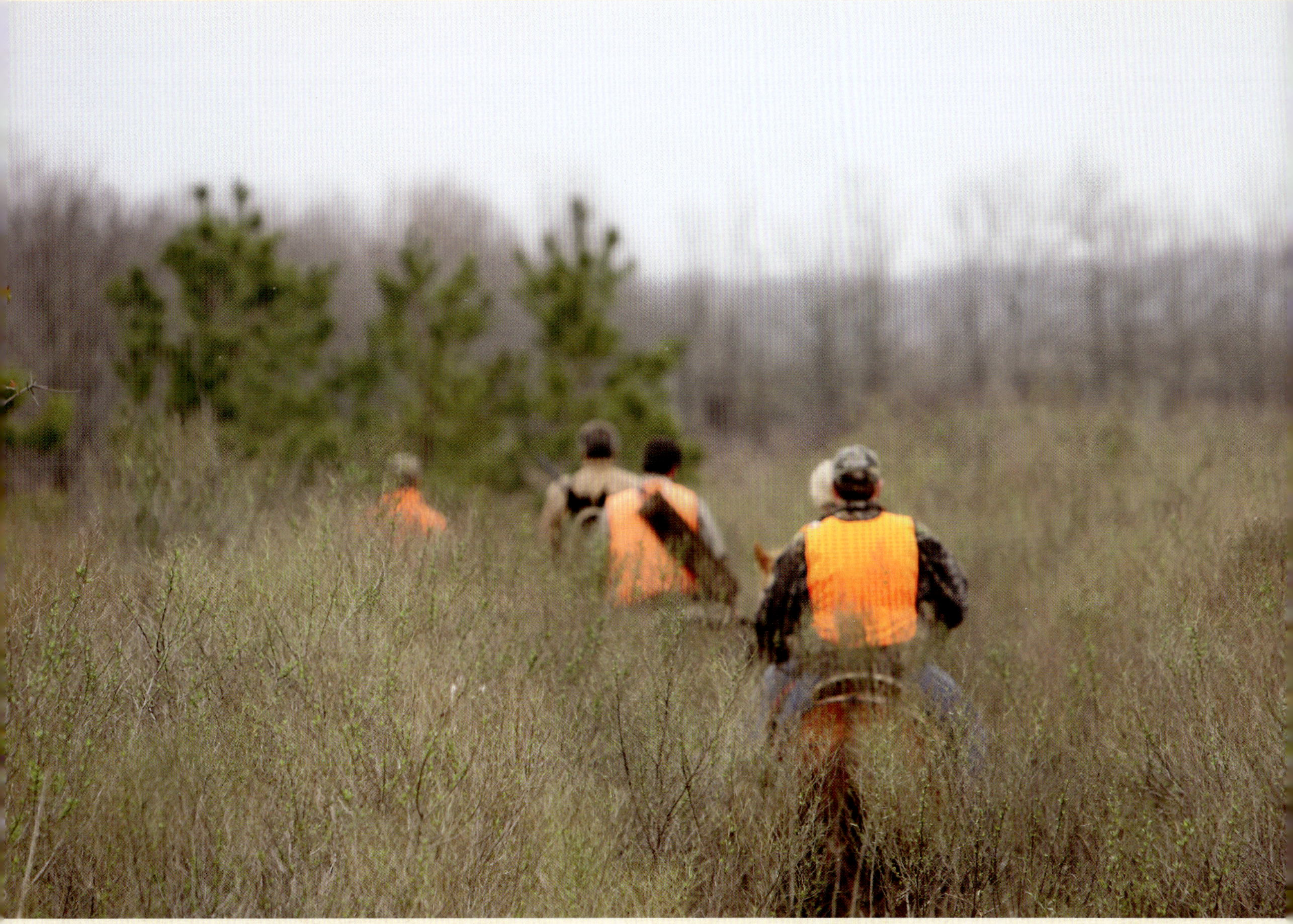

Tall brush.

‹ Road through planted trees for conservation.

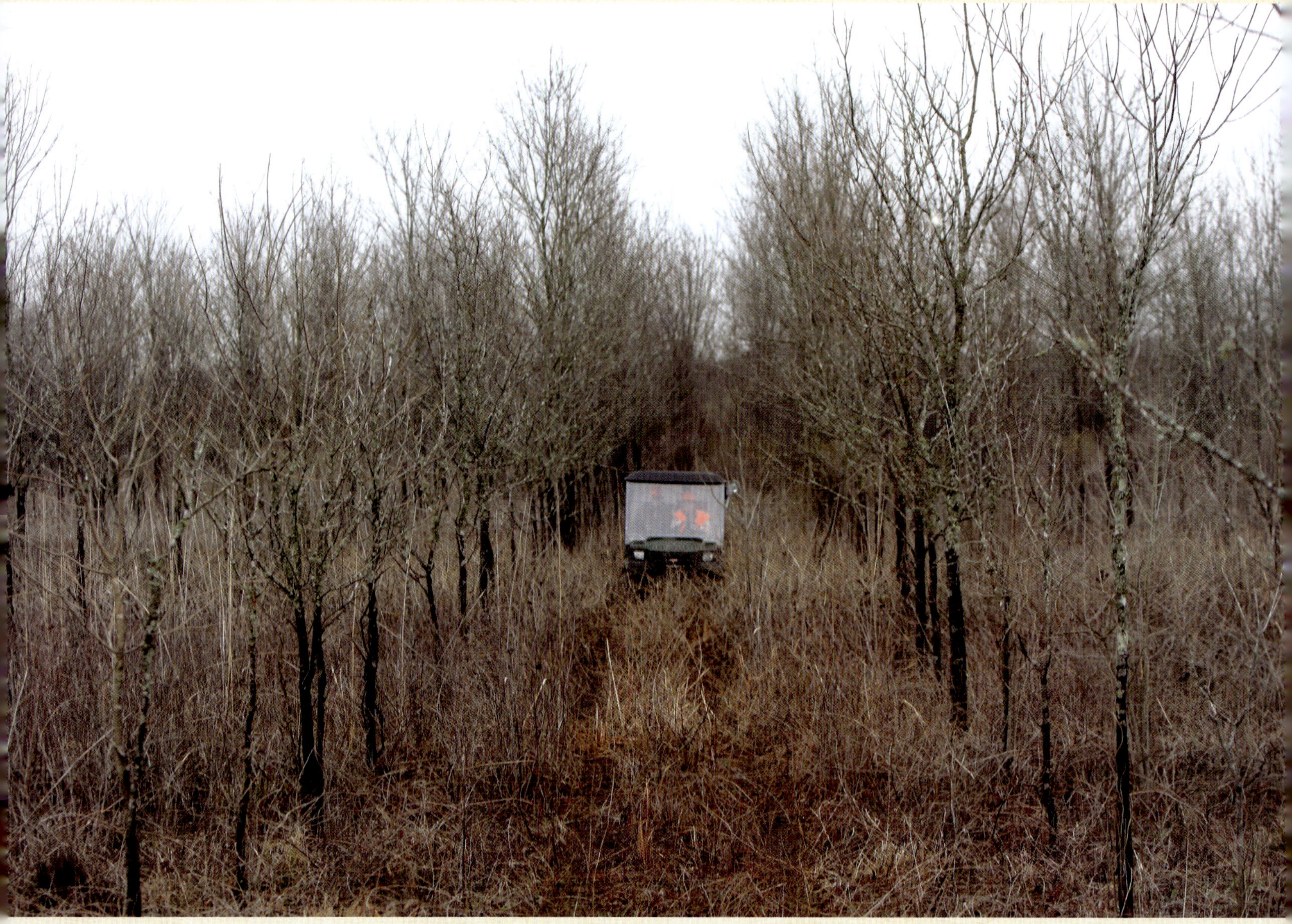

Hunting in a Ranger through planted hardwoods.

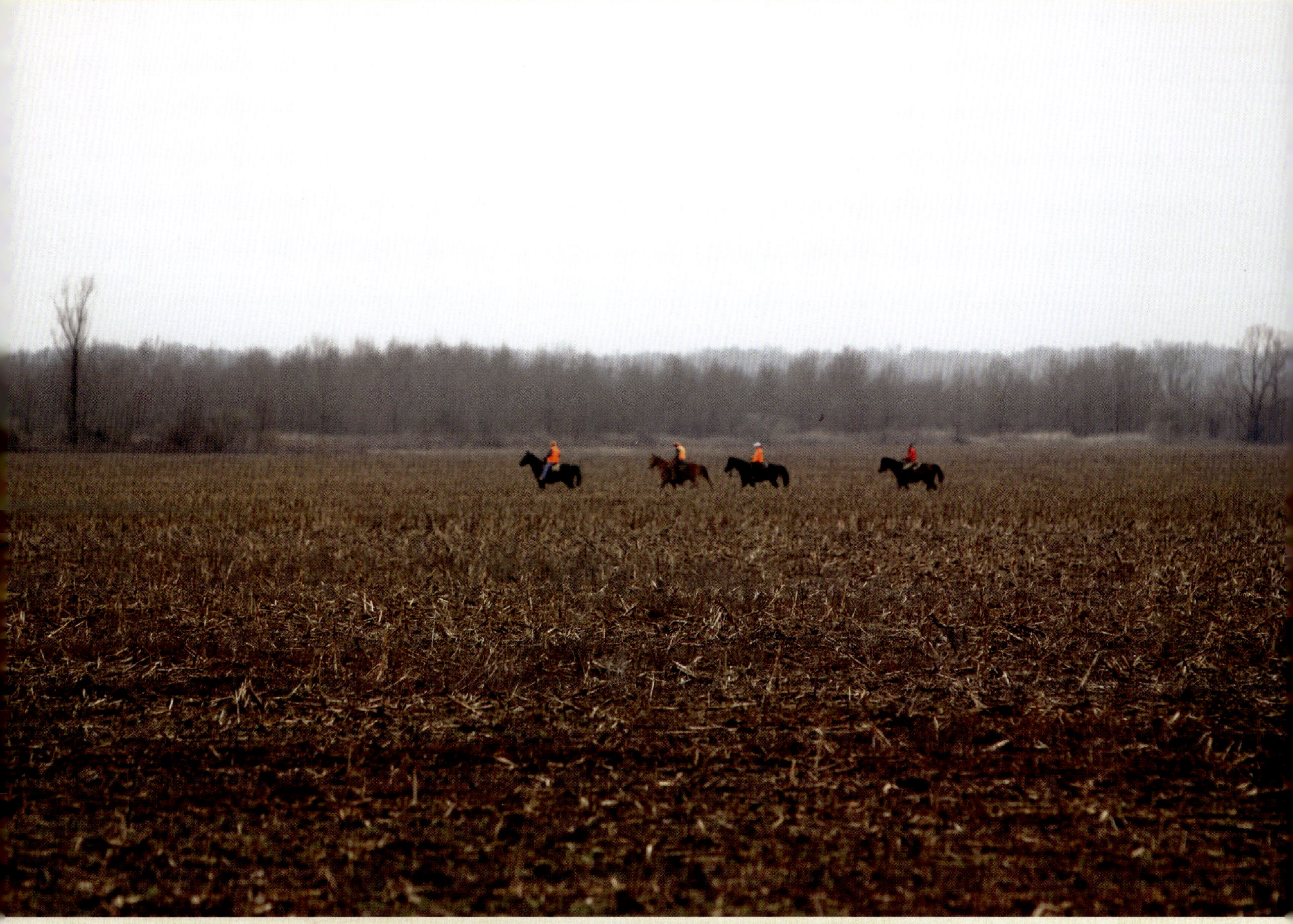

Riding through the fields.

Overleaf left: A Tennessee Walking Horse.
Overleaf right: Clark Gordin and Stumpy, a Shetland pony.

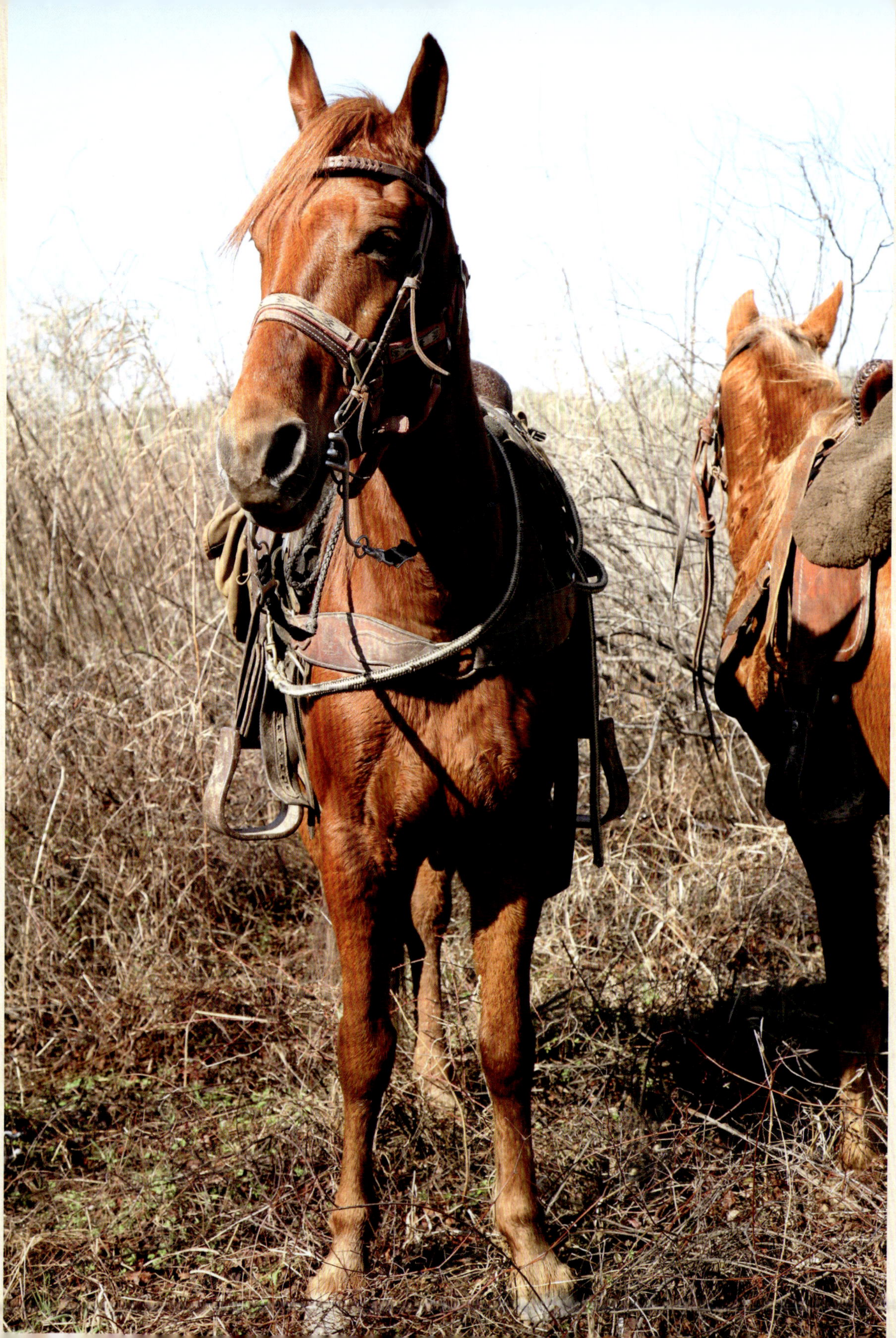

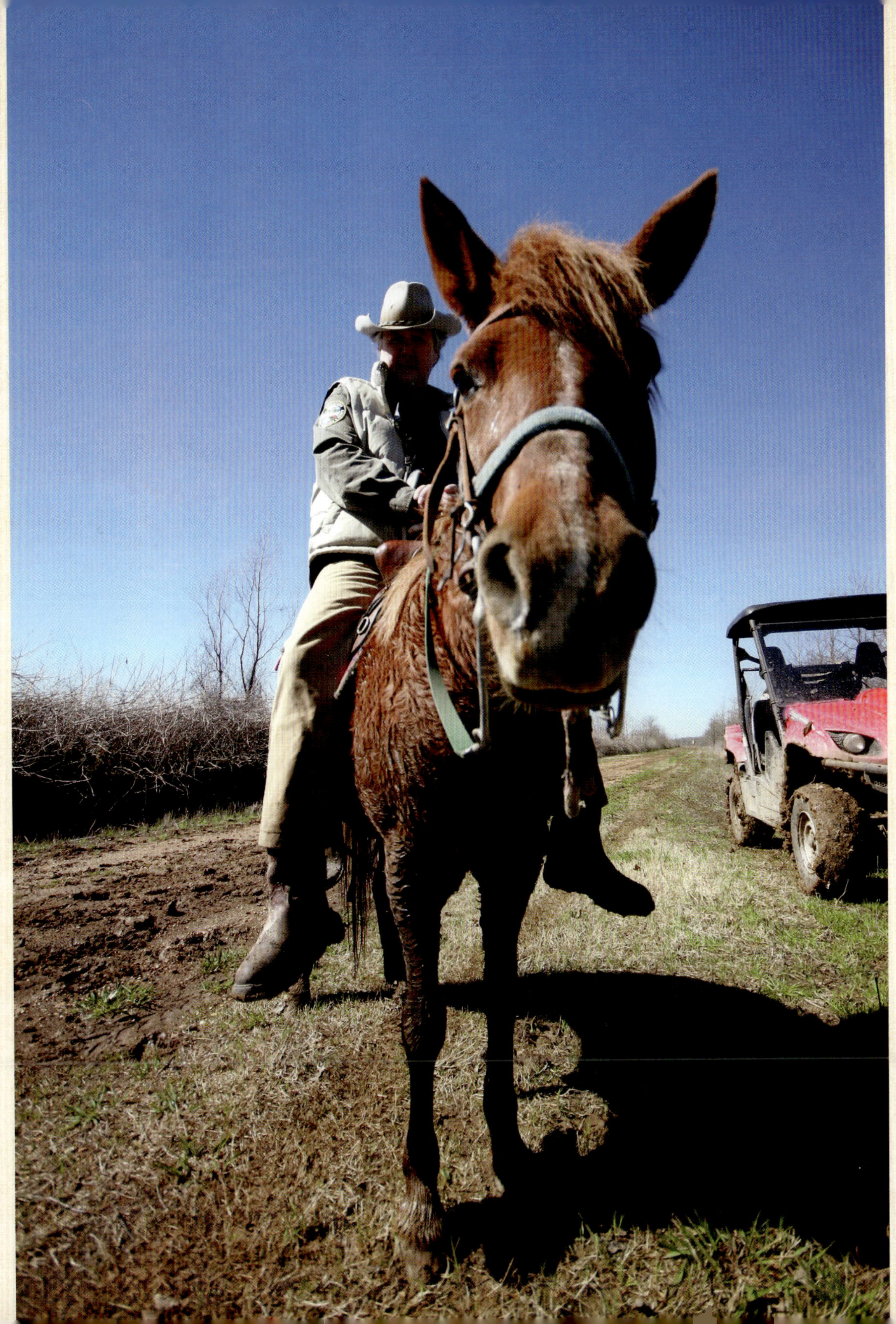

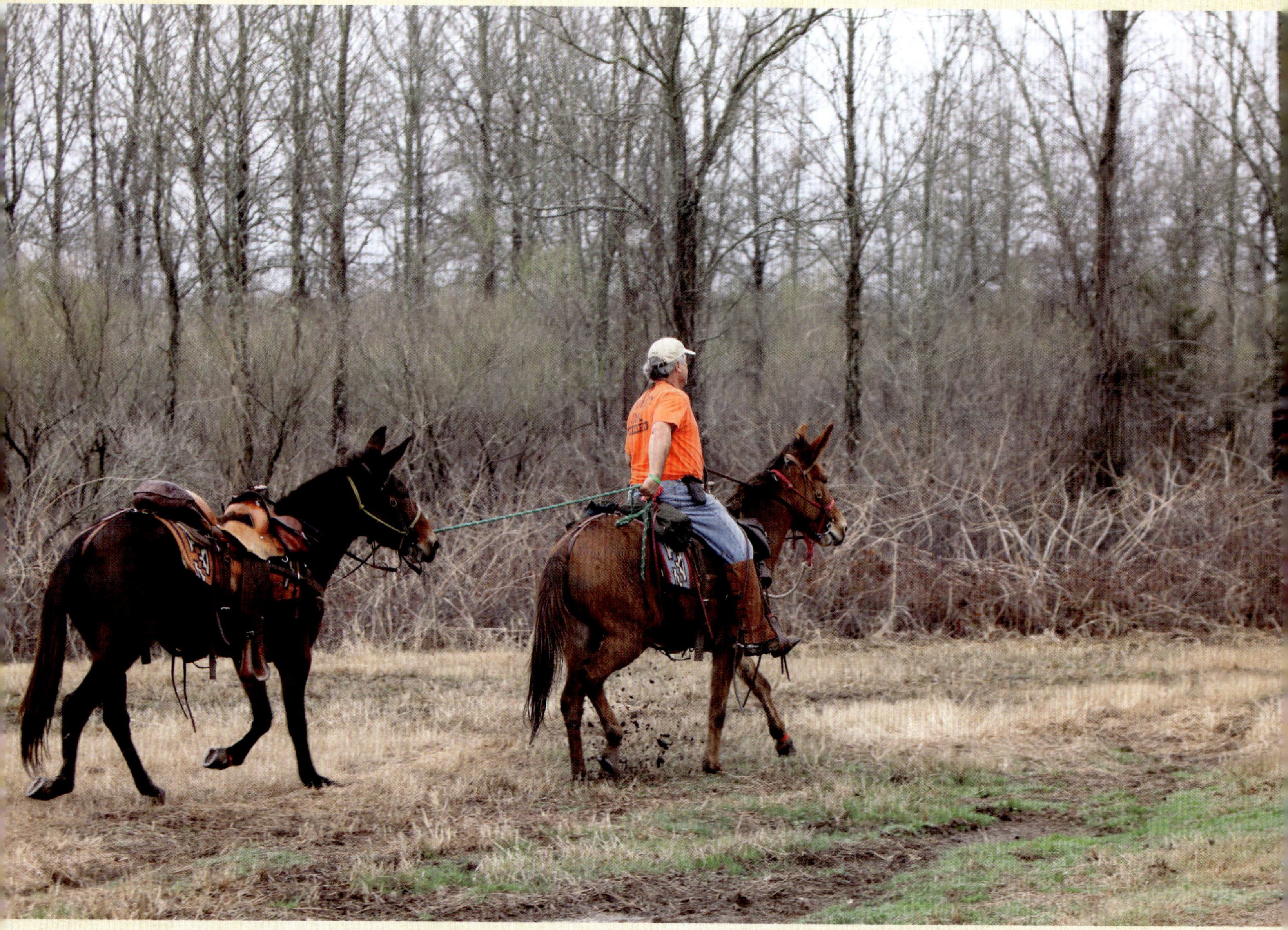

Doc Merideth and his mules, Gin and Tonic and Tootsie.

Wick Eatherly and his horse, Skipper, a quarter horse. >

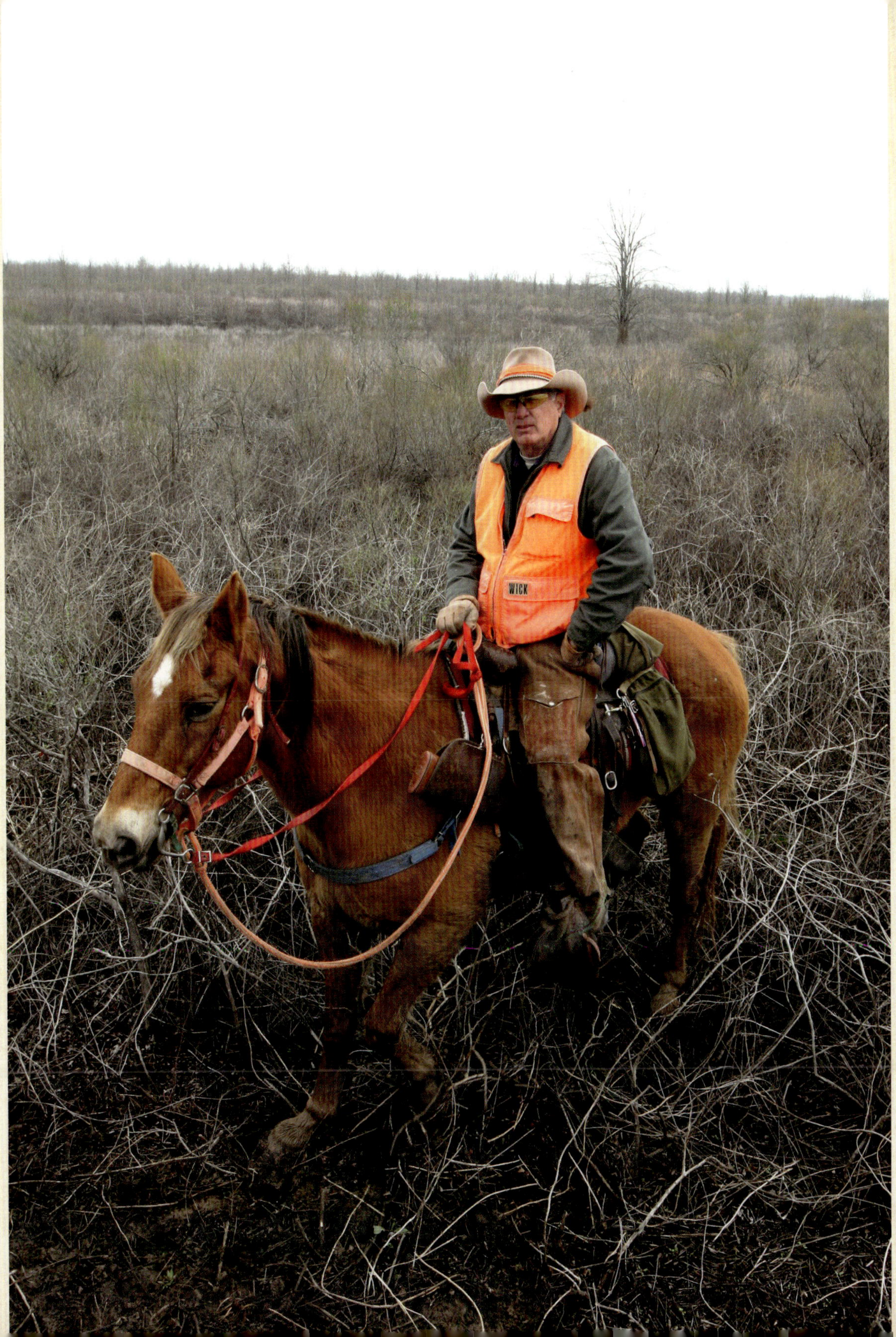
WICK

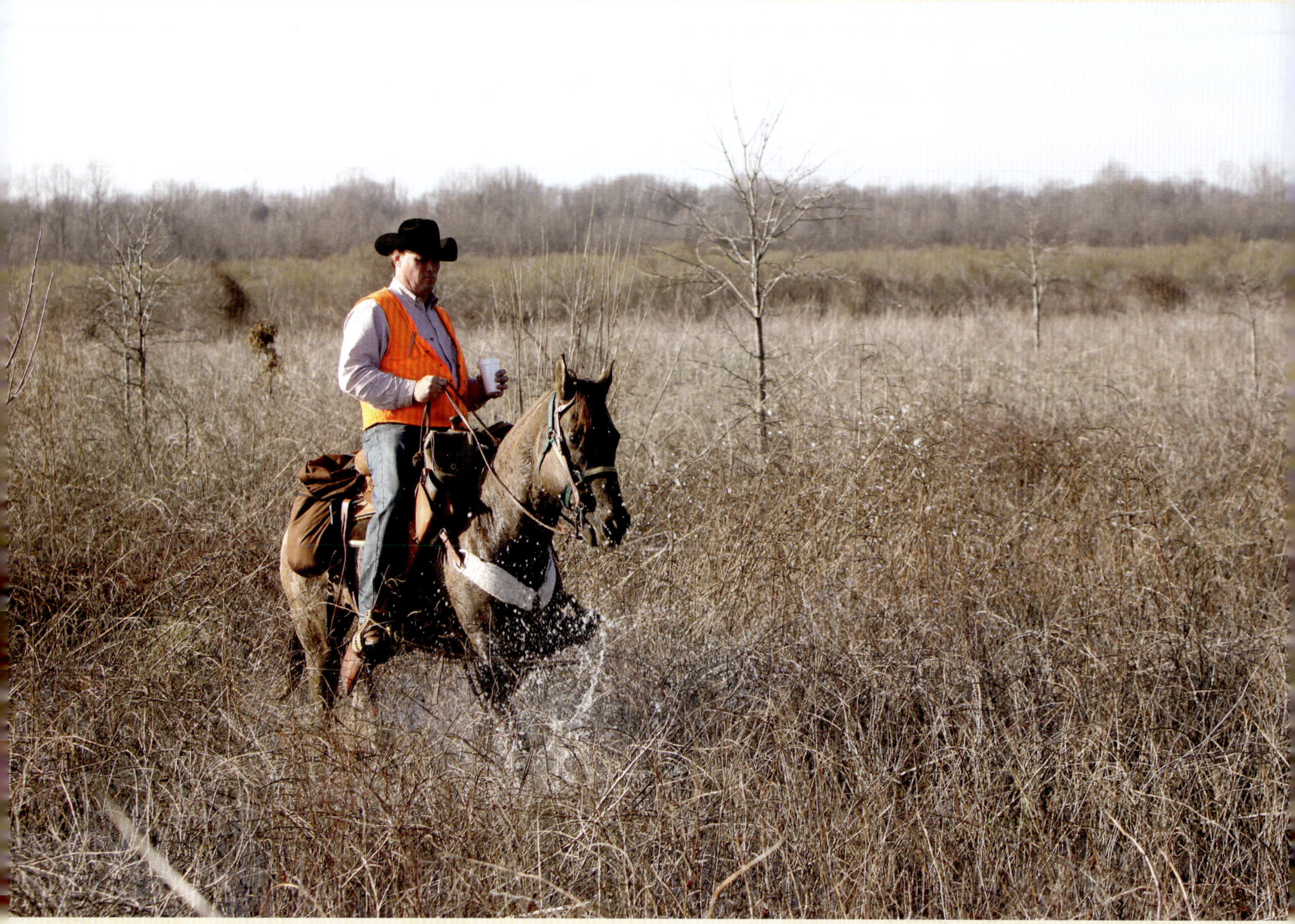

Pete Fulgham and Peanut, a quarter horse.

Bobby Jo and Robert McConnell, Smokey, a Tennessee Walking Horse, and Dusty, a quarter horse. >

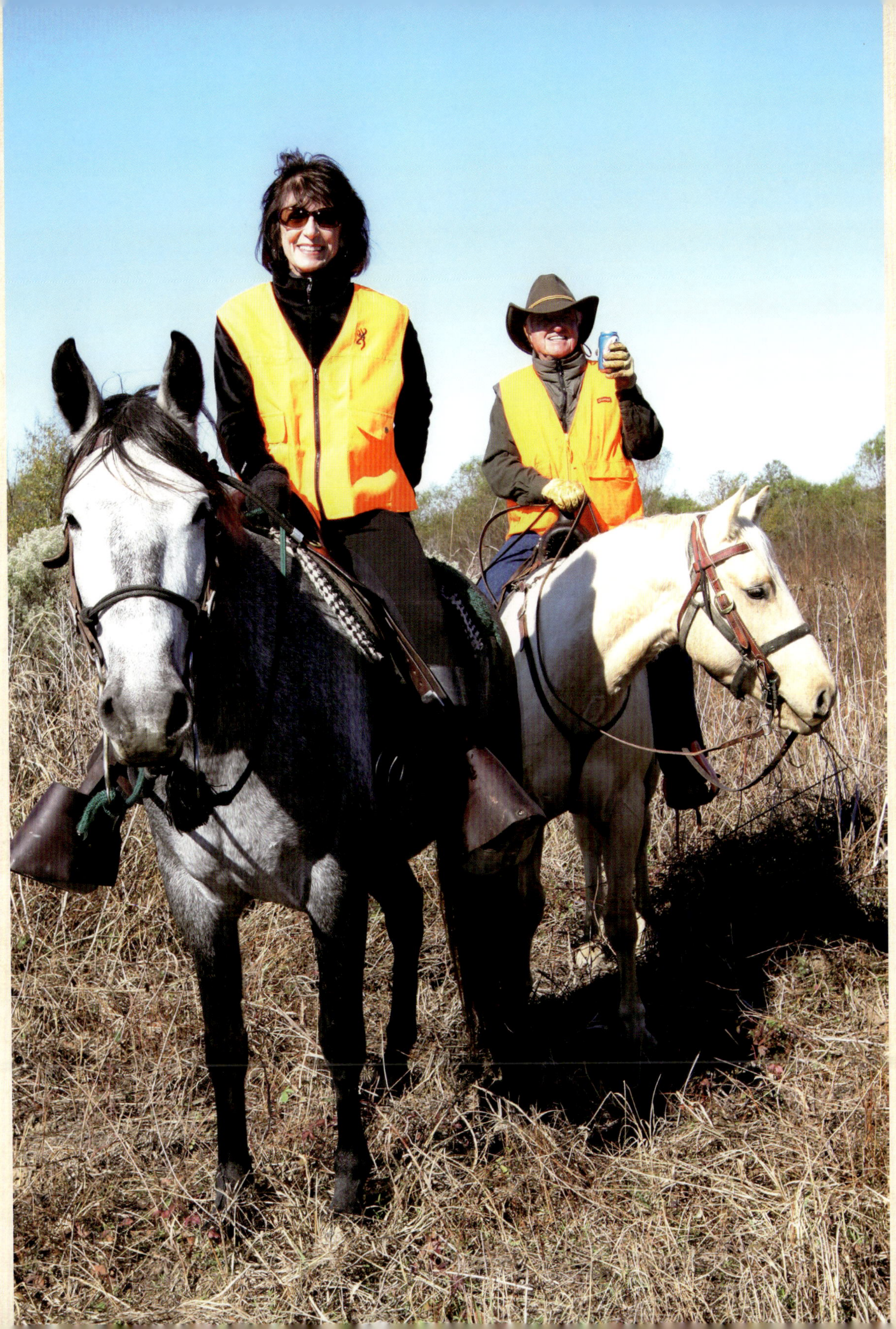

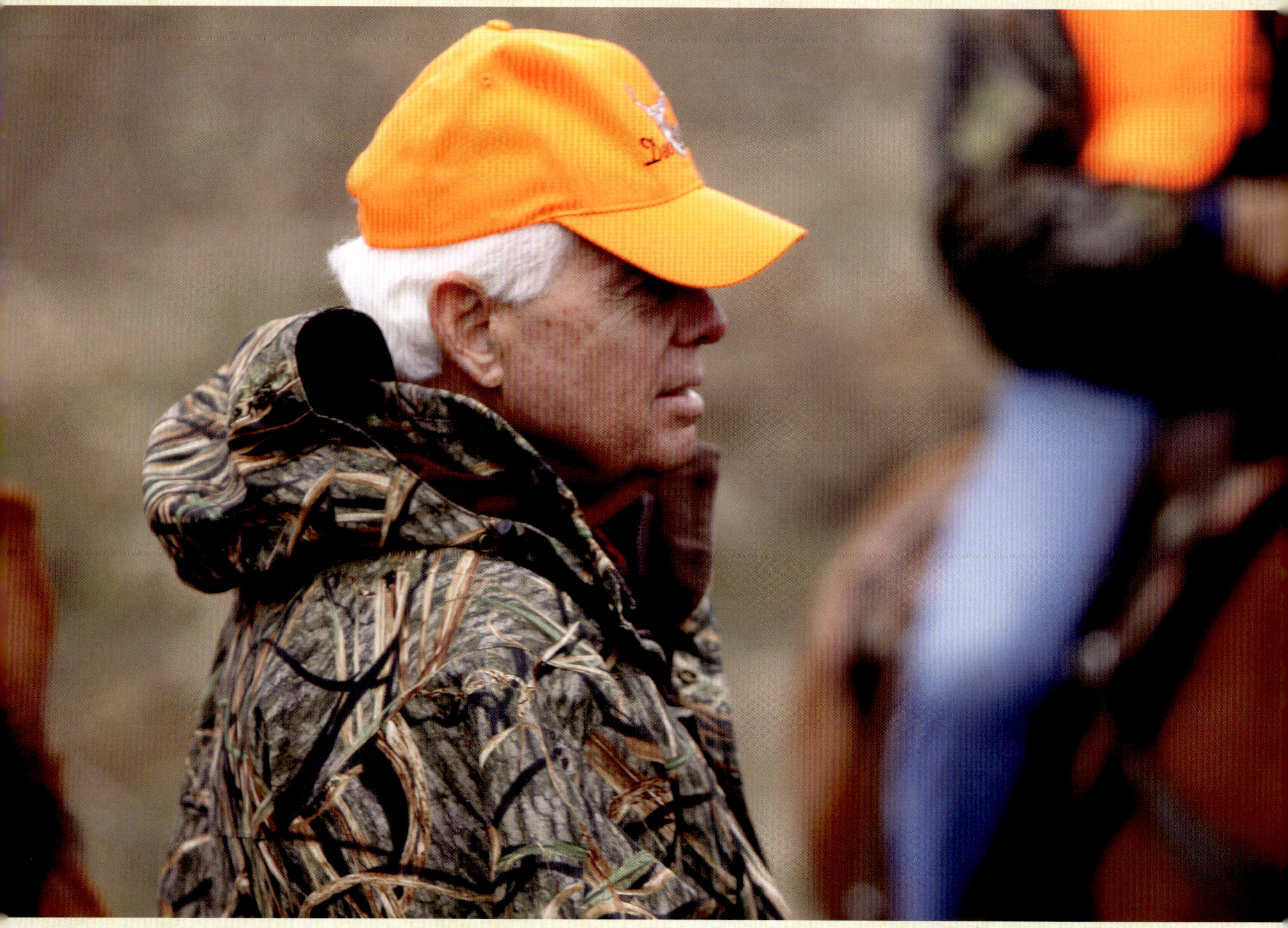

Philip Farr.

David Weeks. >

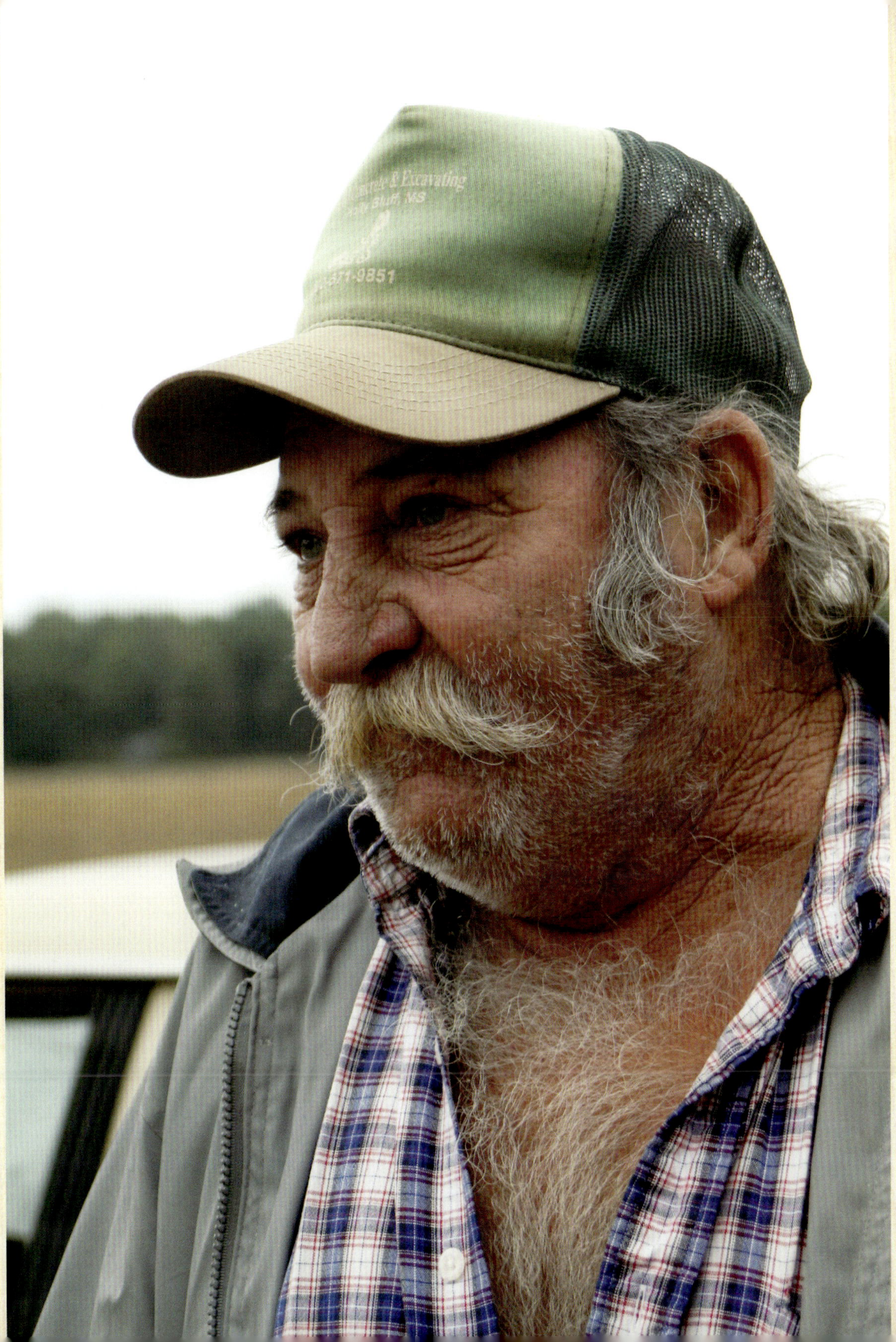

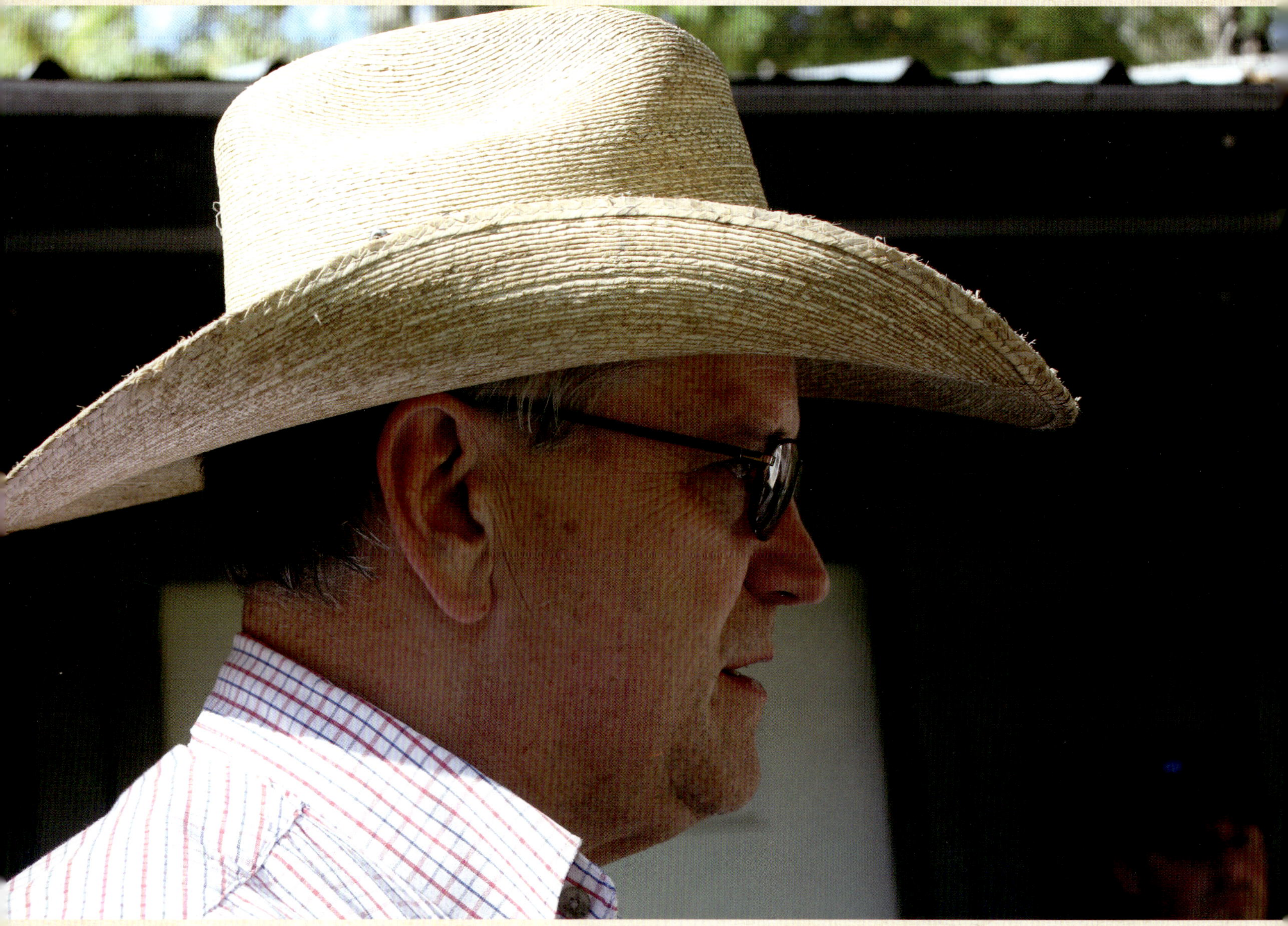

Mike Braswell.

Spitty and 44 Magnum. >

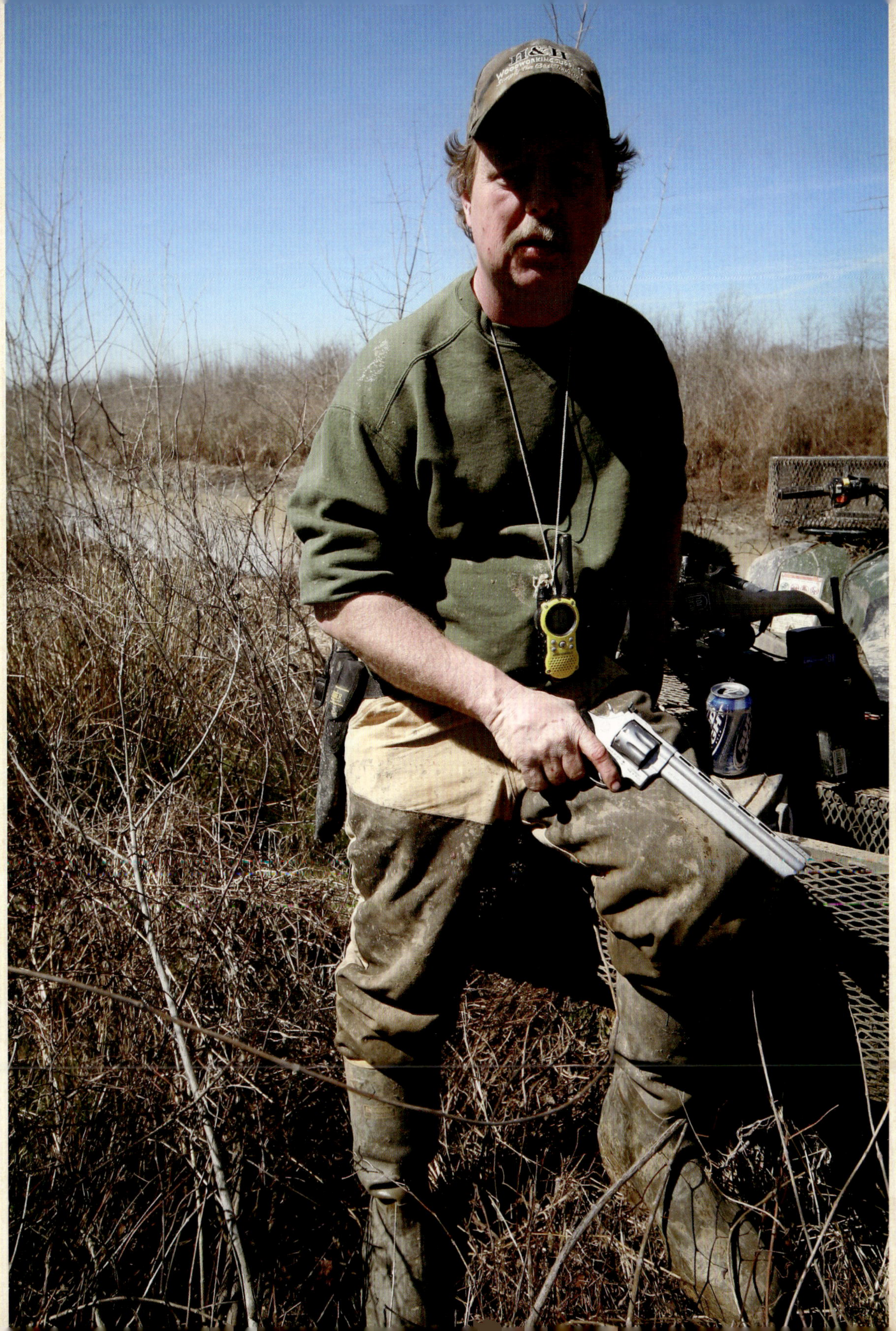

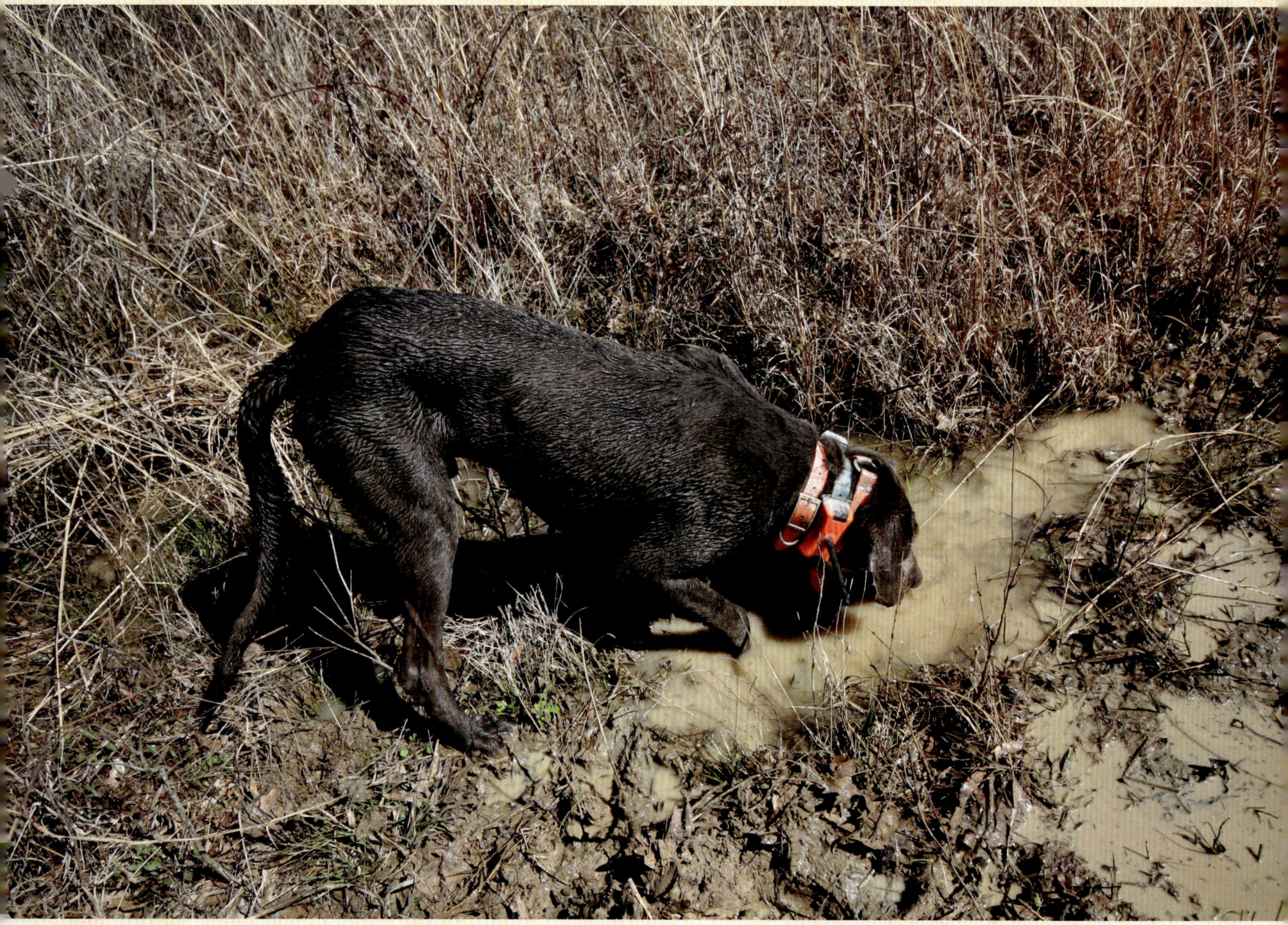

Plott.

Jeff Cremeen. >

Overleaf left: Rickey Lowery, Lea Brent, John Lowery, and Stacey Waites (in back).
Overleaf right: Brandon Yeager and Frog, a bulldog.

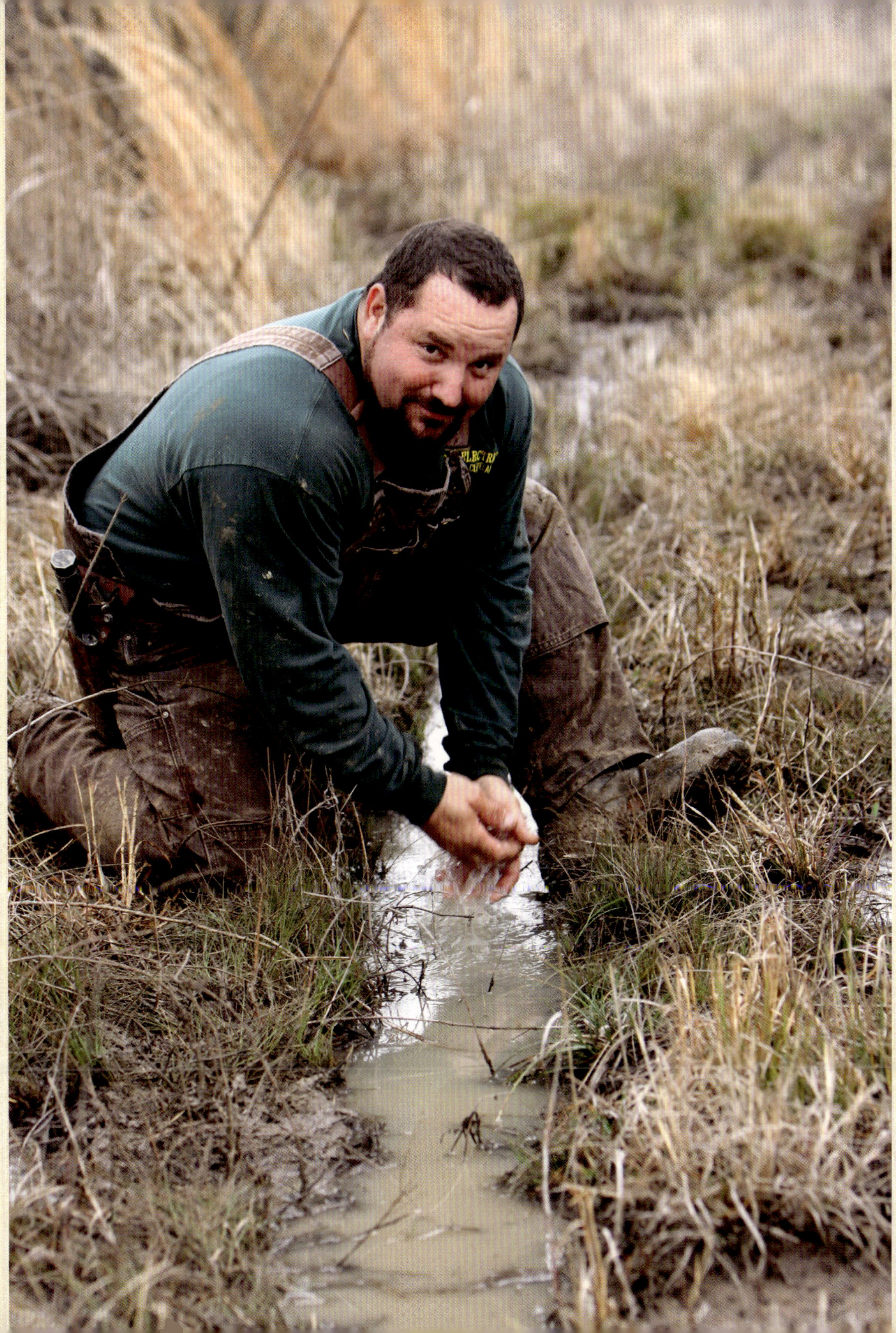

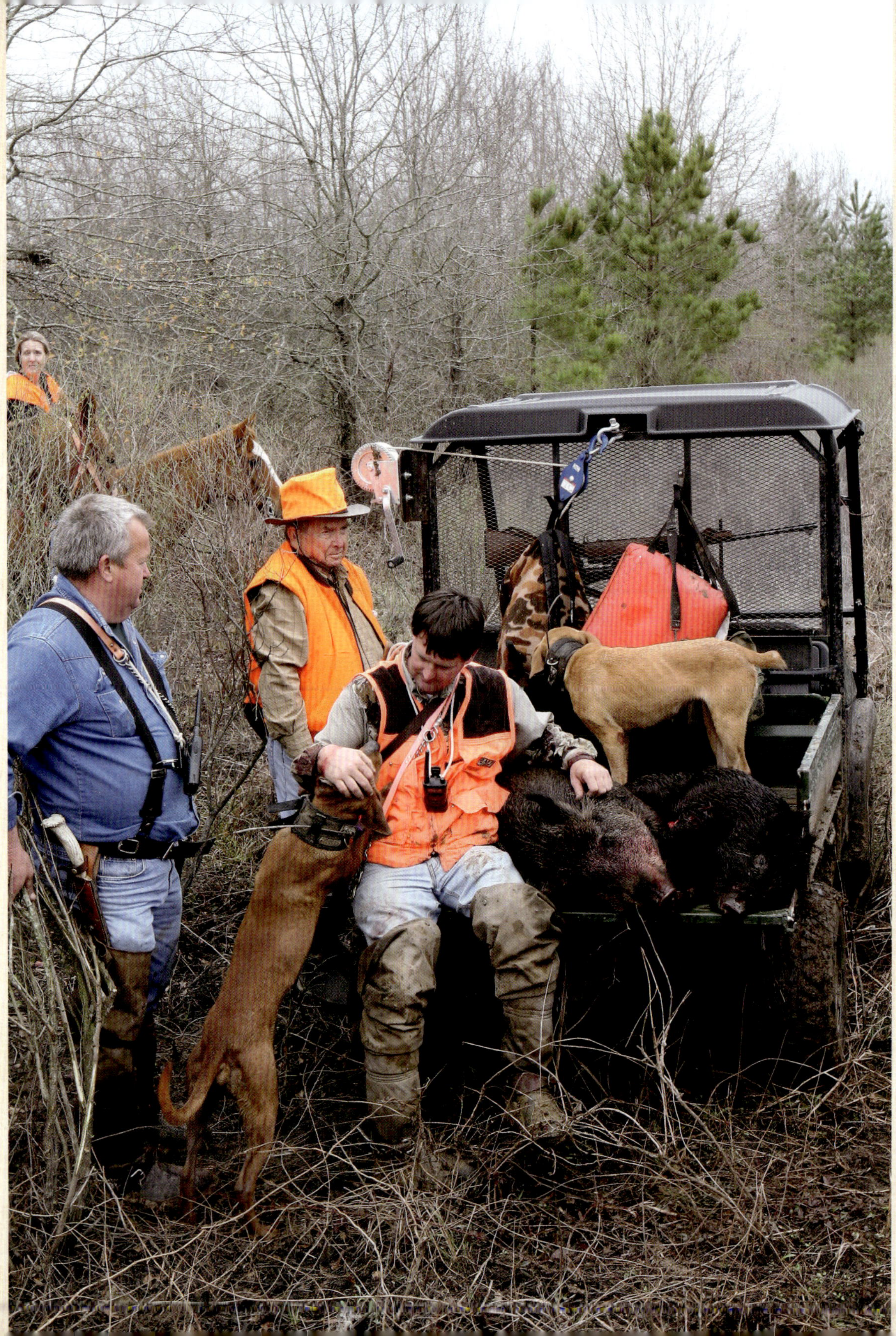

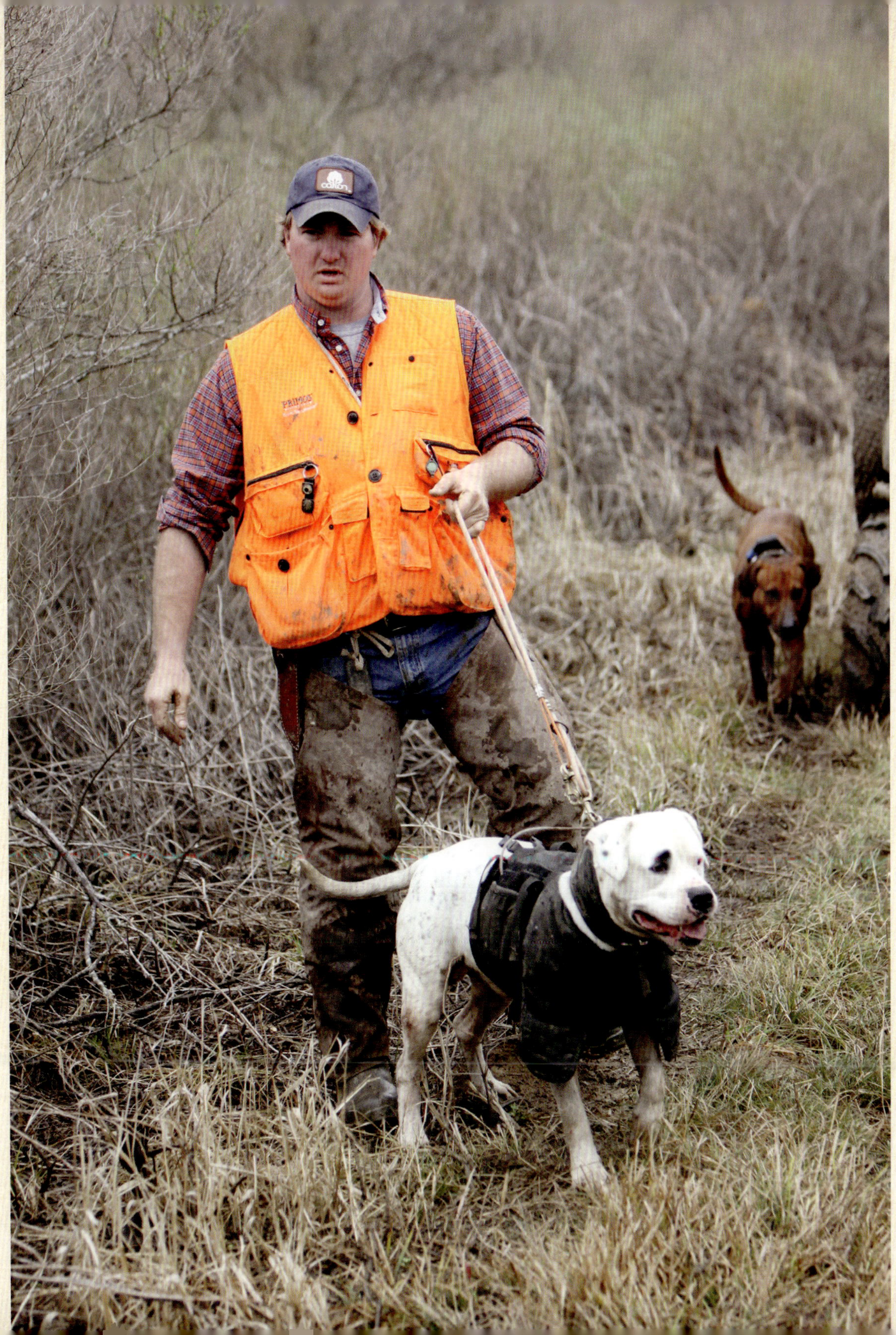

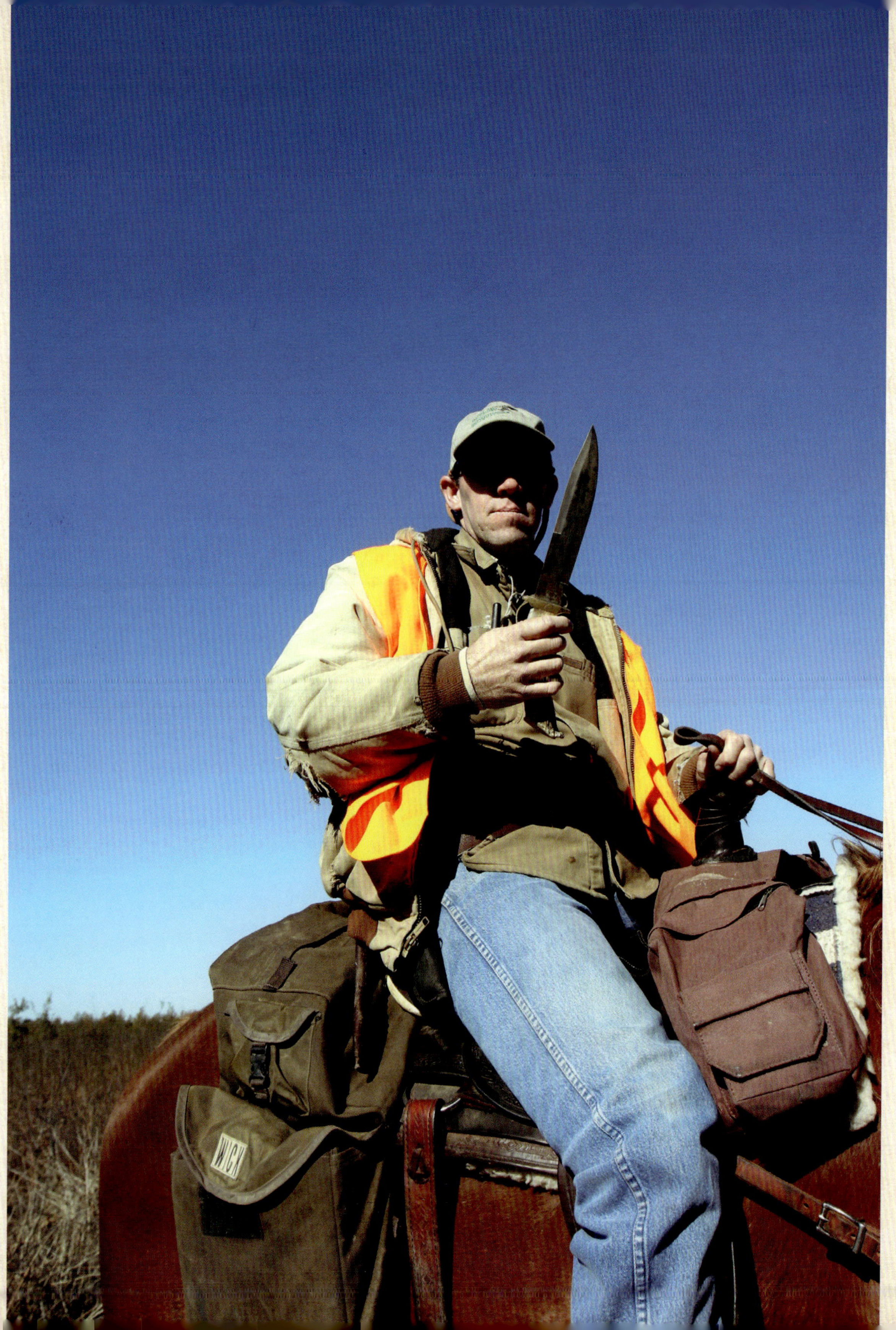
WICK

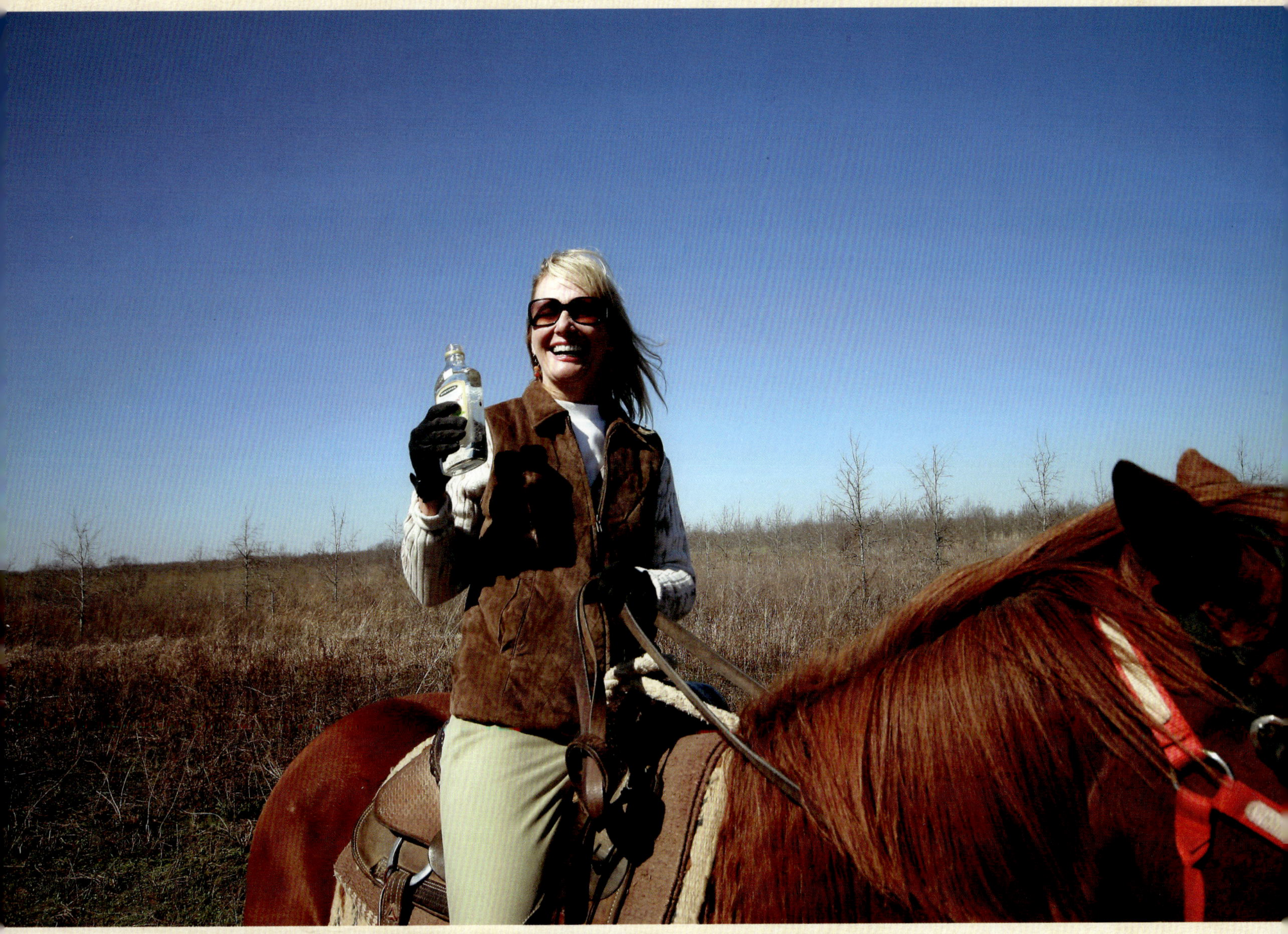

Nancy Bancroft, social observer, with hunters' antifreeze and Baron, a thoroughbred.

< Hines Outlaw.

Overleaf left: Rickey Lowery and Brandon Yeager, tracking dogs with Quick Track tracking systems.
Overleaf right: Steve Ratcliff, Molly, mountain cur, and downed boar.

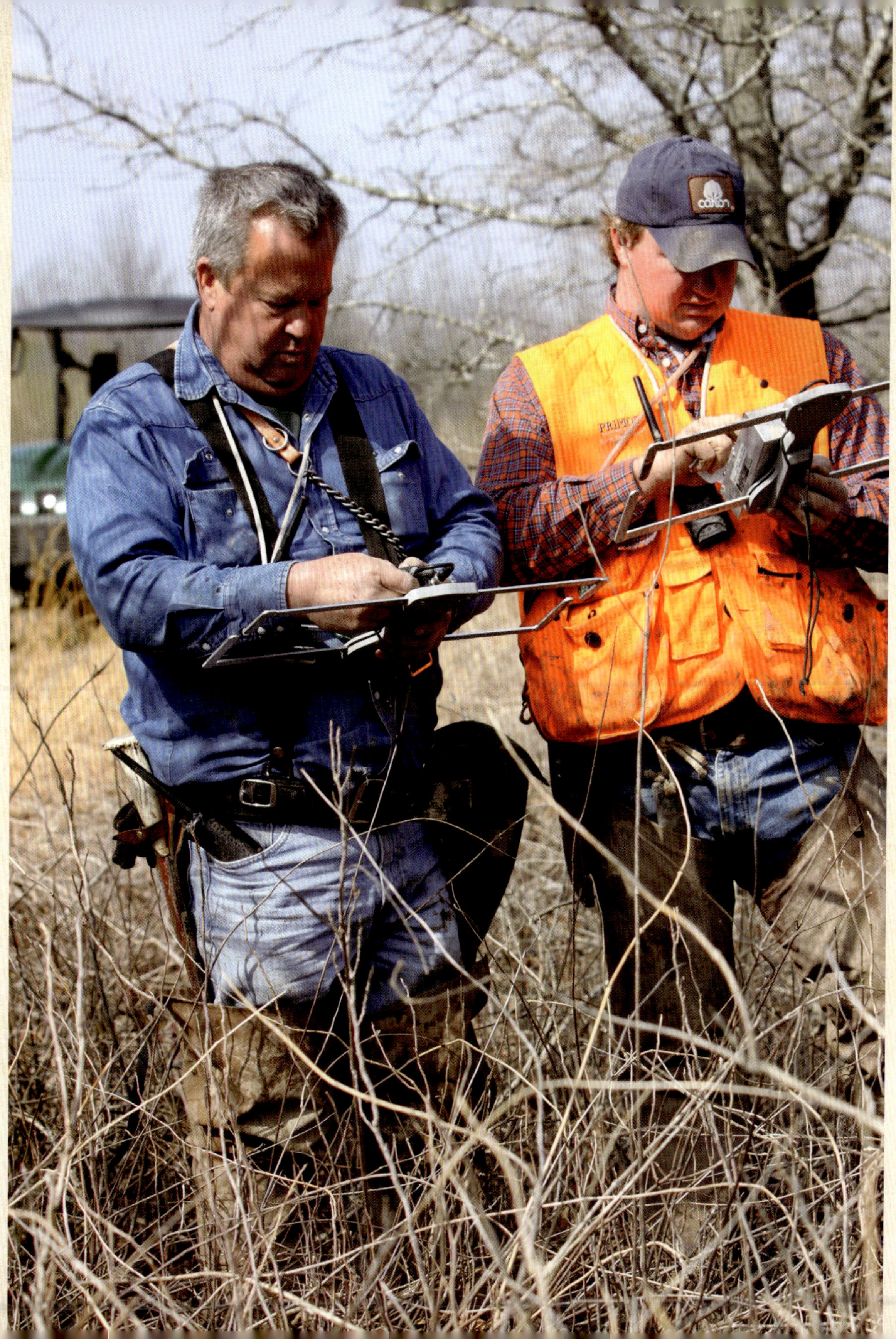

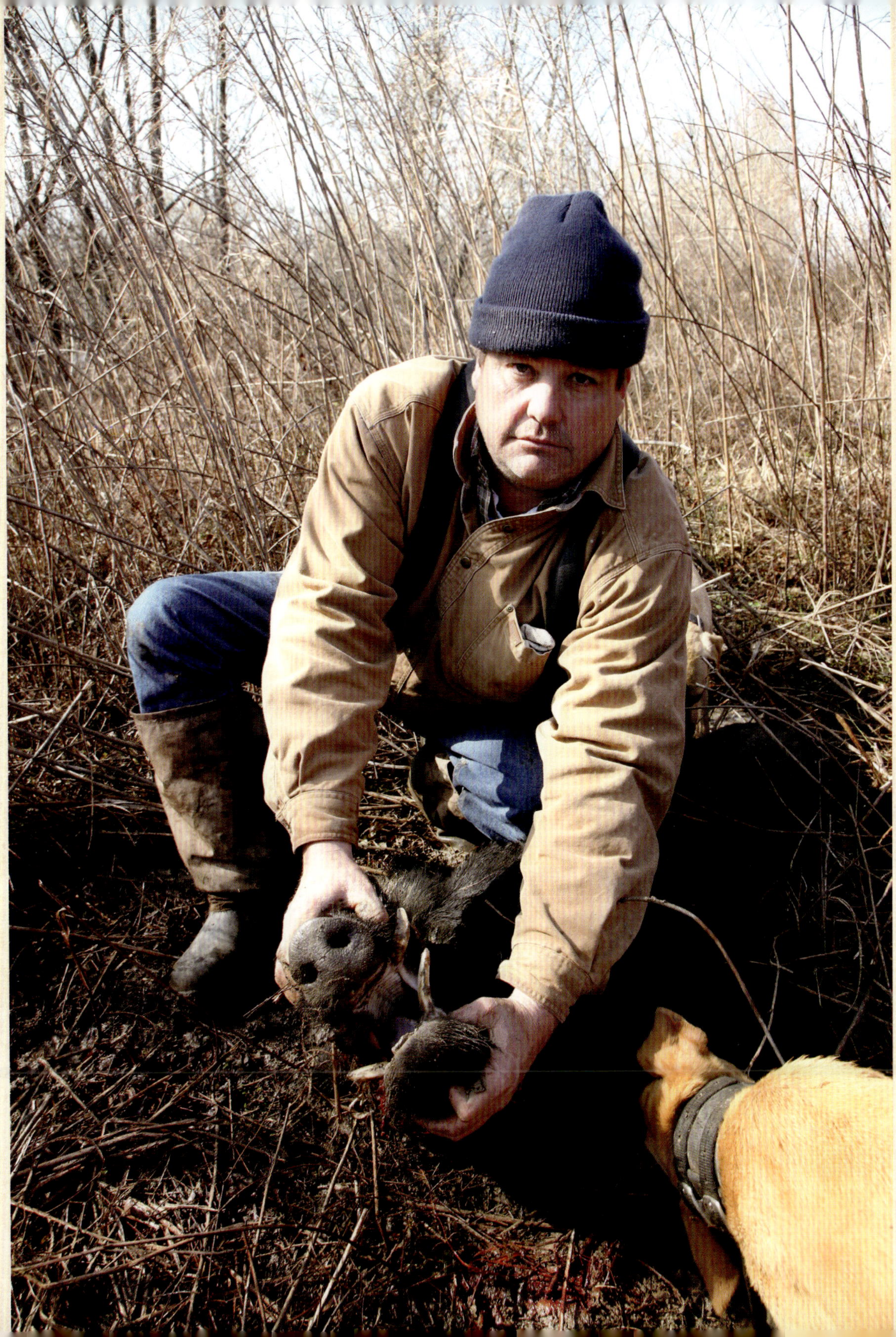

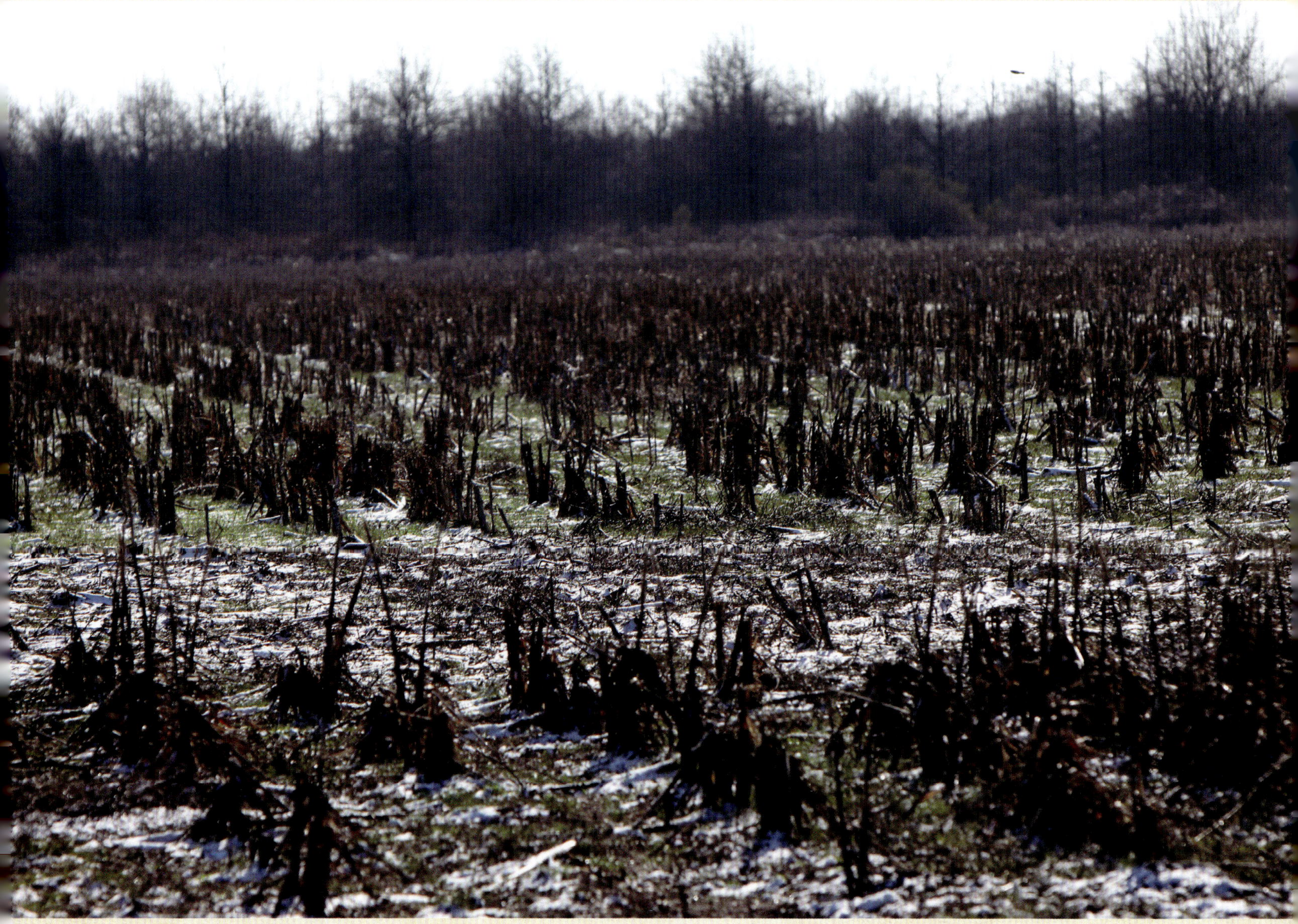

Corn field with freshly fallen snow.

Dan, Yeller, and Rooster, yellow black mouth curs. >

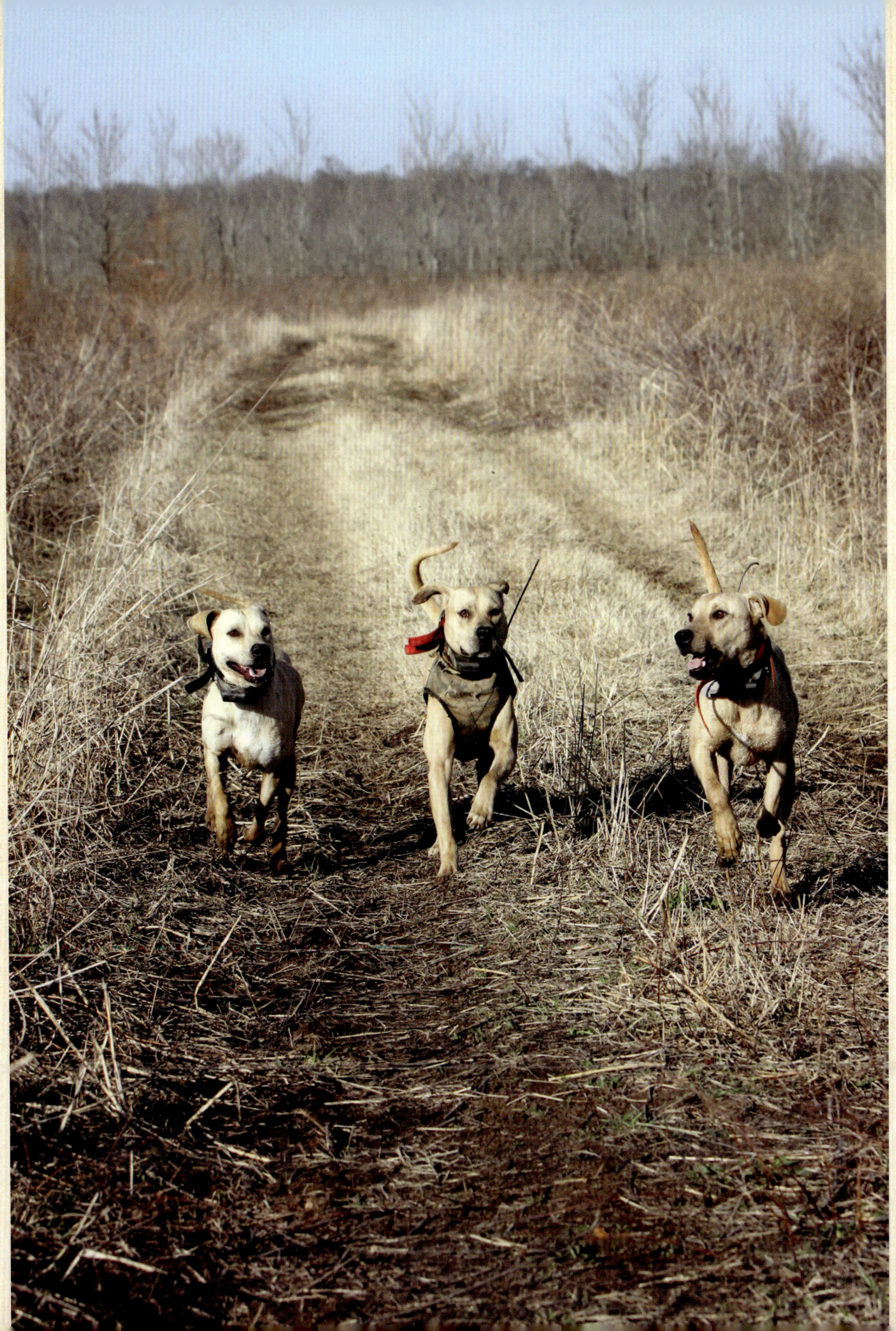

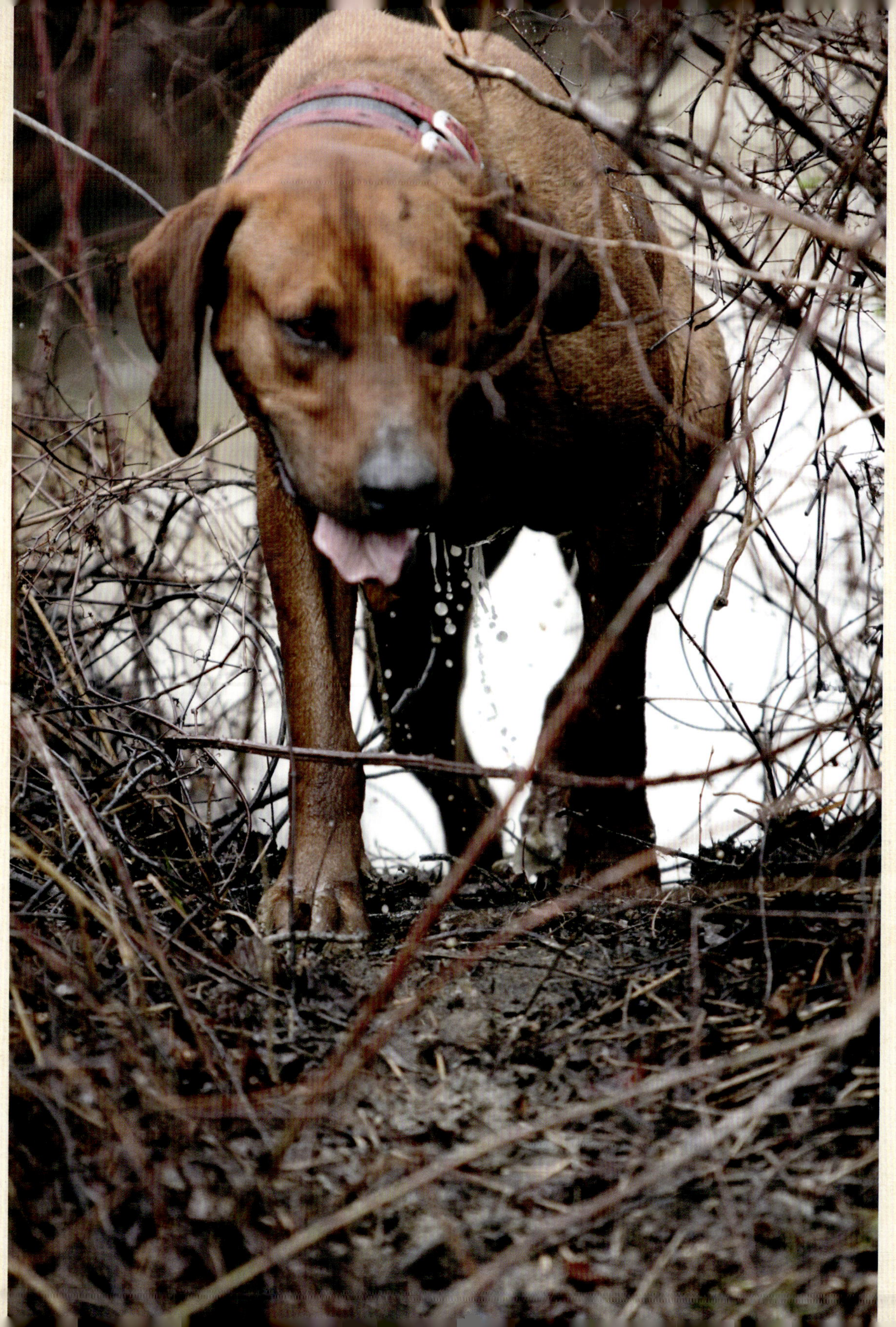

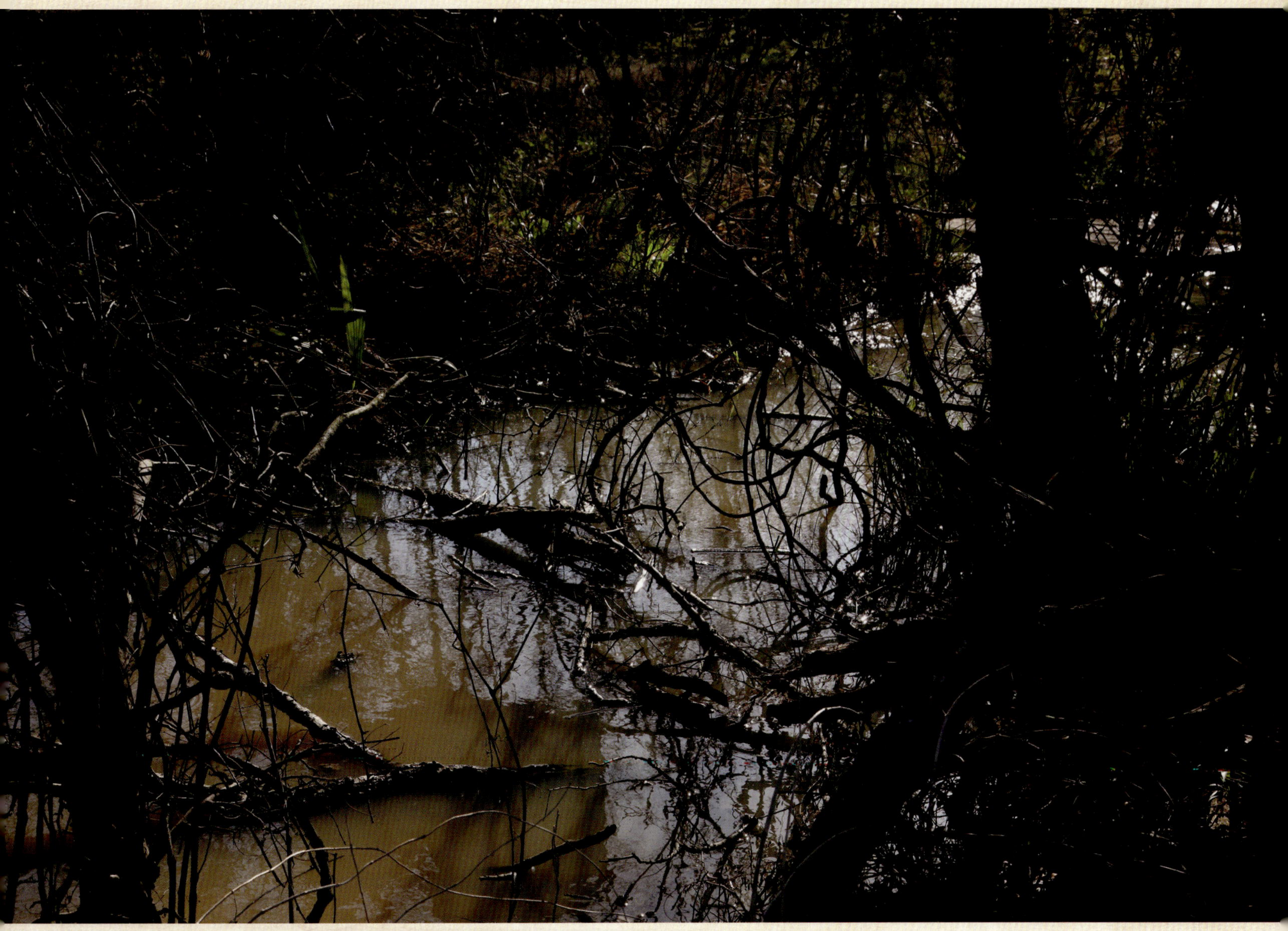

Muddy slough.

< Levi, redbone black mouth cur.

Overleaf left: Slough.
Overleaf right: Lake Discovery.

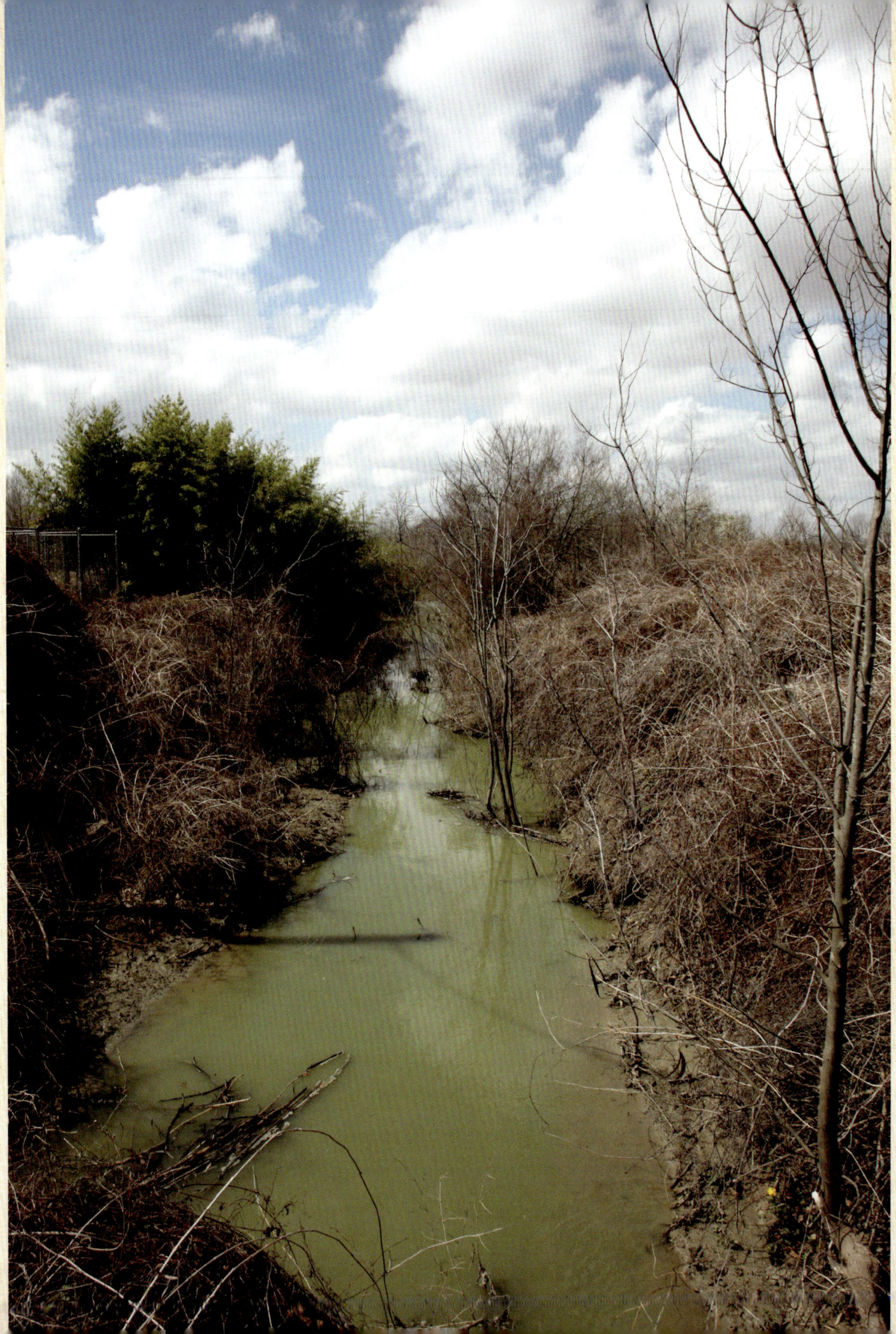

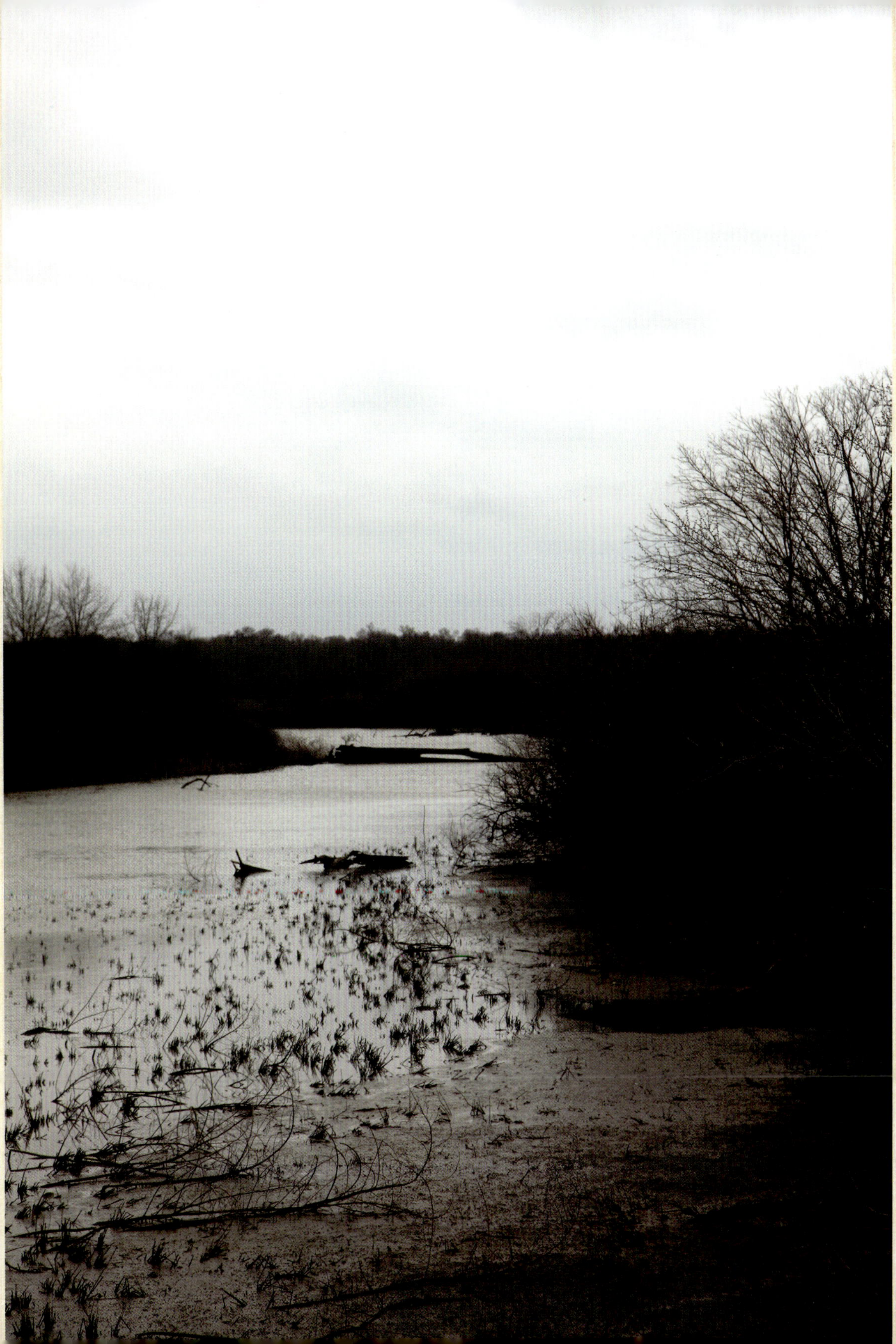

Oscar Meyer, Smith & Wesson, Lennon & McCartney

BARRY BERMAN

Shotgun loaded, gripped by my side, I move through the darkness. I'm in my bare feet, soles soft as ducks, my toes tracking laserlike through the pile carpet. Silence is crucial. So when my knee slams into the corner of the coffee table, I do not say ouch, but I do think: gaaaaaaahaaaaa. In this life-and-death moment, I must utilize my pain—process it differently from how I would the pain of, say, nicking myself while shaving. Here, now, in the wilds of my apartment, in the middle of the night, I convert my pain to strength, as calories convert to energy. My pain becomes my edge, my ally, my meditation—my portal to a precise and perfect presence. I hold my pain in sanctified silence, for the slightest outward display might alert my target and trigger a defensive attack—which I can ill afford. Because I am a hog hunter.

I hear a movement and swiftly react by pinning myself against the wall. My head is two inches from the frame containing a sepia photograph of my maternal grandparents (Esther and Shlomo, two unsmiling Polish immigrants—and unsmiling why?—could it be because they've never gone hog hunting?). I control my breathing, lower my pulse, and thrust my ear toward the blackness. The sound I hear is either a wild boar foraging for the remote control or the elevator across the hall moving to a higher floor. I surmise that it's the elevator and wonder if it's carrying that cute girl who lives in 916. I wonder if she has the slightest inkling as to the thrill and danger currently taking place in apartment 802. I make a mental note to tell her later of my heroics. If there are any.

I press on, leaping like Geronimo over the inch-and-a-half aluminum threshold between the living room carpet and the linoleum kitchen floor. There is no turning back now: the hunter, once engaged, must proceed.

It all happens in the blink of an eye: I lunge forward, throw open the refrigerator—the glaring light revealing a jar of mayonnaise (a decoy)—and the butt of my shotgun knocks backward, against the toaster. I love that toaster—makes perfect toast ("dark" is dark, "light" is light—there is a glimmer of truth in my complex world—and when you add cream cheese to that truth, well . . .)—but what comforts one on a new morning can wreak havoc in the lonely night of the hunter.

With no time for sentiment, I shoot the toaster, then turn back to the refrigerator, wrestle open the meat drawer, and train my gun on the shrink-wrapped package of bacon. Oscar Meyer is no match for my Smith & Wesson. I grip steadily the trigger of my shotgun and—

—wake up.

That was the dream I had the night after I'd read the preface to Melody Golding's anthology of hog hunting stories (the very book you're reading now). While real hunters prowl the forests in blood sport, I'm in my kitchen, shooting a rasher of bacon.

The fact is I have never hunted and never will. I'm not cut out for it. I grew up in Washington, D.C., where guns were tools of human destruction and meat came from the grocery store.

But now I am older and have seen other worlds.

Many years ago, while residing in Los Angeles, I had an

idea for a movie. The story would require the protagonist, caught in a dire circumstance, to return to her native home in the Deep South after an absence of fifteen years. At that time, my experience with the South was limited to a trip to Miami (which, to genuine southerners, is actually Cuba) and to a few summers on the beaches of Virginia (which again, to real southerners, may as well be Alaska or some other great neighbor of Russia). I also had my impressions of the South, courtesy of movies and TV shows I'd seen while growing up: *To Kill a Mockingbird*, *In the Heat of the Night*, *The Dukes of Hazzard*.

I did, however, know enough to know that if I was going to write of a place central to my story I must be familiar with that place. I located a small town on the Louisiana side of the Mississippi River with an alluring name: Waterproof. I doubted anyone had ever heard of the town, save those who lived there, and decided that I must venture there to see what it was all about.

Days before I was scheduled to leave LA—and still with no contacts on the ground in the South—I was on the phone, chatting with some friends in New York, who, a couple of years earlier, had filmed a Budweiser commercial on a Mississippi River barge. They'd made some friends while on the job and offered them up, saying, "You should call Steve and Melody. They live in Vicksburg and helped us on the shoot. They know everything about the South." I instantly phoned the number they gave me. Melody answered and our conversation went basically like this:

Melody: Hello-oo.

Me: Hi, you don't know me, but I'm a friend of—

Melody: Great, come on down. We'll take care of you and give you our house.

They'll give me their house? Just like that? For once I wished I was a cat burglar. As easy as she'd made it, I could feel like a really competent thief.

But then I got concerned: precisely because Steve and Melody were about to make my path so smooth, I'd have only myself to blame if I wrote a really garbage screenplay. My suspicion grew: so this is what's meant by "southern hospitality" . . . give your house to a complete stranger with no questions asked and make it open season for his deepest inadequacies to hunt him down and kill him.

Bags packed, flaw in the operation noted, I flew to New Orleans, rented a car, and drove four hours north.

It was nighttime when I reached Tensas Parish and, following the directions Melody had given me, I turned right at that egret, veered left at that Spanish moss, made a sharp right at that big ol' rock, and found myself in front of a spectacular home on the banks of Lake Bruin, just across the river from Vicksburg, Mississippi.

Steve and Melody welcomed me warmly to their "camp."

To me, a "camp" meant a simple tent erected in the desert, easily disassembled when it came time to grab your unleavened bread and flee. But their "camp" was three thousand square feet of homespun luxury: floor-to-ceiling windows looking out onto a peaceful lake; a dock and boathouse down by the water, heaven to any painter, poet, or fisherman; a fully stocked kitchen with a big, convivial table and lots of chairs; couches, TVs, air conditioning, board games, and bedrooms the size of most monasteries.

They offered me the largest of the guest bedrooms and said they'd show me around in the morning, introduce me to a bunch of people, and answer all my questions.

Aha! Step two of the old southern hospitality con: give the writer nothing to complain about. Give him everything he could possibly need to facilitate his work, including a hammock next to his very own lake in which to drown himself when he gets writer's block.

My logic was airtight except for one thing: Steve and Melody meant everything they said. They really did want me to learn about the South and to have access to the people and places I might want to write about. They really did want me to write the best movie I could, whatever it was, as long as it was written from my soon-to-be southern heart.

Don't get me wrong: I've known many nice and hospitable people in my life. In fact, some very nice and hospitable people have been humiliating me for decades. But Steve and Melody gave me their house, of all things—and all I did was mention some folks from New York. Imagine if I'd mentioned some people from Mississippi. I could probably have had power of attorney.

By the middle of the next day, after meeting a number of Steve and Melody's friends—the Vizards, Watsons, and Hardwicks (all of whom quickly became my friends, as well)—I reached the conclusion that my suspicions were baseless. There was no cunning or subterfuge behind the storied southern hospitality. It was not a myth, gimmick, nor ploy. It was real—a fable come true—like the full red moon I saw the night we boated across Lake Bruin.

For the next seven years, I returned frequently to Mississippi and Louisiana, spending great stretches of time hosted by my southern friends, soaking in the elements and writing my script. Eventually, the film got made. For me, it was a labor of love. For my friends, the love they showed me seemed no labor at all.

So what does this have to do with hog hunting?

All the time I've spent with Steve and Melody has been on their turf. They have repeatedly extended to me their southern hospitality and I have thus far been unable to show them any of my Washington–Los Angeles–South African hospitality.

But then my chance arrived when Melody sent me the preface to *Panther Tract: Wild Boar Hunting in the Mississippi Delta*. While reading it, I saw an opening to offer a tiny gesture of repayment for the enormous generosity she'd unfailingly shown me. After years of her offering houses, cars, friends, meals, and trips to the Mississippi Delta Blues Festival, I could finally bring something substantial to the table: I could offer my own short story for the book.

While reading Melody's work, I learned a little bit about hog hunting, an activity I never even knew existed, and a bit about those who endeavor it. In addition to the preface, I read some of the hunters' stories. Notwithstanding all the guns, knives, and blood, much of what I read reminded me of the South that I had come to know and feel somewhat akin to—the South that I have failed to revisit for too long now.

And then I got to thinking: what if hunting, even in the twenty-first century, continues to answer a basic and natural calling in mankind? What if man's instincts for survival and dominion are so bound to the psyche that the impulse to hunt (within legal and logical boundaries, of course) was less a luxury than a necessity? What if someone like me—not and never a hunter—by reading the words of hunters, were to find a window of curiosity into something previously anathematic? And what if—just what if the Beatles were right?

What if happiness is a warm gun?

A Lifetime of Hunting

MIKE BRASWELL

I am from Lexington, Mississippi, and I have been hunting all of my life. I've been hunting since I was five or six years old—big enough to go to deer camp with my daddy up at Steele Bayou between Valley Park and over here by Redwood.

I started hog hunting, let's see, probably in about 1980. I was catching cows and breaking horses for a living and had a bunch of Catahoula curs—I always had a place in Mississippi where somebody wanted me to go catch cows with my dogs. When the cow market sort of went bust, my daddy-in-law, Vernon Jones, who hunted with Bubba Cunningham, Wick Eatherly's uncle, said, "You ought to start hog hunting." They told me that they used to hog hunt down here. I started going to Delta National Forest because I wasn't a member at Delta Wildlife. So, I couldn't get anybody to go with me. Wick wouldn't go with me, he said, until I got my pack right. Well, when I went to Delta National and turned my dogs out they'd go looking for cows. They didn't know we were hunting for hogs. One day about two o'clock in the afternoon, I was coming in and a big boar hog crossed the road. My Catahoula cur named Blue was staying right beside my stirrup because that's the way I trained them, but you put too much control on a dog, it takes the hunt right out of them. He saw the hog cross the road and I sicked him on the hog and all the other dogs followed him when he took off and bayed the hog up. I had about six dogs with me at the time. I found the hogs with Ole Blue and the rest of the pack! After that, I called Wick and said, "You better get your horse ready."

So, I started hunting and finding hogs. I had a big yellow Texas cur that I was hunting with one time. There was a twenty-acre cutover you couldn't hardly walk through and we went walking up through there and we were duck walking and saw that big ole hog come and I said, "Shoot him, Wick! Shoot him!" Because I didn't have a gun on me. I don't like to kill a sow—I like to kill a big ole boar hog. The killing is not the fun; I like to see the horses and dogs work.

I had a little dog named Pete who used to sit on the saddle and ride with me and I used him for a catch dog. He didn't catch until I sicked him on the hog. He was a great dog.

My two best dogs I ever had were a dog called Ox and one called Barney. Barney was Ox's puppy and he weighed close to a hundred pounds. I could tell when we went hog hunting if they bayed a sow or a boar. A sow, he only barked a couple of times, but a boar he would chop bark and not stop.

I was hunting down in Natchez one time where it is just hills and gullies, and the dogs had bayed up a big ole boar hog down in a ditch. I was walking up in there because it was so thick you couldn't get through there with a horse. Well, a three-hundred-pound boar hog came after me and I couldn't run from him—there is no way to run in that stuff. I waited until he came right up to me and I shot him in the head and he slid right up to where I was standing. I shot him with a Winchester .375.

At Caulk Island one time a big spotted boar hog caught up in a sand field. I was riding the best horse I ever owned—Woody—he was my hunting horse. I rode up there and the hog came and got up under my horse because he was trying to get the dogs off him. I took my rifle out and stuck it down by my right stirrup and shot that boar hog in the back and killed him right there. Woody, my horse, never even moved. I killed several hogs that way with Woody. We've had a lot of close calls.

At Catfish Point a hog was feeding on a ten-point buck that somebody had shot and didn't find. The hog was bayed up and the Brutin boys I was hunting with were going to take pictures of him and film him. Well, that hog looked straight up at us and away from the dogs. When hogs look at you something's going to happen. He started toward one of the Brutin boys and like to have got him and when his brother killed that hog he slid up between that boy's legs, but that was a close call. According to where we hunt sometimes we use four-wheelers but I prefer hunting on the horse because we can get to it and I can hear my dogs.

I've been hunting in Natchez, Caulk Island, Catfish Point, Delta National, Delta Wildlife, Morgan Break, Belzoni, Slaughter, Charleston, Oxford, Tupelo, Tilapo Plains in Tennessee, West Memphis, Arkansas, Little Rock, Willow Run, Morgan City, Alabama, Georgia, Udawah, Tennessee (a suburb of Chattanooga), Oklahoma, Merigold, Grenada Lake, and more.

Caulk Island has a lot of hogs. Merigold and Caulk join each other, and it's one big thicket. They cut so much timber at Merigold that hogs have plenty of places to hide. Thickets are bad for dogs—hogs turn around and wait on them in a thicket and try to get them. It's bad stuff to hunt in.

We've put horses on barges and carried them across the Mississippi River to Big Island. When water gets up you got to put your horse on a barge and cross at Caulk. You can see how broke your horse is when you put them on a barge. Bubba Cunningham had a horse that had been trained by Melvin Weeks and he fell off the barge one time. He was on a little pontoon barge, so I just cut him loose and he swam to the other side and didn't get hurt.

There's not a season on hogs. If they have a place to breed and hide you are always going to have hogs. Farm hogs gone wild are feral hogs. They are spotted, red, and white. Russian boar hogs have hair that splits on the end; it's a European-type hog and that's how you can tell them. Wild hog meat is not as fatty as a tame hog. Wild hogs eat acorns, which makes the meat sweet—real good. I don't ever waste anything I kill. There is always somebody that wants that meat. I'll give it to somebody.

Wick and I've been hunting about forty years together. I know I can always depend on Wick and he's always going to help me. Wick is always going to get there.

People don't pay me to come get hogs off of their property. It's just not sport if you get payed for it. People know I like to hunt and that I'll come if they need me. We used to start hunting at early daylight, but not so much anymore. You look for signs and you know your dogs will find those hogs. Our dogs love to hunt! I put trackers on my dogs' collars. I don't leave a dog in the woods. For me, I don't use Kevlar vests. I've had dogs get caught in tangles and vines and such and I just don't like to use them. I'm kind of scared of them. We just have too many sloughs, and breaks, and swamps. I've had to swim and get them loose before. According to where you are, sometimes if I don't have any other choice, I've had to swim after the hog. I've swum lakes and across the Pearl River down at Port Gibson, Big Bayou, our deer camp we've swum across several times. But if I know it's a big hog, I'm going to get there for my dog! I don't like to swim—but I will. But I don't like it. My horse can swim too.

I'll blow a goat horn and those dogs will come back to me—it's just like whistling for them. They will hunt out. When a dog is about seven or eight months old, I will start them out hunting with their mama. They have to have a good experience or they won't make good hunting dogs. If they get hurt when they're puppies or on their first time it scares them too bad and it's not good.

I love my dogs. If they don't eat, I don't eat. I always treat my animals like I want to be treated. I love those dogs and I'm going to take good care of them. My best dogs have been Catahoula, that's why I'm so partial to them—they were the first dogs I started with.

I don't want a dog that will bark on the track. A hog will hear the dog and stay just ahead of them. A hog will run up to five or six hours. You got to respect hogs. It takes a lot to kill a hog. We have killed them before that have had buckshot in their guard plate, some with arrows and one even with another tusk embedded in his hide.

The biggest hog I have ever killed weighed 870 pounds! I've never seen anything that big—it was unbelievable! He was a tame hog that got loose and couldn't run fast. He was white. We shot it with a rifle and weighed him on the cotton scales. We gave it to the people on the place and it probably fed five or six families at least. But 250 pounds is big for a "woods" hog.

Wild Boar Hunting Along the Mississippi River

PEMBLE DAVIS

The context of these true stories is on and around Indian Point, which encompasses the hunting clubs of Merigold, Beulah Island, and Caulk Island, all located in Bolivar County, Mississippi, and Desha County, Arkansas. These hunting clubs are located on the east side of the Mississippi River south of Rosedale, Mississippi.

In the 1800s there were two main settlements along the river in this area. One was Prentiss in Mississippi, and Napoleon was across the river in Arkansas. These small towns acted as trading posts and stopovers for the paddle wheel boats along the river.

During this period of time all of the deer and turkey vanished because of no game restrictions. These animals were a target for food and were sold to frequent travelers along the Mississippi River. Prentiss disappeared during the Civil War and the inhabitants moved to Beulah and Rosedale. The settlers left behind the hogs that lived off the land and became wild after a few years. They were called feral hogs. Their numbers increased and they became a sporting animal because they were wild.

I remember quite well cypress log hog traps on Hog Pen Slough and Buffalo Bayou. These traps were built by the settlement of Prentiss and were used for commercial selling of hog meat.

The only permanent inhabitant on Indian Point was Harry Lampard. Harry worked as a light tender from Arkansas City to Rosedale. Harry had constructed a fence on one-half of Indian Point to hold his hogs. The fence ran five miles from the Mississippi River to the upper end of Lake Whittington. Red Wherry used the other one-half for his hogs. The acreage for each amounted to about five thousand acres. Harry moved to Rosedale in 1950 but returned to stay in his shack to sell hogs for a few years.

Hamilton Rials raised hogs on Caulk Island during the same time so this was another twelve thousand acres for these animals to roam. Beulah Island got the overflow from Indian Point and its land totals six thousand acres. This is approximately thirty thousand acres of hog hunting in one area all along the Mississippi River. Neither Harry nor Red could catch all their hogs each year so many grew to enormous size without any hunting pressure.

In the late 1920s, hunters arrived mainly for ducks and geese. There were very few deer and no turkey. These men had hunted deer in south Mississippi before moving to the Delta. They pooled their finances and purchased twenty deer that were turned loose on Indian Point around 1938. The hunters formed Merigold Hunting Club and would not hunt deer for seven to ten years. In order to keep the poachers out of the area, these men became avid hog hunters and stayed in the woods during the winter.

Hog hunting with horses and dogs became a sport in order for someone to always be in the woods to protect the deer. In 1945, deer season was opened and each member was allowed one buck per year. These men knew the hog population should be reduced and since baiting was banned, the sport for hog hunting took off. The hog dog became a valuable asset. It was a true test of the quality of the dog if it would not run a deer. The hog dogs were in great demand and the owner had no problem finding a place to hunt.

In 1954 when I was old enough to mount a horse, I began to hog hunt. I would ride on the back of the horse with my granddad for hours. The hog hounds would be in front of us and when they bayed I would jump off the horse and run as close as I could to the dogs and carefully pick my shot at the largest male hog. Sometime we would shoot three or four hogs. We never shot a sow with pigs as this was not sporting. After the shooting, we would rope the hog's hind legs and pull it to the closest road. Sometimes this took hours. These hogs weighed between 250 and 350 pounds and were hard to drag through the woods.

The members of Merigold Hunting Club got so good at hunting hogs, the population drastically declined after ten years of hunting. These same men decided to put a limit on the number of hogs taken from the woods, and made the hogs one of their game animals.

During this period deer season lasted only three weeks, with a split season during November and December. Placing a limit on the hogs gave hunters an opportunity to hunt longer during the winter.

Mr. L. L. Vance, from Drew, Mississippi, was a very good hounds man. Mr. Vance had deer dogs and separate hog dogs that would not run a deer. Most of the good dogs came from Catahoula Parish in Louisiana. It was nice to know that the dog was on the trail of a hog and not a deer. I shot many hogs with Mr. Vance and learned a great deal from him about ethical hunting and about the woods.

One year Hilliard Lawyer and Frank Wynn brought a man named Ben Passmore hunting. Ben was from the Ozarks and had ten of the best dogs I ever hunted behind. Ben would not ride a horse and went barefoot most of the time. When his dogs bayed, it was a good hog. The largest hog ever taken from the Merigold Hunting Club was in January 1964, on YanKum Bayou near Lake Whittington. C. L. Beckham, Happy Pleasant, my brother, and I were hunting with a dog named Ring and another dog named Kelly. These dogs got on a hog track around 9:00 a.m. and could not hold the hog until we arrived on foot. We were exhausted by 1:00 and started the two-to-three mile walk back to the truck. The four of us had almost made it back to the truck when we heard dogs barking in the distance. My brother and I were a lot younger than the others so we took off in a slow run and, after thirty minutes, we got into a thick cane patch on our hands and knees. I was crawling in front toward the bayed hog. The hog ran in our direction and I shouldered my Winchester Model 100 .308. I pulled the trigger to an empty chamber. My clip had fallen out during the crawl. My brother shot the hog at seven steps. We both stood up and looked at the hog and did not believe what we saw. It was large—very large! Mr. Beckham and Mr. Pleasant finally arrived and they could not believe what they saw either. It took us three hours to make a road for the truck to carry the hog out. We had to winch the hog into the bed of the truck. Mr. Beckham drove directly to Dossett Gin in Beulah to weigh the hog. Buck Frazier, the gin's manager, gave us a weight receipt that showed 527 pounds! I am sure someone has topped this but I can prove the weight on this trophy hog.

I have had the good fortune to hunt with quite a few good hog men over the years. There are many ways to hunt hogs. My favorite is to slip hunt into the woods in thick cover just before daylight. Hogs are a lot smarter than any deer. Shooting a hog from a tree stand is like shooting a squirrel on a limb. If a hunter stalks a hog in the thick cover and consistently kills a large boar hog, he or she is a true hunter and just not a shooter.

Dr. Jon Meyer has always been a master of hog hunting on the ground with a bow. He will not shoot a big boar because most of the time the angle of the shot is wrong and the hog will get away and die later. Dr. Meyer shot his first hog forty-five years ago with a Herters Ram X May Arrow. He is a very ethical hunter.

The late Bailey Peyton, from Hollandale, Mississippi, and I hunted together for forty years. We chased a hog for four hours one day. The hog swam Lake Whittington so we decided to go and get a boat, which took us three hours. We returned and paddled across the lake. I could hear the dog bayed on the high bank. We climbed the sandbar bank and we shot a 275-pound boar. The dog had been on this hog for a good eight hours. The hunt lasted all day for one good boar hog.

I have had the opportunity to take many trophy hogs and hunt with many fine men and dogs. These men who love the chase and love good dogs and horses are getting fewer in number. At the same time, there are a few left that have what it takes to enjoy the hunting and not just the shooting.

Mike Braswell, Jimmy Bruton, and Dr. Don Blackwood are among the few men who enjoy the hunting of hogs. Mike and Jimmy have good horses and great dogs and they use horns to call the dogs away from the sows with the little pigs. We were hunting in February 2007 and had not crossed the trail of a big boar. The time was getting close to lunch but I said to Mike, "Let's make a pass across the sandbar to the river." Most of our party decided to go back to the truck for a break. Mike knew I had seen a very big hog there during deer season and I would not shoot him because it was too difficult to get him out. The party of five horses and three dogs started off on the forty-minute ride to the Mississippi River and then we headed south toward a large log pile. I knew the big hog would be sleeping there if he was anywhere around. The dogs were fifty yards ahead and all of a sudden all hell broke loose. The boar came out after the dogs and the chase was on. Jeffrey Bruton shot the hog in full gallop with a .30-30 at twenty-five yards. This was the largest hog taken at Merigold Hunting Club that year.

Hog hunters are a different breed. They don't hunt in stands, they don't use mineral licks or bait, they don't use trail cameras, and they don't shoot sows with little pigs.

The area that I have hunted for the past fifty-two years probably has some of the best wild boar hunting in the U.S.A. I am very fortunate to have had the opportunity to share my experiences with other sportsmen and sportswomen.

Two Hogs, One Bullet

DYLAN CHRISTMAN (AGE 7)

It was Monday, July 13, at about 10:30. My dad, my papaw, and I were at Caulk Island riding around in the yellow hunting vehicle. There are high seats in the back and I was in the middle seat. I had a .308 bolt action in the gun holder by me. I also had a knife and a .357 pistol. We were riding around looking for deer in the big field where I killed my eleven-point buck in December. We drove in the big grassy field; the ground was hard and the grass was yellow. I told my daddy to look to the left, and he did, but I took my gun and aimed to my right where I saw two hogs and I shot one time. Two hogs went down. We drove closer to them and found that my one bullet had killed both of them! They were almost full grown, about seventy pounds each and the tusks were one inch each. The bullet went into the first one's stomach and then into the second hog's jaw and windpipe. My daddy picked up both hogs and said, "These jokers are *heavy*!" We took them to the skinning shed and kept the meat for summer sausage and breakfast sausage. It was a *great* hunt!

Horns and Hogs

WICK EATHERLY

I've been hog hunting since 1979, about thirty years. It's too much trouble to hunt with a knife; you have to get too close. So, I always use a gun. Mike Braswell is who I always hog hunt with. I started hunting with my uncle and Mike's daddy-in-law, who had been hog hunting since the sixties, and we took it up from them. We wanted to hunt hogs. It's fun! We went through a lot of dogs before we got a good pack, and we use all kinds of dogs. We don't use catch dogs, per se. They get killed too easy. Our dogs are bay dogs—no bulldogs. Our dogs are silent trailers and they can get up on the hog quicker. The hog will turn on you and you gotta stand your ground—you can't run. I always ride the horse. On the horse you can hear better and cover more terrain. When I hunt I like to go with only three or four people. I want to see everybody, especially when I shoot. I use a .375 Winchester rifle and a pistol, just in case I have to get off my horse. Hogs have cut my chaps before when I got too close. You got to kill the hog before he gets to you. If the hog turns and looks at you instead of the dogs you gotta do something quick. A hog is way quicker than most folks think and they can run, jump higher, and even swim faster than most people think and the thick stuff doesn't bother them. Some hogs have a very strong odor. Some male hogs are too strong especially when breeding. But whatever I kill, I bring out. I never leave anything I kill in the woods. Somebody gets the meat. Lots of people who live around me and know I am going hunting say, "Bring me one, bring me one." You can use and eat everything on the hog.

One time we were hunting on the Mississippi River with some boys from Hollandale. We got down off the horses because it was so thick and we couldn't see. Well, the hog broke and came charging after one of the boys. He started running and tripped and fell. The hog started coming right after him fast. This boy's brother ran after the hog and shot and killed him, and he and the hog both fell on top of the boy who fell.

A hog is ten times smarter than a deer and can smell twenty times better than a deer. It won't take the pressure from a lot of hunting. The hogs will leave the area if it's hunted a lot. They won't stay until they starve to death. They are gonna find something to eat.

A smaller hog, say, 100–150 pounds, is better to eat. A good hog to eat is about a year and a half old. There is not a hunting season for hogs. You can hunt hogs any time of the year—on private land only.

Mike Braswell, who I hog hunt with, has twelve dogs that wear tracking collars so we can keep up with them and find them so they won't get lost. The range is up to five miles out in the open and, in the woods where it is dense, two miles. Every now and then we'll have a dog get cut by a hog. Sometimes thick briars keep a dog from being able to get away. Most dogs know to stay back and don't have that catching tendency, but a catch dog will catch the hog and hold on to him. The hog's hide is too tough for the dog to hurt the hog. A .22 caliber bullet won't go through the shield on a hog's shoulders. You can't stick a knife in the guard plate on a hog and so you have to go underneath the hog and stab it in the heart under its front leg.

My horse stands still no matter what. I can shoot off of him all day long. His name is Slick. One time, a hog came up under my horse and knocked my feet out of my stirrups! The hog came so fast and I shot him twice but it didn't do anything; he just kept on coming. My horse didn't do anything but just stood there. That hog got so close that I couldn't shoot him again and so the other boys I was with shot him and killed him.

I try not to get off my horse; I shoot off my horse because it's safer than being on the ground. Before everybody got sophisticated we used horns to communicate. Mike Braswell has a goat horn and I have a cow horn—each has its own distinctive sound and that's what we used before we had radios. We would blow the horn for different things. Blow three times: killed something. Blow two times: missed. Blow a lot: come quick, bring yourself over here, we need some help! That's the way we used to communicate. Horns were all we ever used. Now, we use radios, which is more convenient.

I always have hog meat in the freezer. We hunted on two or three good spots this year and got some good hogs. Cooked eight hindquarters weekend before last.

One time, I had been deer hunting all day and had my shotgun. The deer dogs bayed up the hogs and I was duckwalking through the thicket. I looked up and saw a huge hog twenty feet away popping his teeth at me. When they pop their teeth they are sharpening them and I saw another hog just ten feet away. I shot and killed one and shot at the other but he ran off. A boy called me two weeks later saying he got my hog and I said, "How do you know?" He said the hog he got had triple ooo shot in the shield on his shoulders and he probably weighed 250 pounds. There wasn't anything around I didn't think I couldn't kill with triple ooo buckshot.

Mike Braswell has the hog dogs. I have the squirrel and deer dogs. We have better luck hunting hogs in the morning. The dogs can smell better in the morning and the hogs move at night. The dogs are winding dogs rather than tracking dogs. Everybody's got a certain kind of dog they like to hunt with. One half bird dog, one half bloodhound. One half Plott hound. The Plott hound originated out of the Carolinas and Tennessee. Also, there are redticks and blueticks.

I squirrel hunt more than anything. I carry the children and grandchildren almost every afternoon. For fifty-four years I've been hunting at Greasy Bayou. I love it! I go every chance I get. We have a "large" time!

The farthest north I ever been hog hunting is somewhere, I think, around Rosedale. But I basically hunt in the Delta. Hunting in the hills is so hard because there are no large tracts. Hog hunting needs a lot of land, say, two thousand acres. Farthest south I've been is Natchez and Port Gibson. It's so steep down there though. I don't like it. It's too rough on the horses. But I like to hunt in the Delta. I ride all day; no matter what kind of hunting I do, I am on a horse. I hog hunt all over the Delta, but I'm not saying where.

People will call, farmers that is, and say, "The hogs are in our corn." A herd of hogs will go down a row of corn that's been planted. They will eat all the kernels and go down row after row after row and eat it all. Once the corn comes up they will leave it alone. But hogs will go straight down that row like a tractor does. Lots of people can't plant corn in certain places because of the hogs.

Everyone in the Delta knows one another and they'll call if they need help getting rid of hogs.

Hogs and Eyeballs

ASHLEY HINES

The door of the cabin swung open and the hunter welcomed us into his home. Our eyes adjusted to the dark after the bright May sunlight. The room was small, with a stone fireplace. The stuffed head of a boar, old and grizzled, with grey tips on his hair, hung on the wall next to the fireplace. The air reeked of wood smoke. The hunter was tall, very tall. He wore blue jeans, boots, a green canvas shirt. He showed us to a trailer behind the house where we would spend the night. We took all of our gear into the trailer and divided up the sleeping arrangements.

We ate dinner and passed the evening playing low ante poker.

The next morning, we were awakened at a civilized hour by one of the other hands. He was quite the contrast to the hunter. He was scrawny and malnourished. His hair was greasy and unkempt. He led us outside to our ride. The vehicle was a Jeep. The body was rusted and the grey paint was splotchy. The Jeep had been customized with a school bus seat welded by the legs over the rear fender wells. The tailgate had been removed in favor of a cage door. Underneath the bus seat was a crowded accommodation for four or five dogs.

The hunter walked up from the cabin, followed by a pack of dogs, some hounds, others various members of the terrier family. There was one scrawny little bulldog, very small, with a bad eye and a deep furrow in the top of his skull. The hunter told us the dog had been hit in the back of the head by a careless client. The bullet had been deflected by the dog's collar, so it had only gouged a groove in the dog's head, taking the eye with it.

We climbed onto the Jeep, sitting high on the back seat. The dogs were loaded underneath and we were off, careening through the woods, dodging low-hanging foliage. The wheels of the Jeep splashed through muddy holes left by the spring rains, spraying fountains of the brown water in the air. We stopped and let the dogs out, to seek out whatever hogs they could find. We drove on through the woods to gain a lead on the pack.

We dismounted the Jeep at a turnout along the road. A path led up a ridge from the road. We followed the path through the woods. The foliage was thick and green. The path was moist and muddy in places. The trail forked after a short distance. Jesse and the guide took the right path, which appeared to continue climbing the ridge. Alone, I walked along the other fork, descending into a hollow. In the distance, I could hear the faint howl of the hounds as they picked up the scent of the pig. A shot rang out from the top of the ridge. I heard a rustling in the underbrush. Suddenly, eight or ten hogs burst over the top of a rise about thirty feet from me. I couldn't shoot, since Jesse and the guide were up the ridge and might be in my field of view. Taking a shot in their direction was too risky. I had never seen a wild pig before, and I didn't know if they would be aggressive or not. I slipped behind the shelter of a large tree and the pigs ran by on either side.

I walked back to the Jeep. Jesse had killed a pig. Due to the warm temperature, we had to take the pig back to the camp quickly, before the heat ruined the meat.

We dropped the pig and headed back into the woods. We had lost the dogs now, so we were on our own. We drove slowly through the woods, keeping our eyes on the forest floor. We came around a bend in the road where a hollow opened up into the flat. Trees arched over the bottom, forming a canopy of shade. The hollow curved away from us into the distance, with a soft mossy carpet underneath. Up the hollow, a hairy black hog had his nose to the ground, rooting up the soft earth. Now was my chance to take care of the hunt from the relative safety of the high-backed seat. I told the driver to pull up slightly, to place the hog a little behind the side of the Jeep. I placed the fore end of the rifle diagonally across the top of the bus seat and found the hog with the telescopic sight of the rifle. I placed the crosshairs behind the pig's shoulder and fired. The pig dropped in a heap. He had never moved. I climbed out of the Jeep and walked toward the carcass. I stopped behind a small tree and glassed him through the scope again. As I said, I had not seen these creatures before, and I was unsure of their propensity for aggressive behavior. I wanted to make sure he was dead. I watched him for a while, and he had still not moved. The guide pulled up behind me with the Jeep. We loaded the hog onto the back. The man asked us if we wanted some eyeball. We were unfamiliar with the term. He pulled a mason jar from a cooler in the floorboard. The jar contained moonshine, flavored with grape drink mix. The eyeballs were grapes resting in the bottom of the jar. We celebrated the successful hunt by polishing off the eyeball.

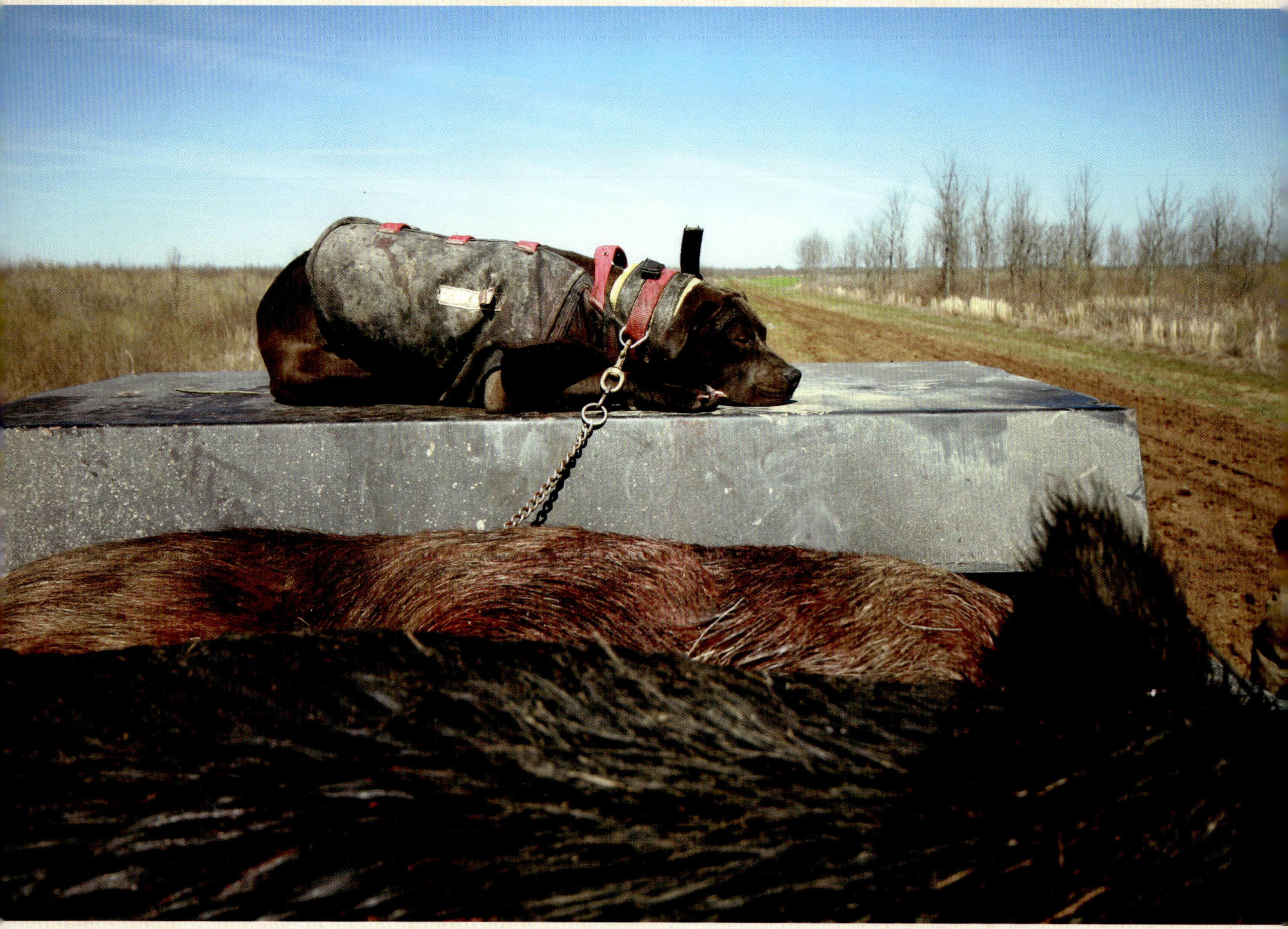

Pit bulldog, resting after hunt, and harvested boars.

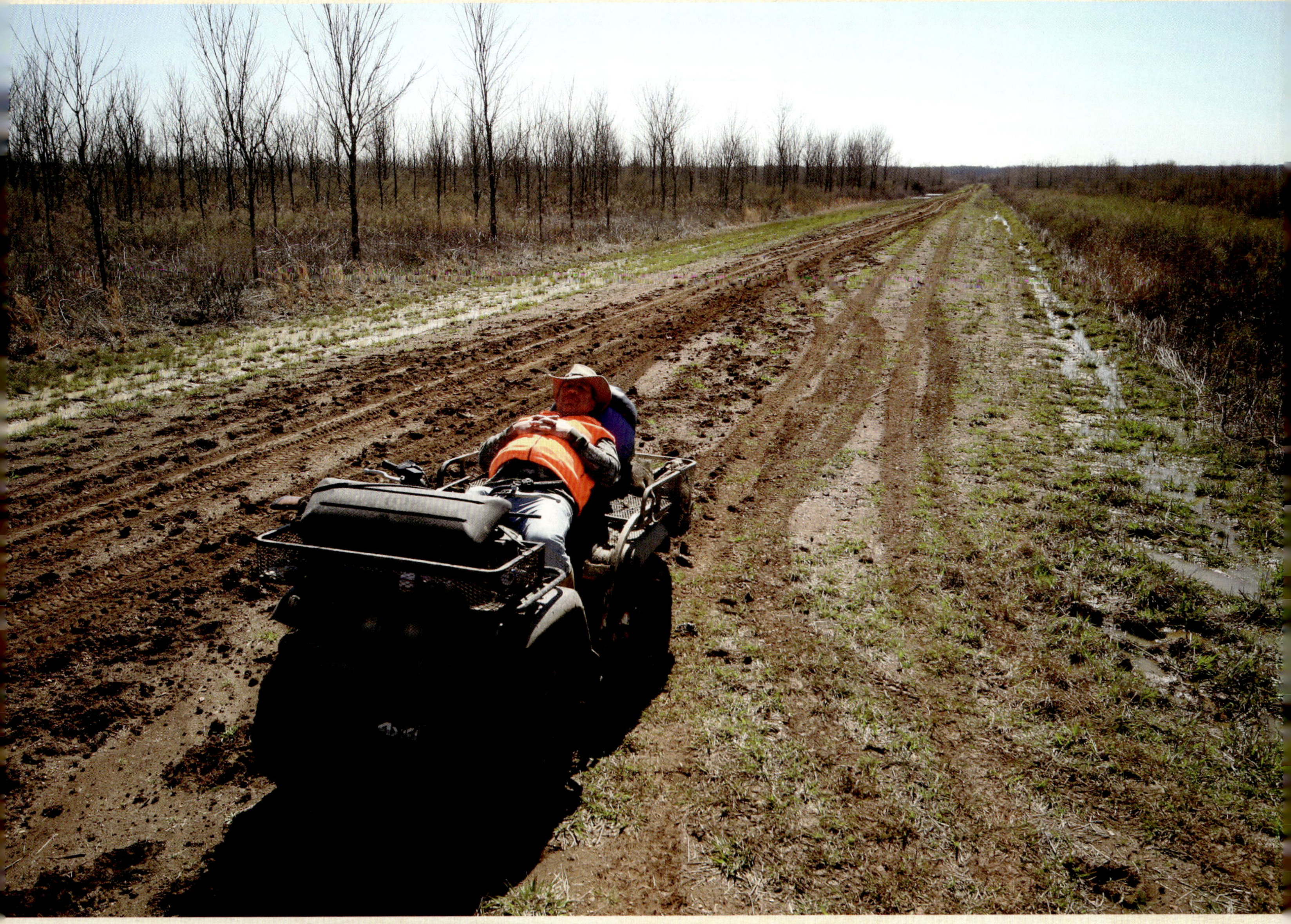

Resting.

Overleaf left: John Lowery and Harley, a Tennessee Walking Horse.
Overleaf right: Junebug, mountain cur.

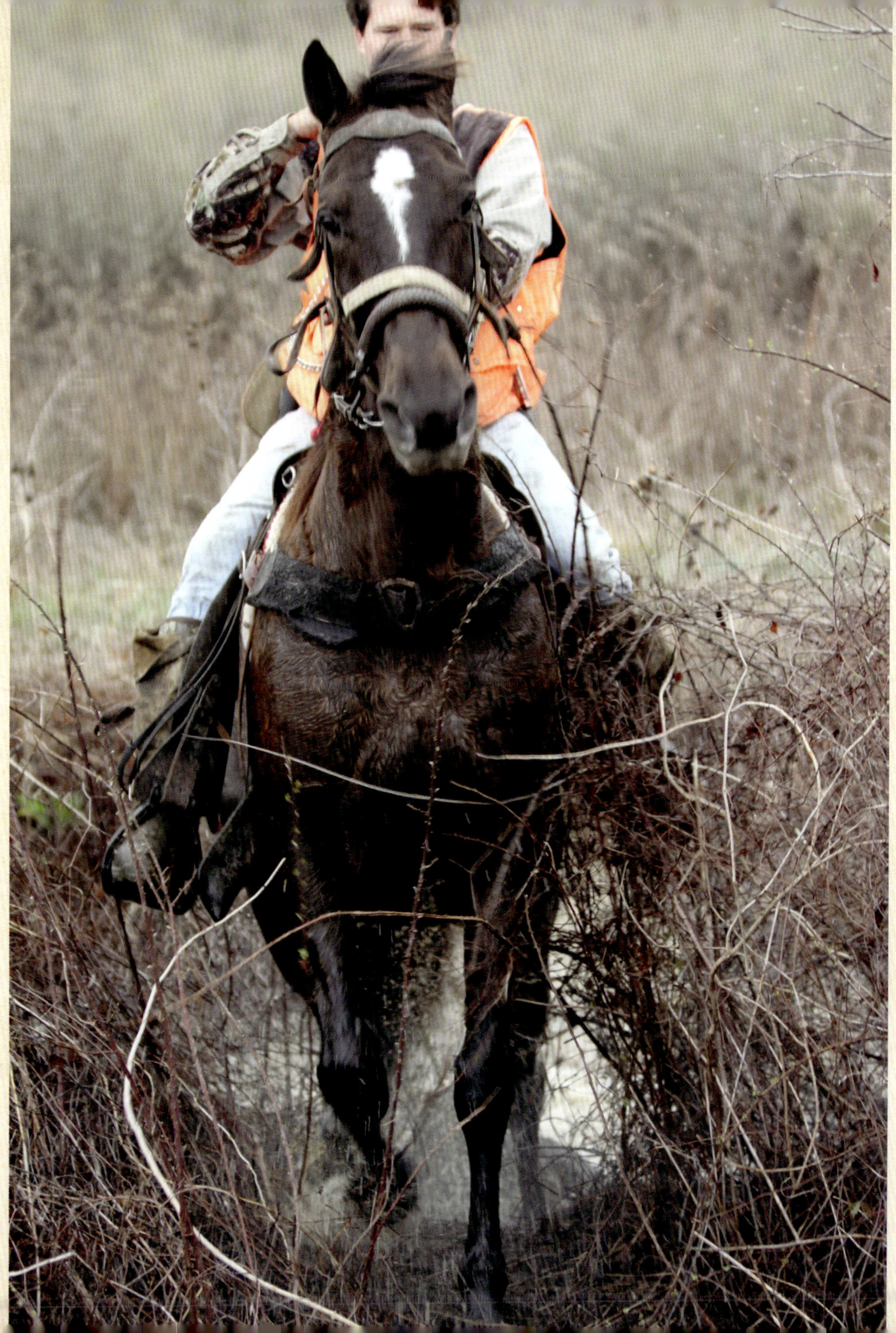

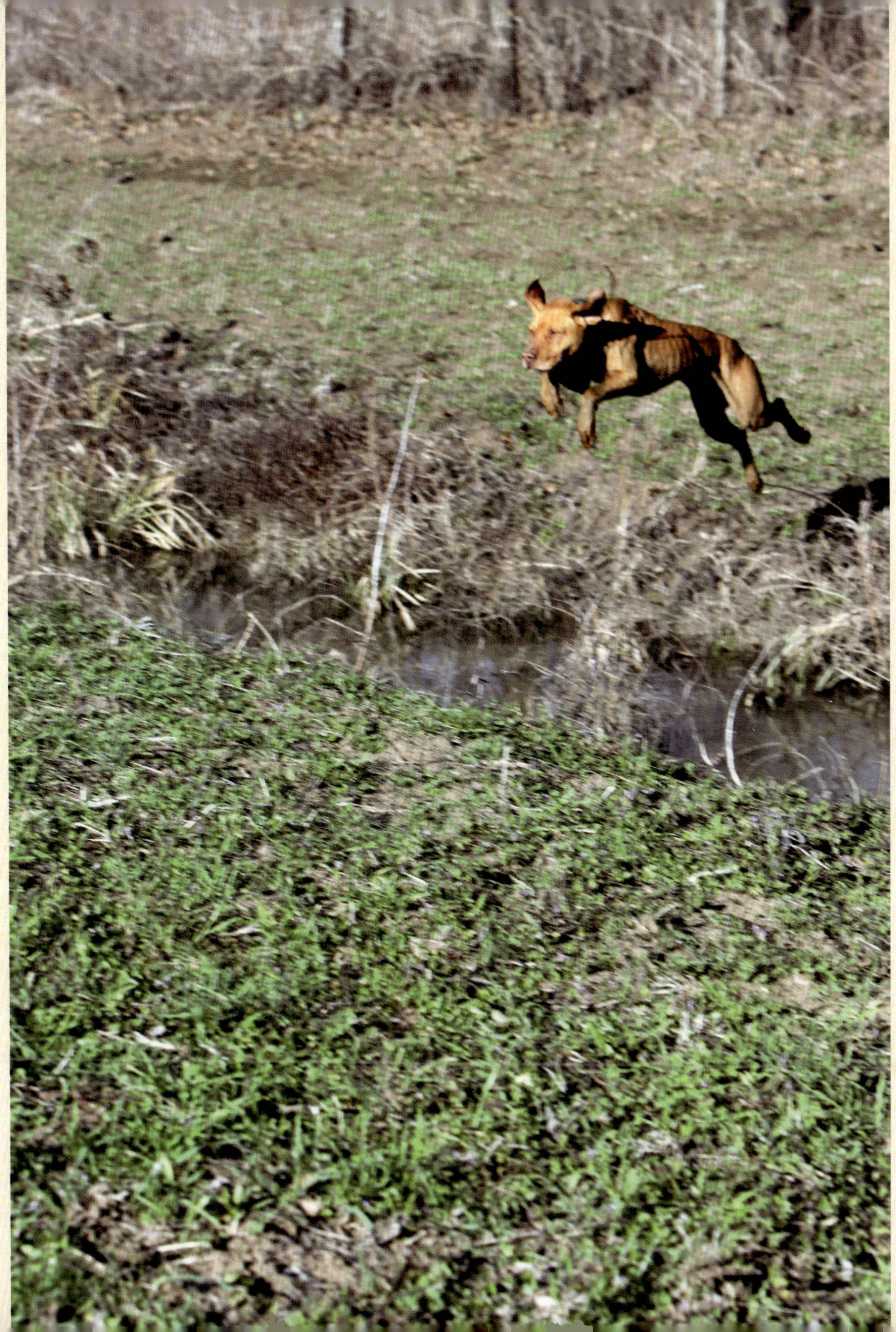

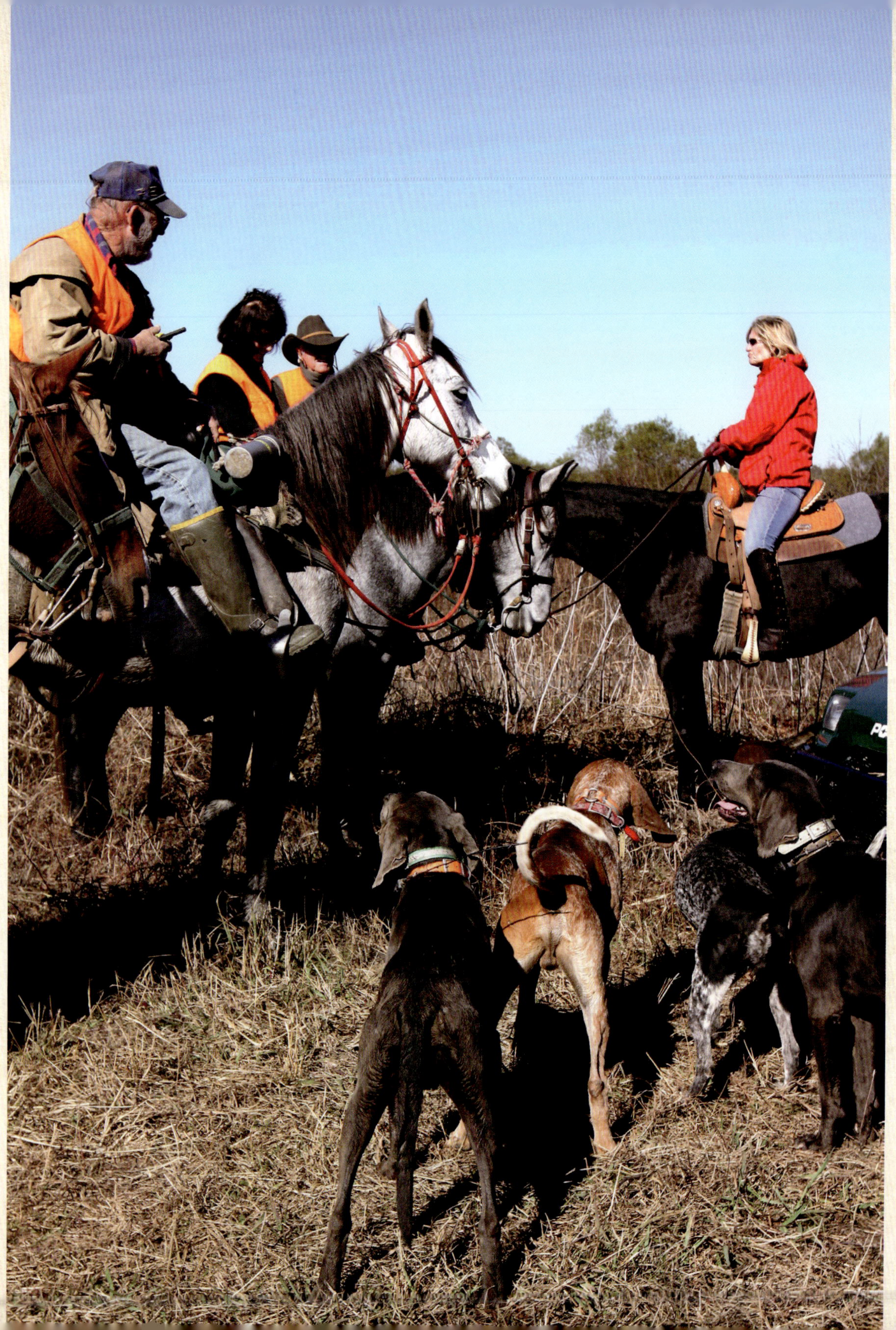

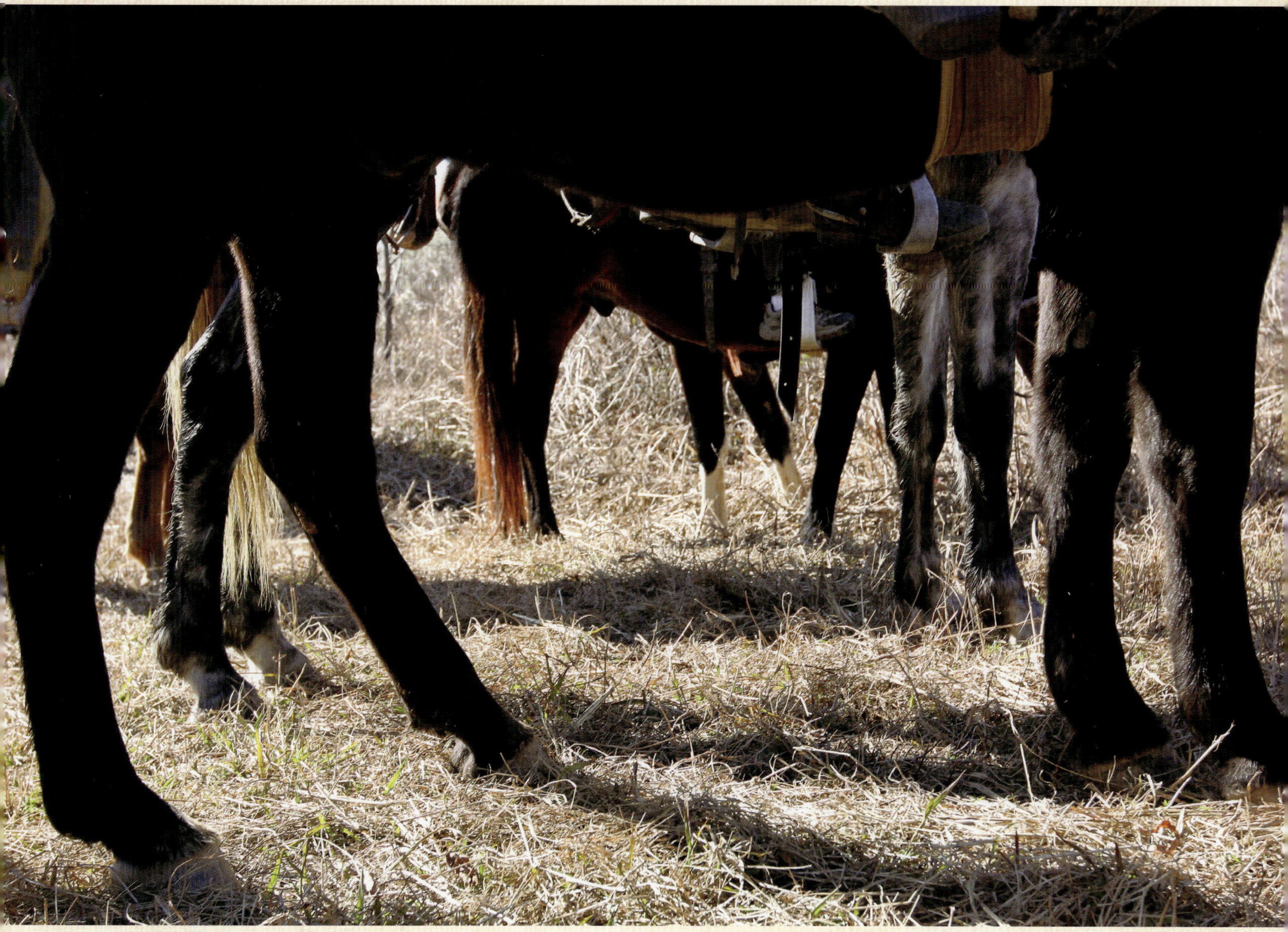

Underbellies.

< Melvin (Bubba) Weeks, Robert and Bobby Jo McConnell, Mary Hardwick, and hounds.

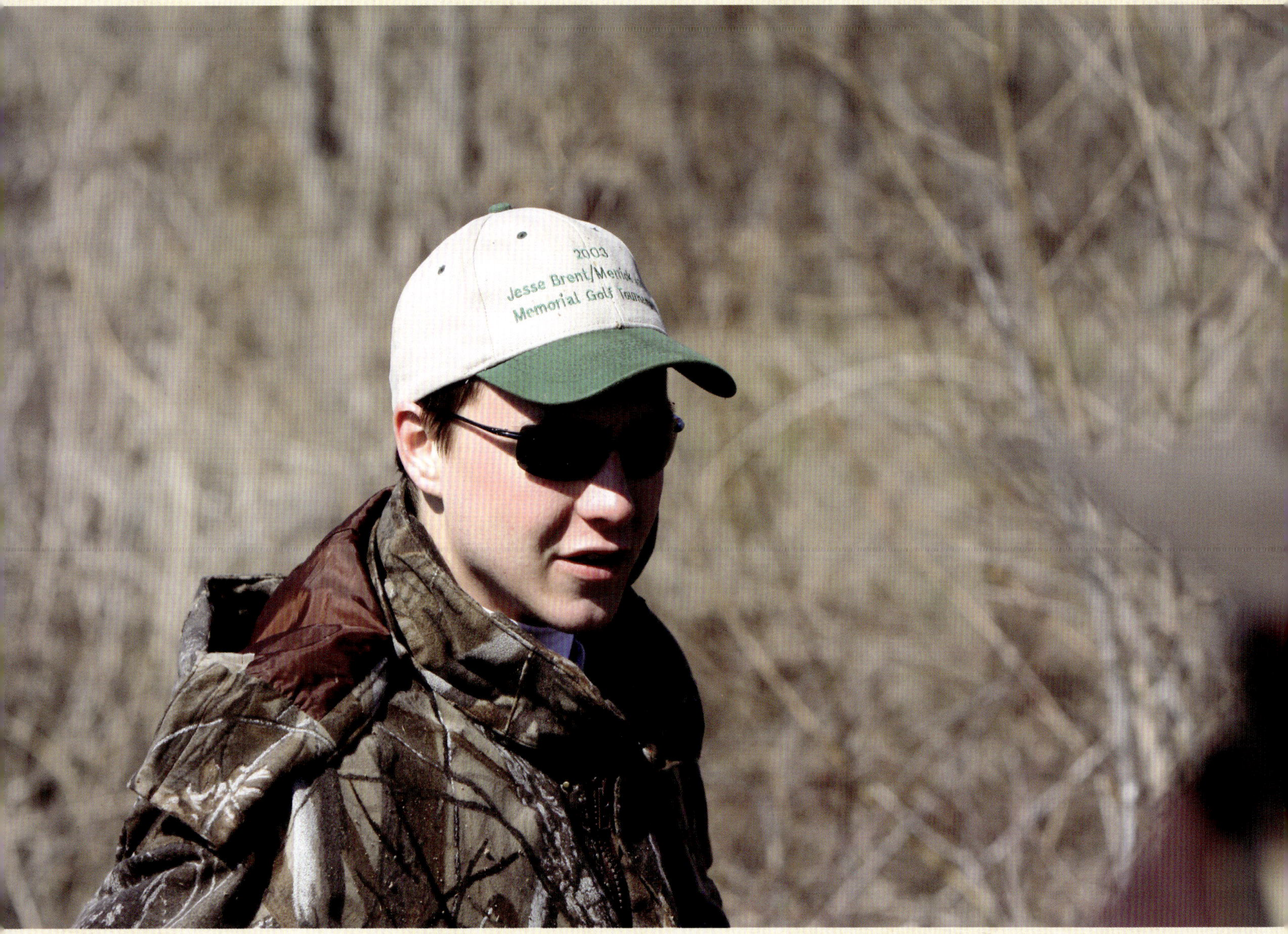

Taylor Haxton.

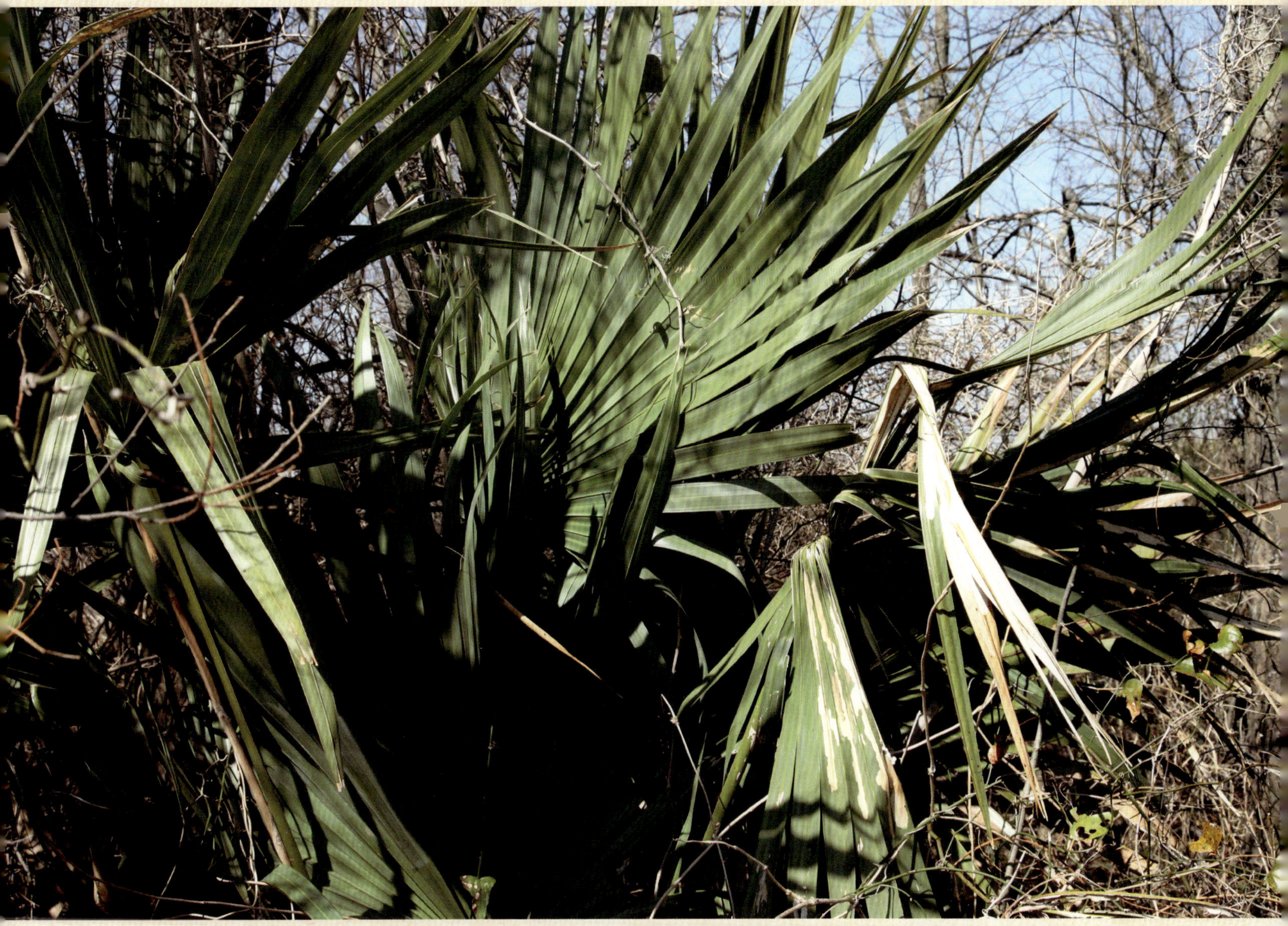

Wild palmettos.

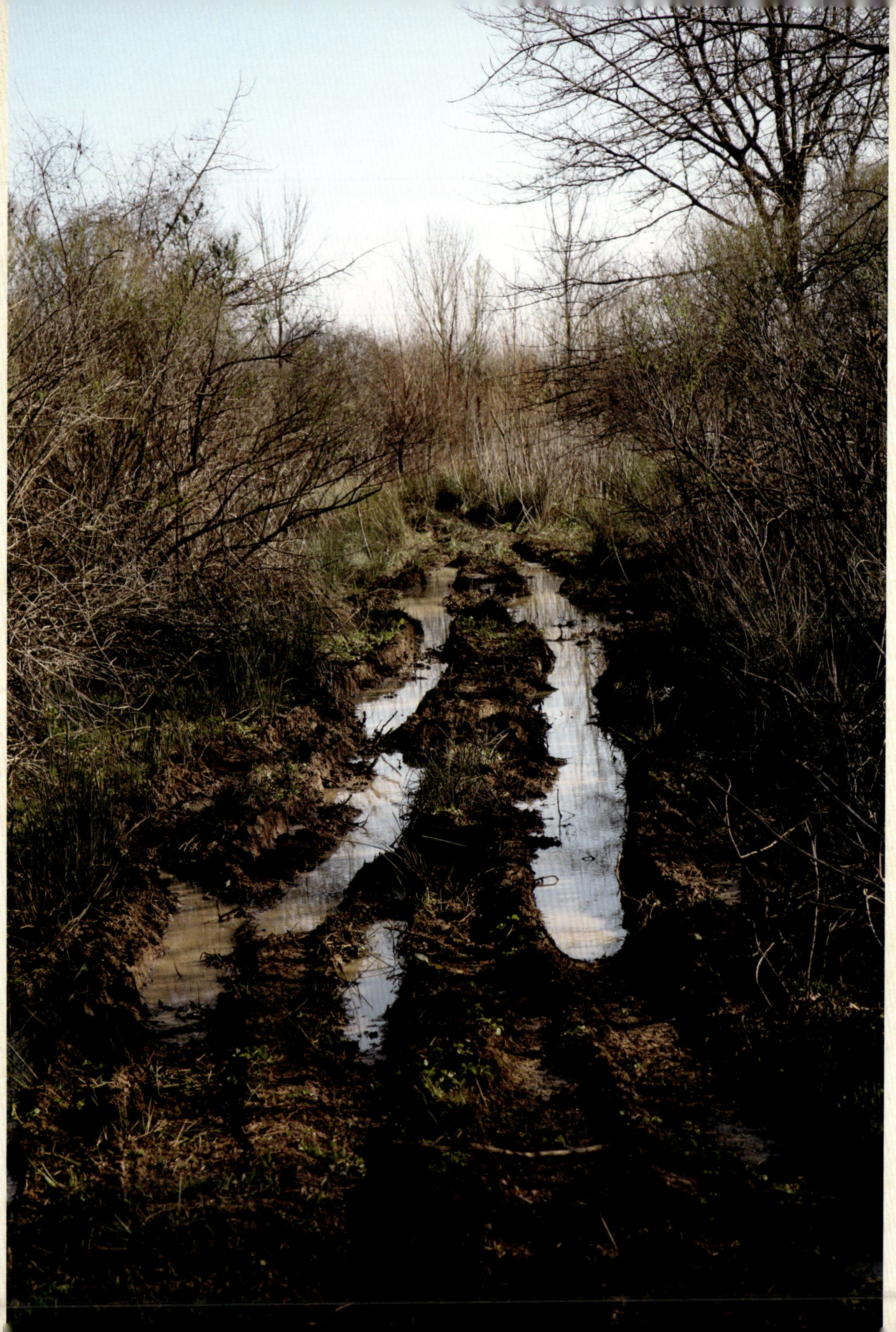

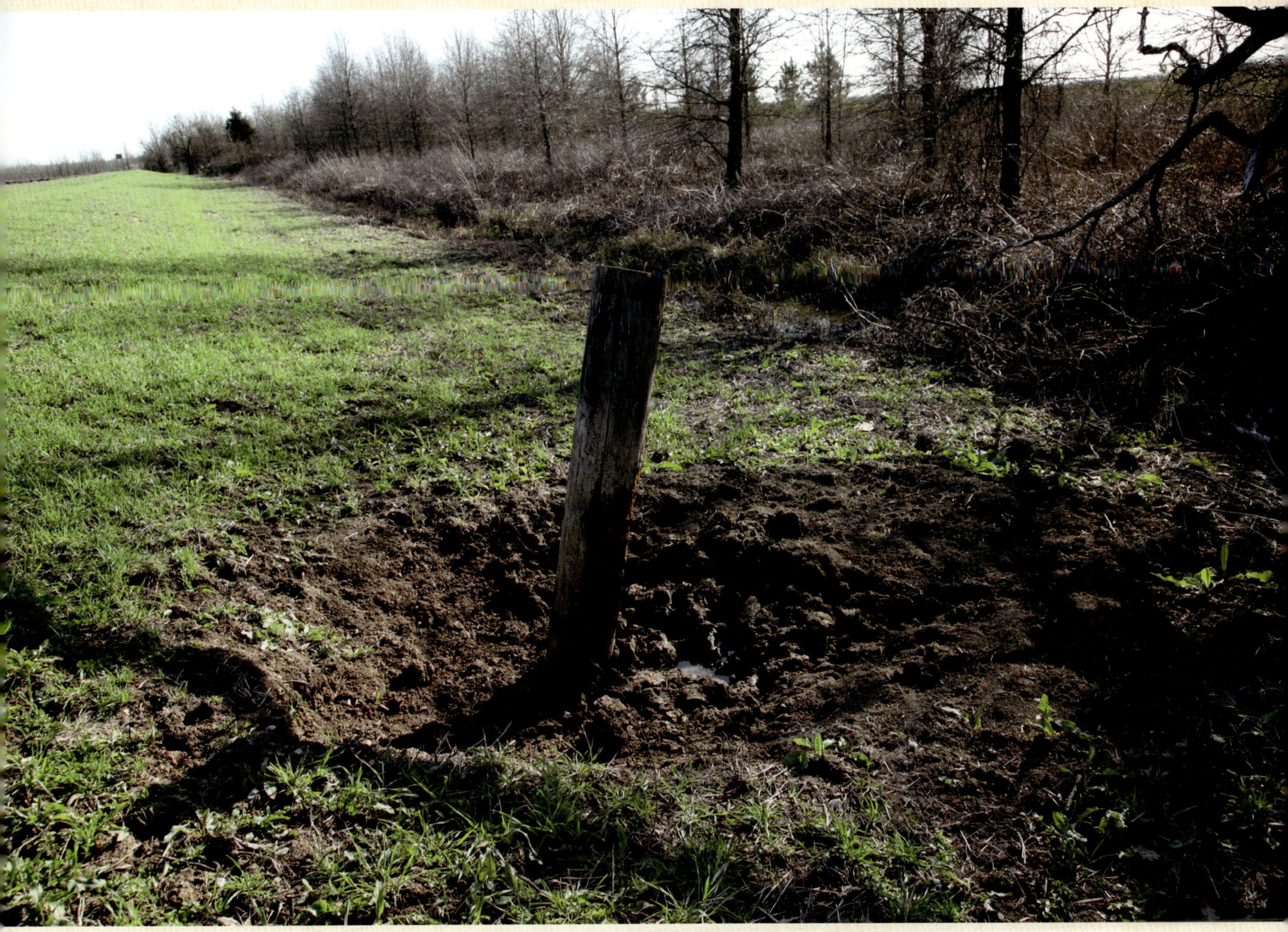

Rubbing post.

< Tracks.

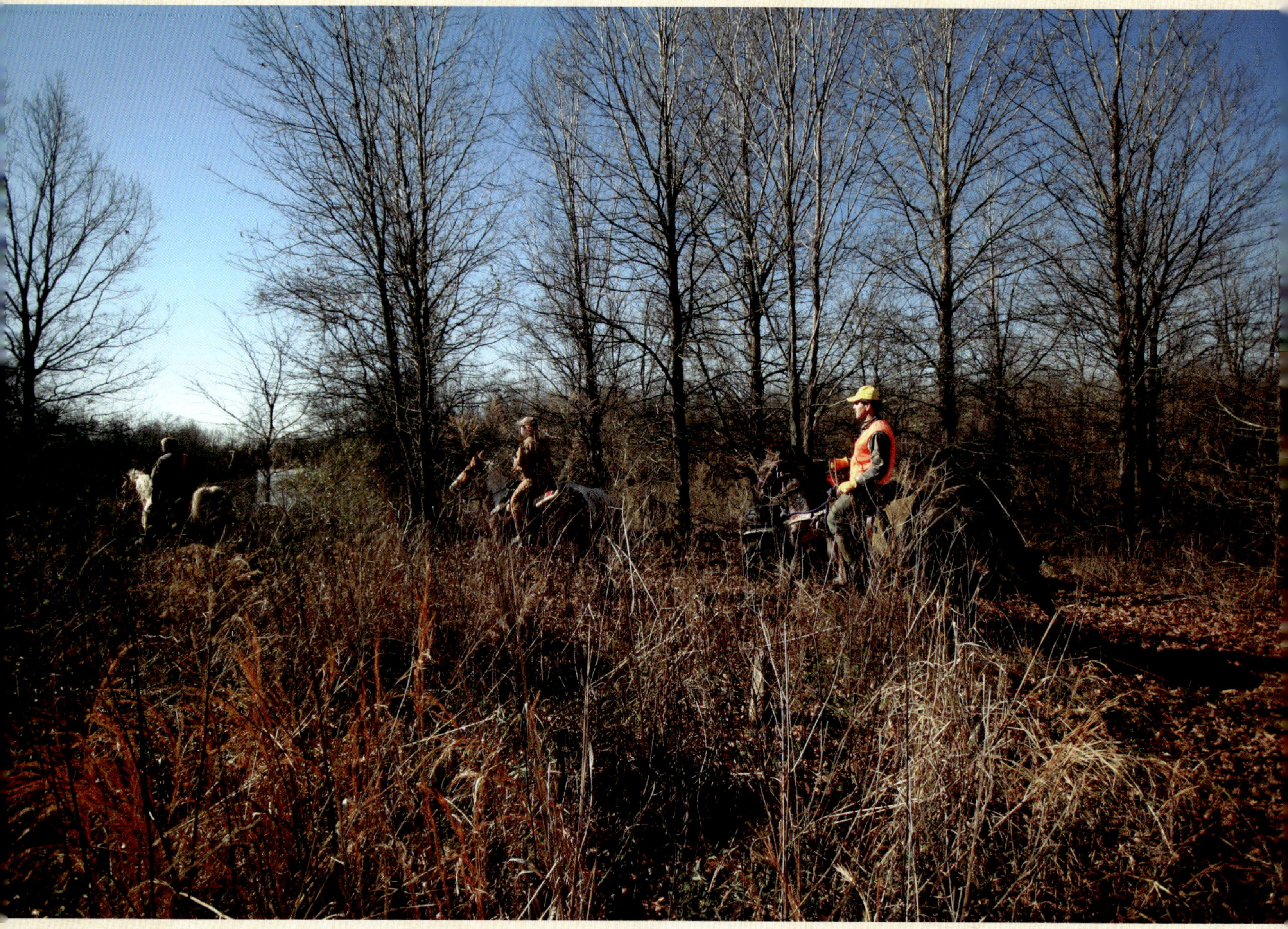

On the hunt.

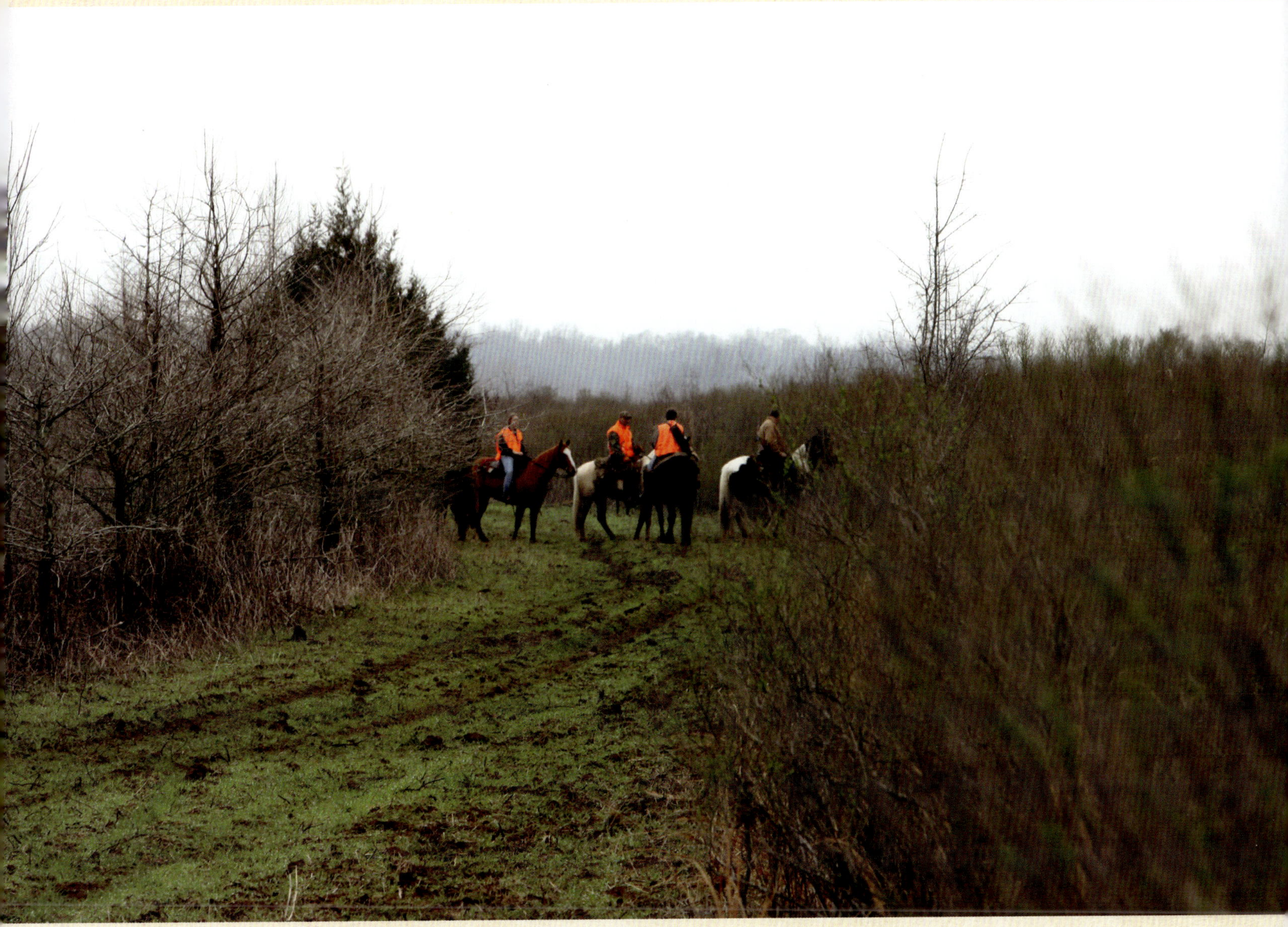

Stacey Waites, Andy Wilkinson, John Lowery, and Steve Ratcliff.

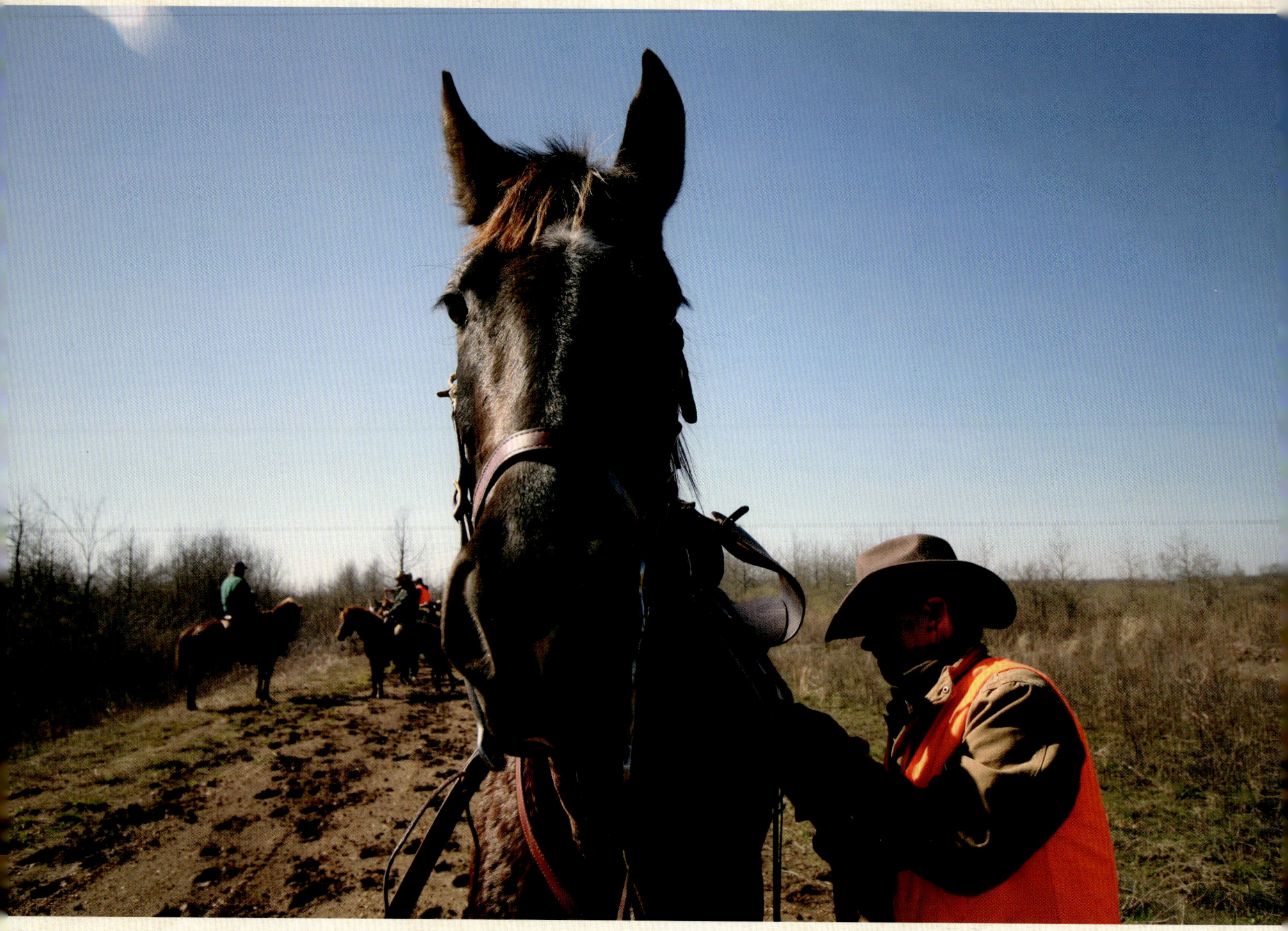

Steve Golding and Star, a Tennessee Walking Horse.

John Lowery, Frank Dantone, Steve Golding, Steve Ratcliff, and Hank Burdine with harvested boar. >

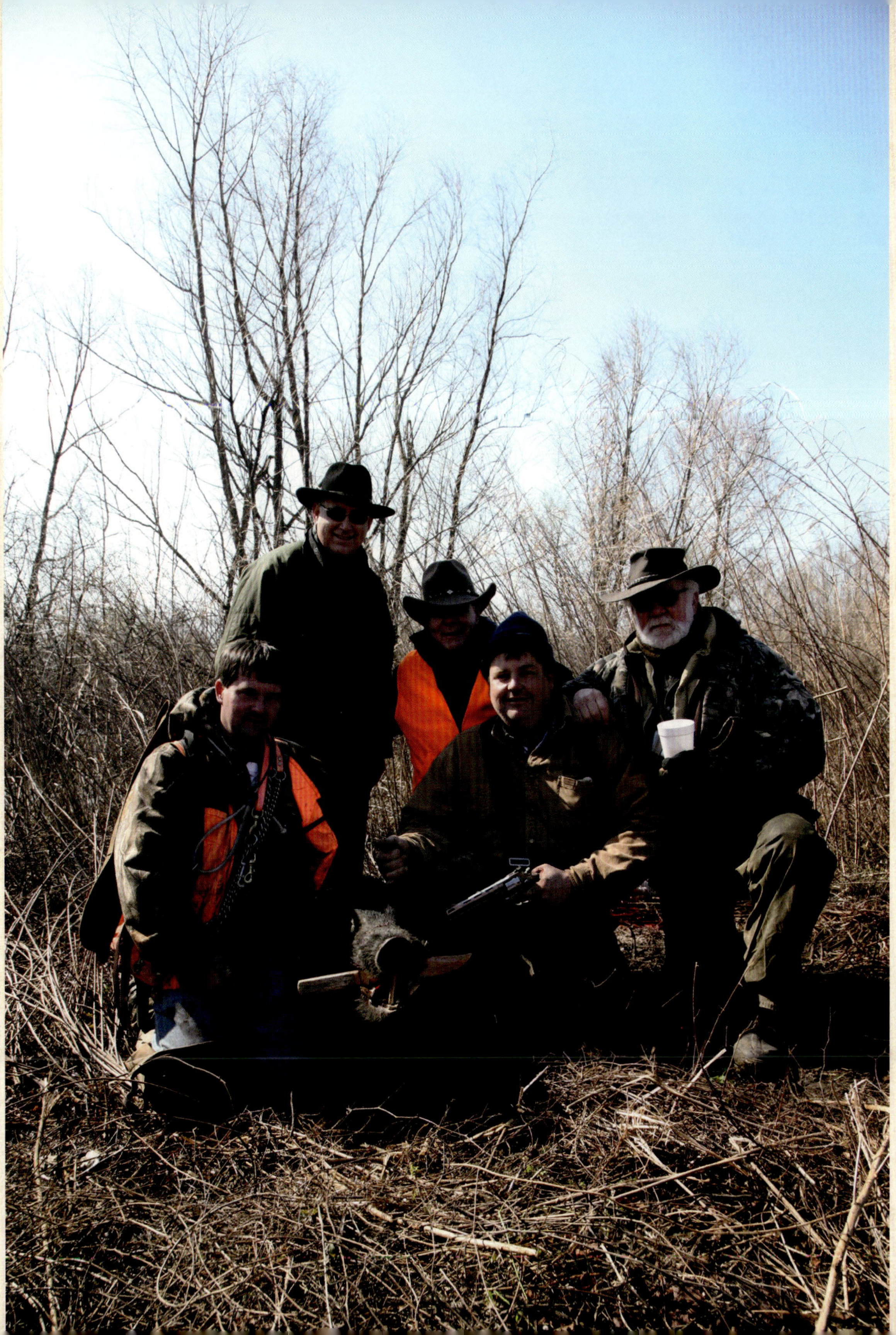

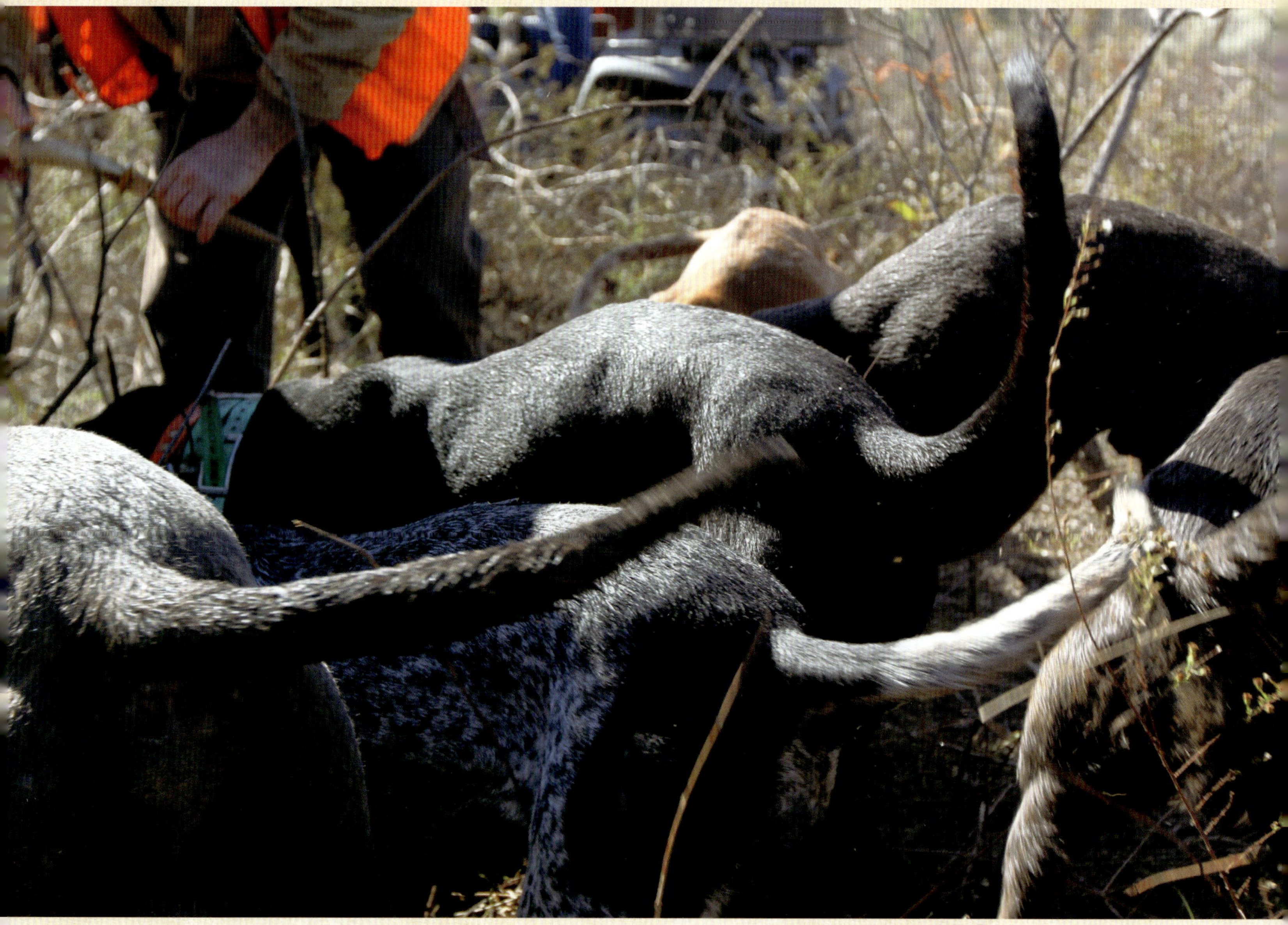

Hound tails.

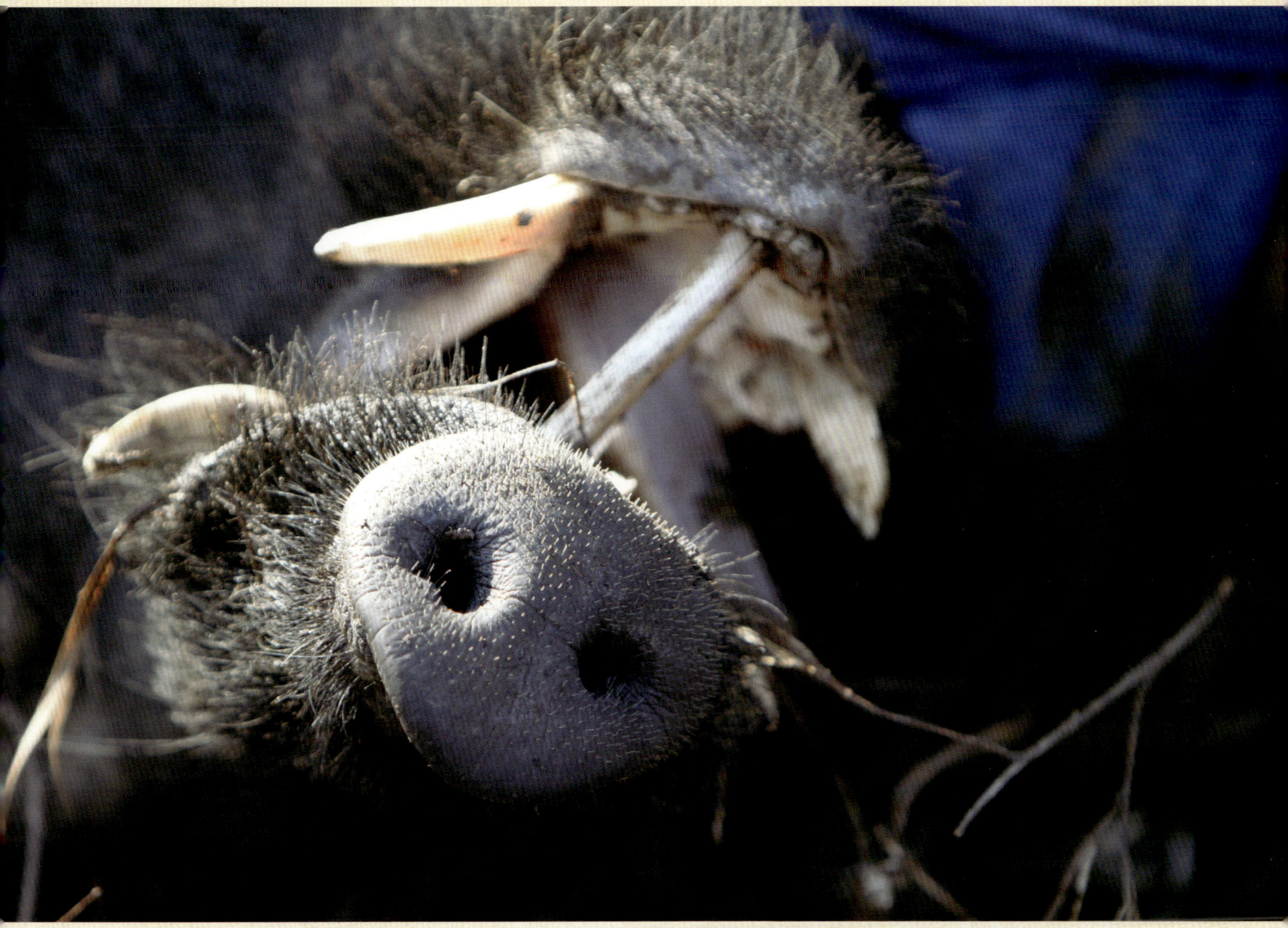

Wild boar with tusk.

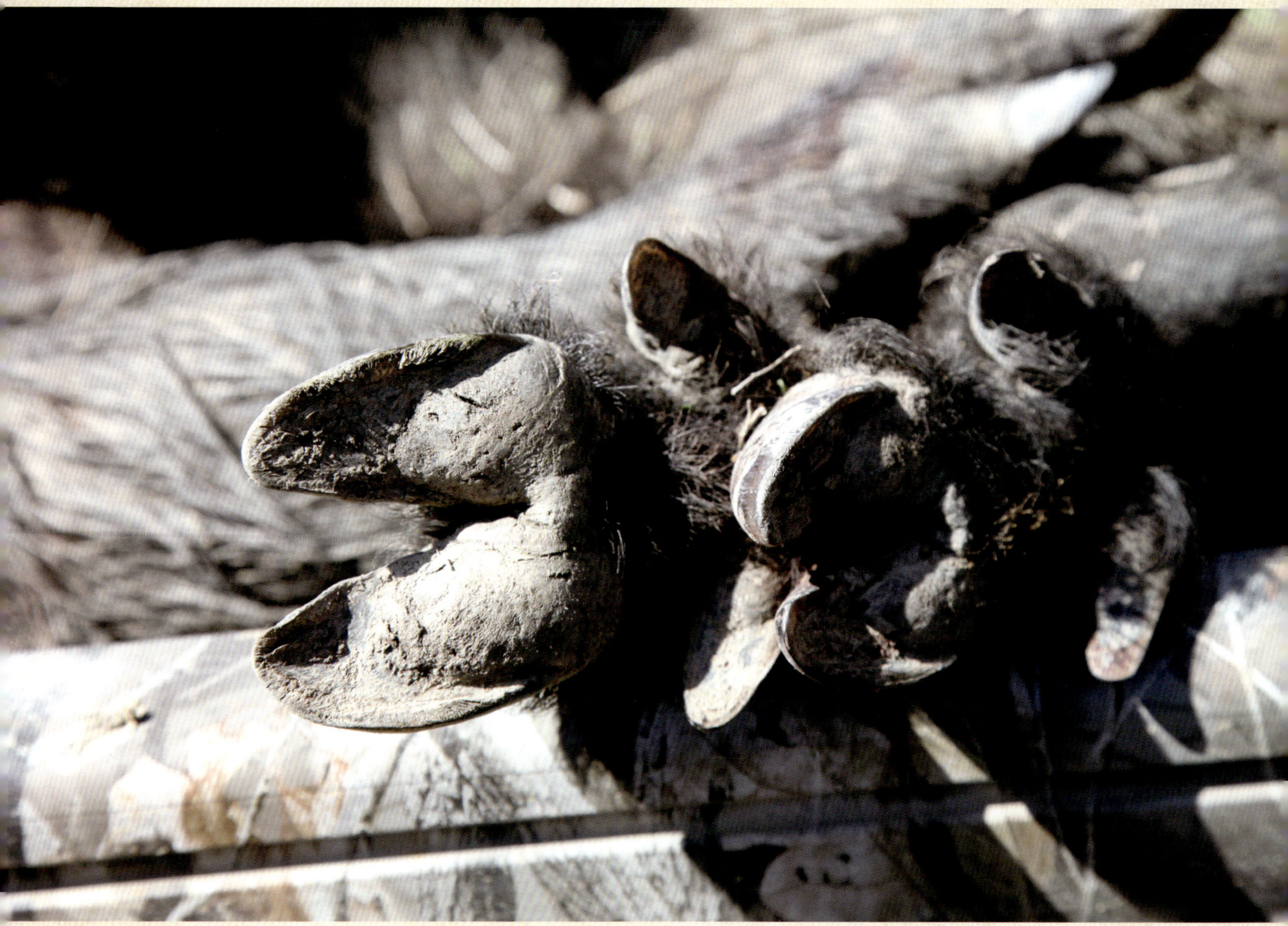

Pig feet.

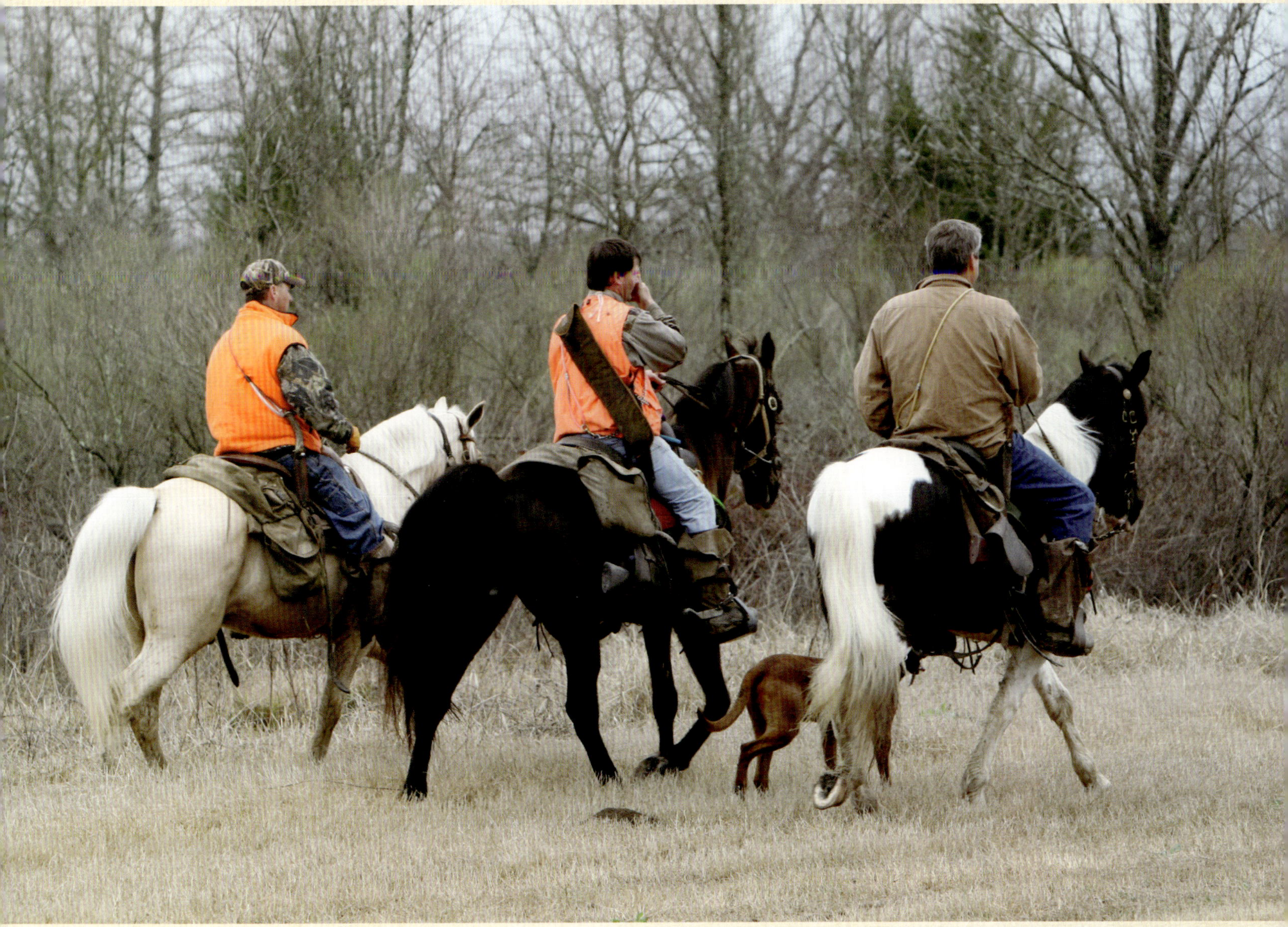

Andy Wilkinson, Vegas, John Lowery, Harley, Steve Ratcliff, Chief, and Levi.

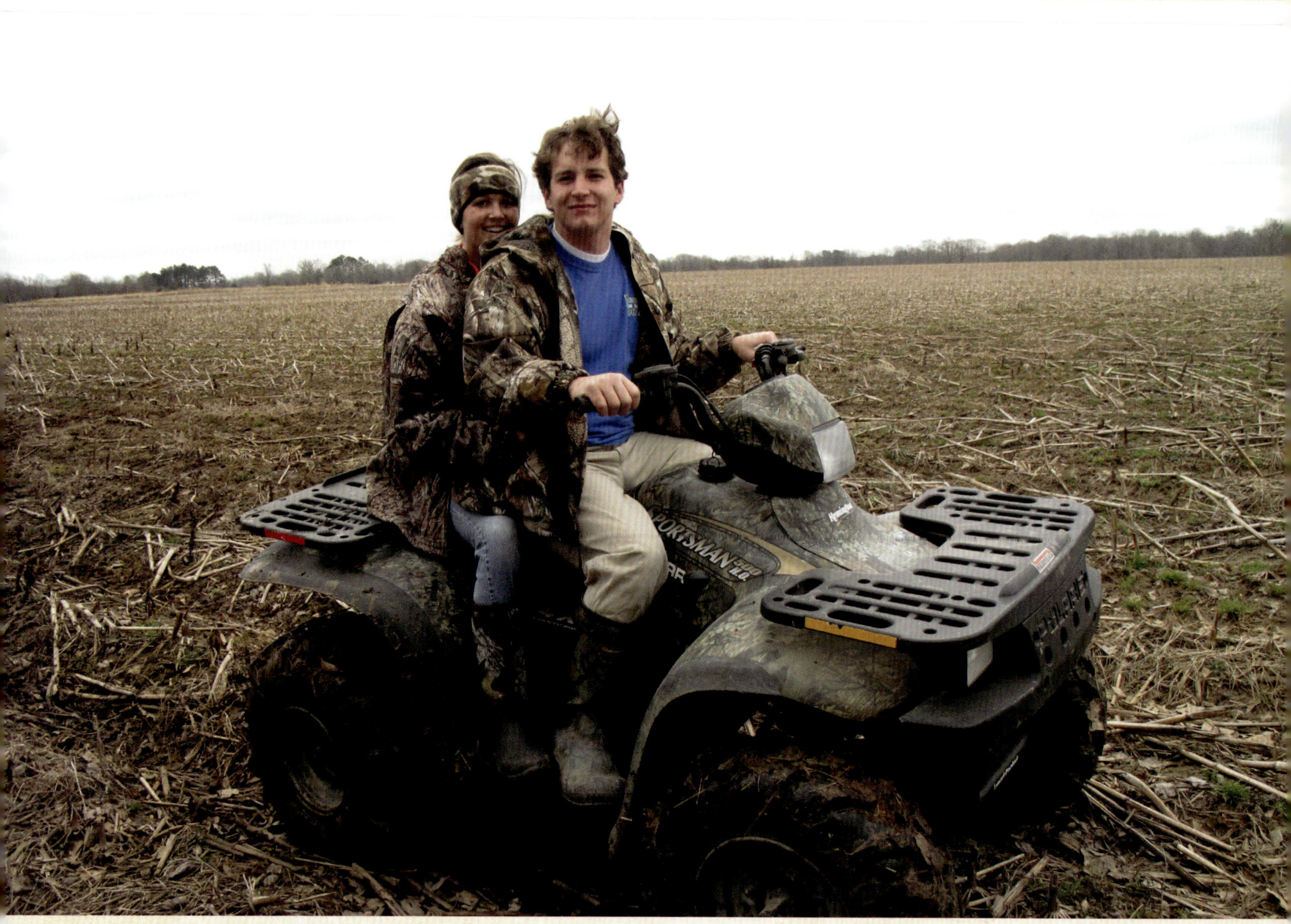

Carley Baker and Austin Golding.

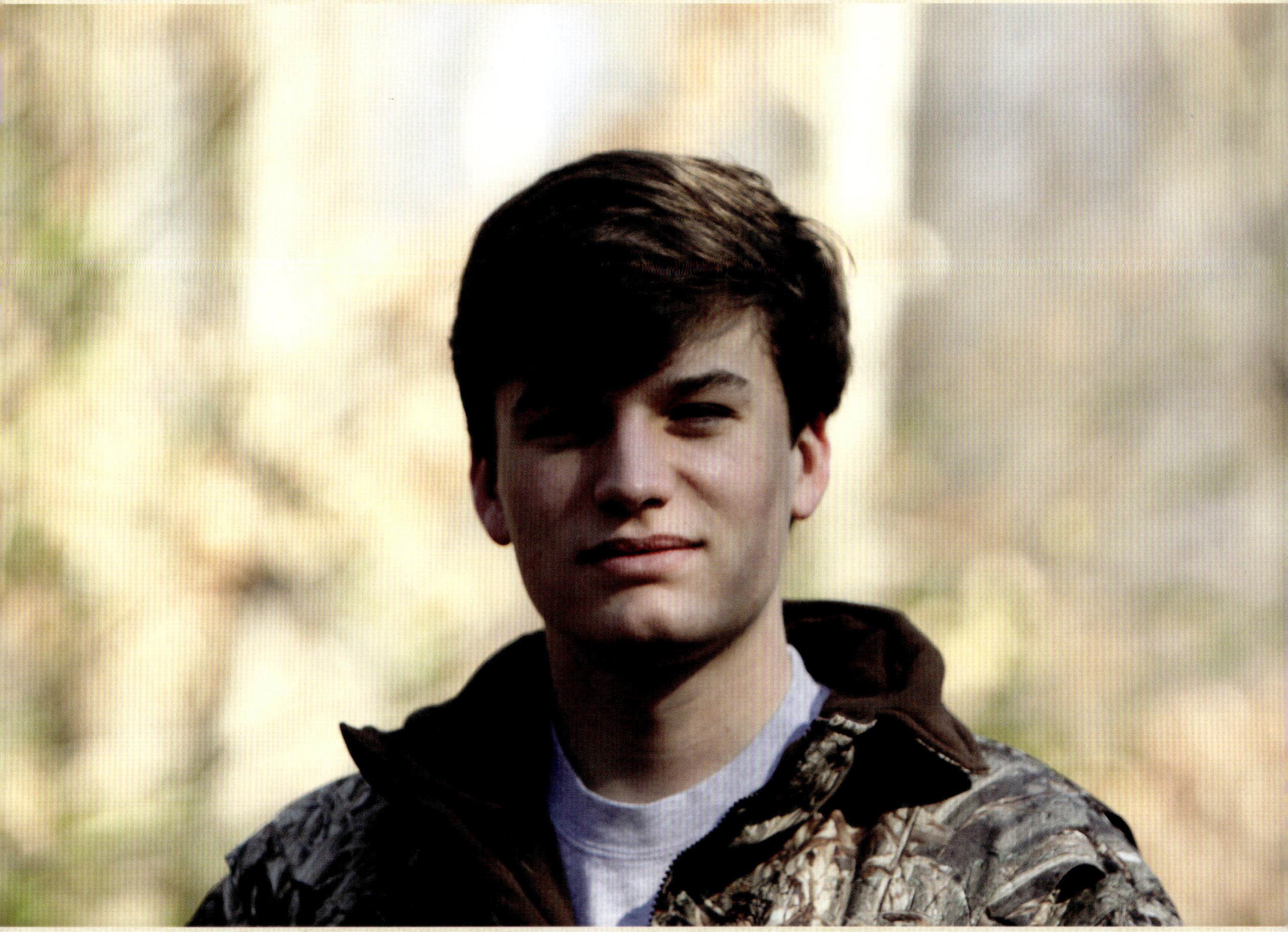

John Reid Golding.

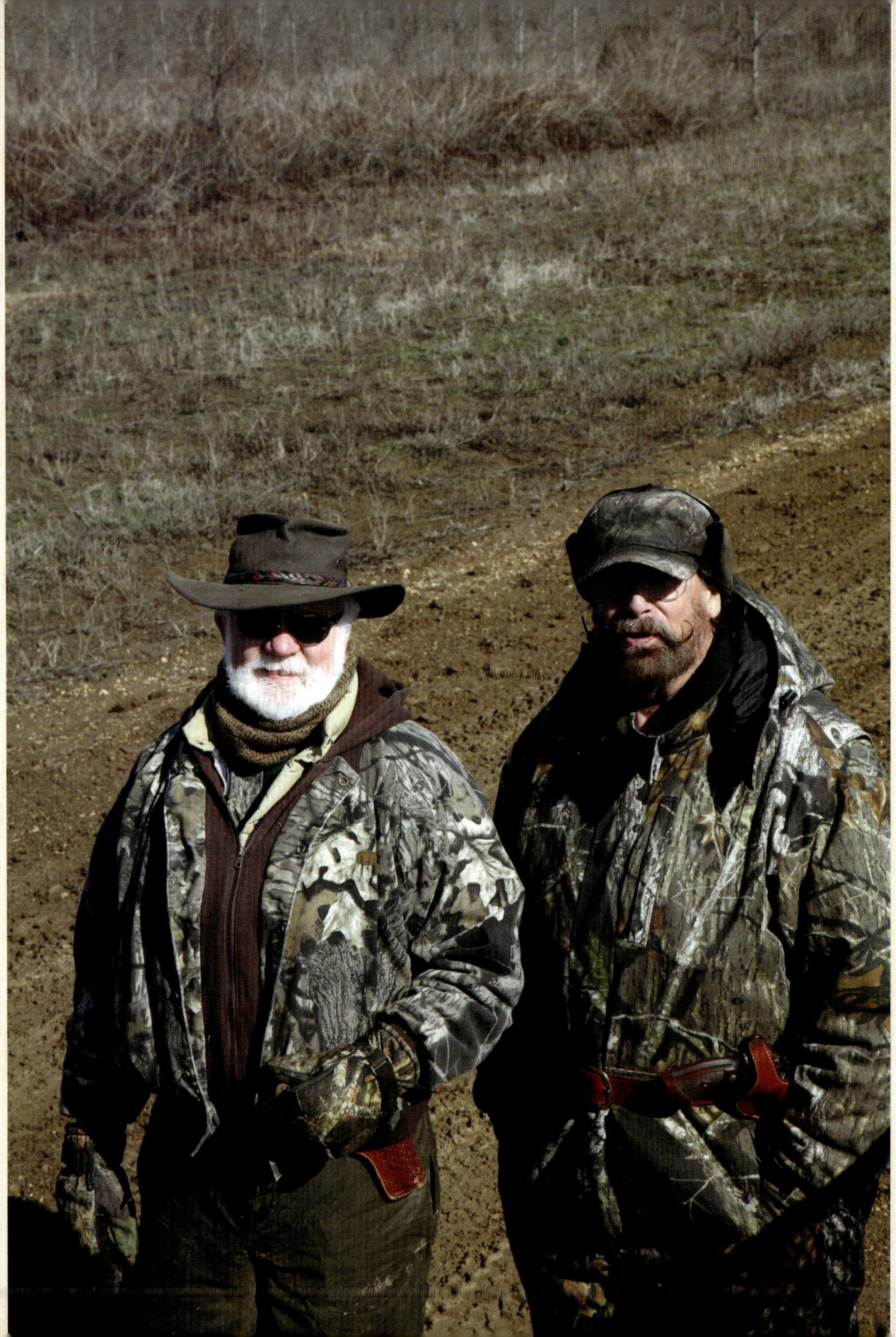

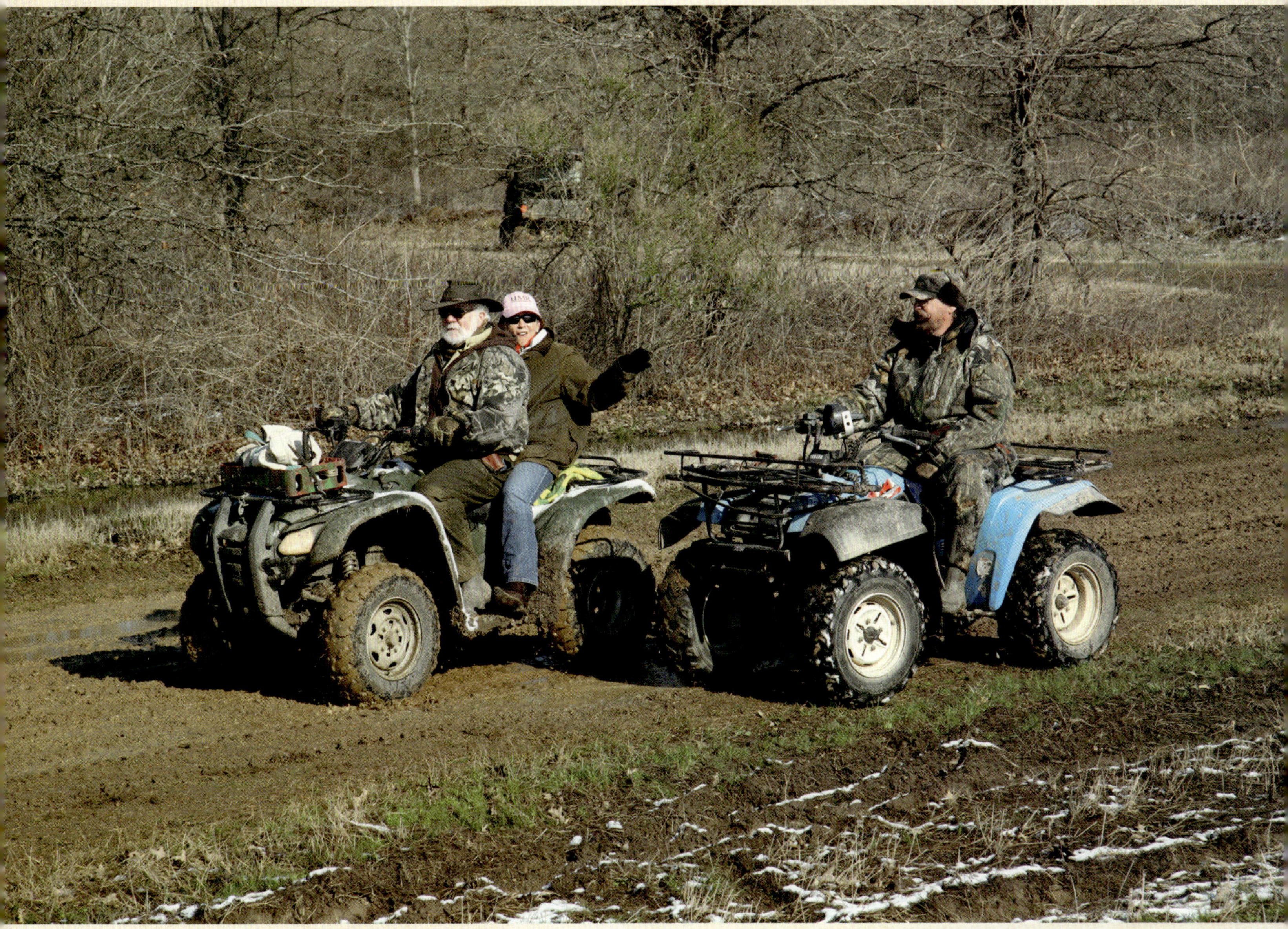

Hank Burdine, Missy Webb, and Sid “Bo Weevil” Law.

< Hank Burdine and Sid “Bo Weevil” Law.

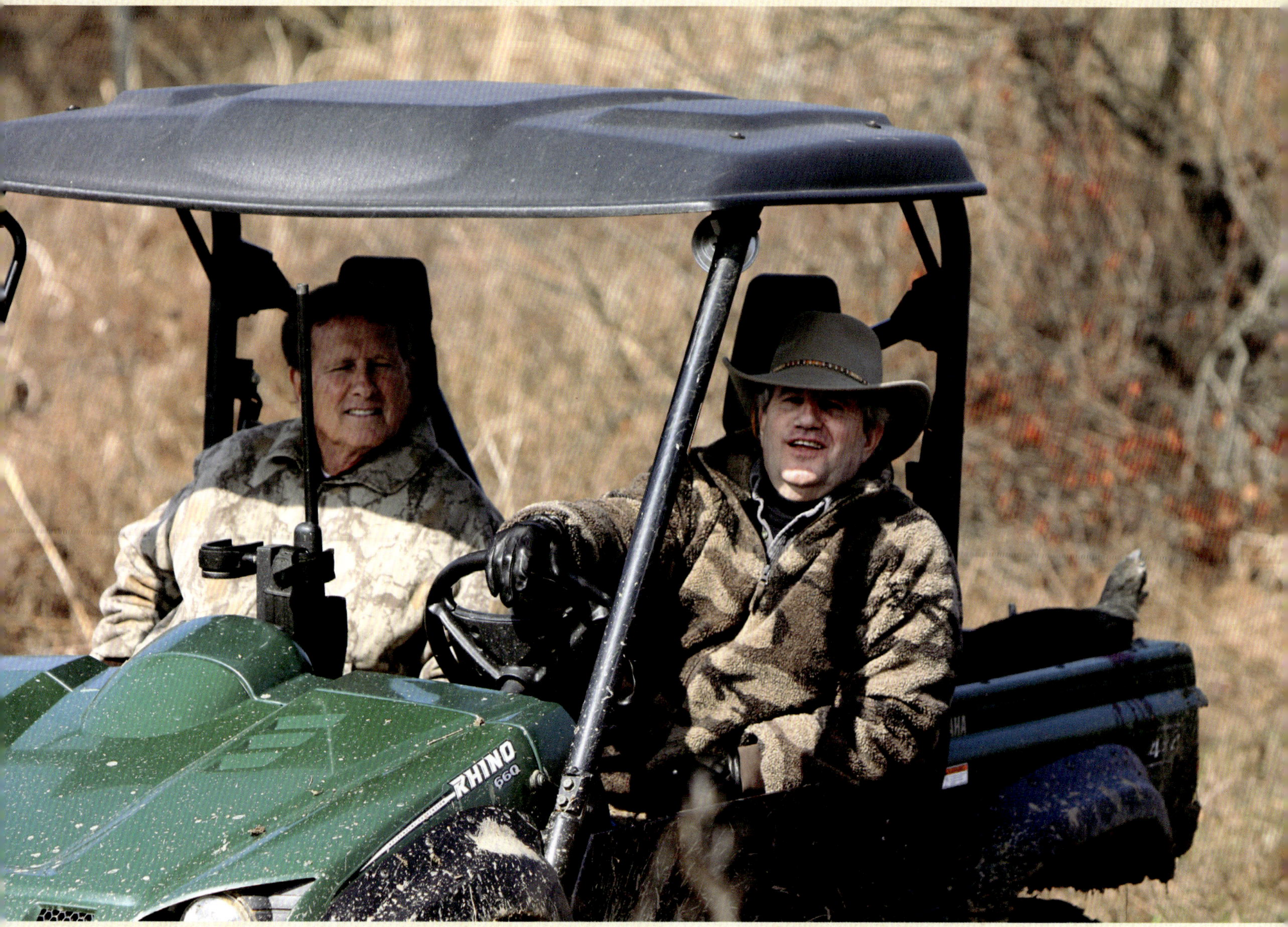

Leroy Frye and Buddy Garrett.

Wick Eatherly, Stacey Waites, and a yellow black mouth cur puppy. ›

Overleaf left: Doc Merideth's mule Gin and Tonic.
Overleaf right: Doc Merideth and Gin and Tonic, his gaited mule.

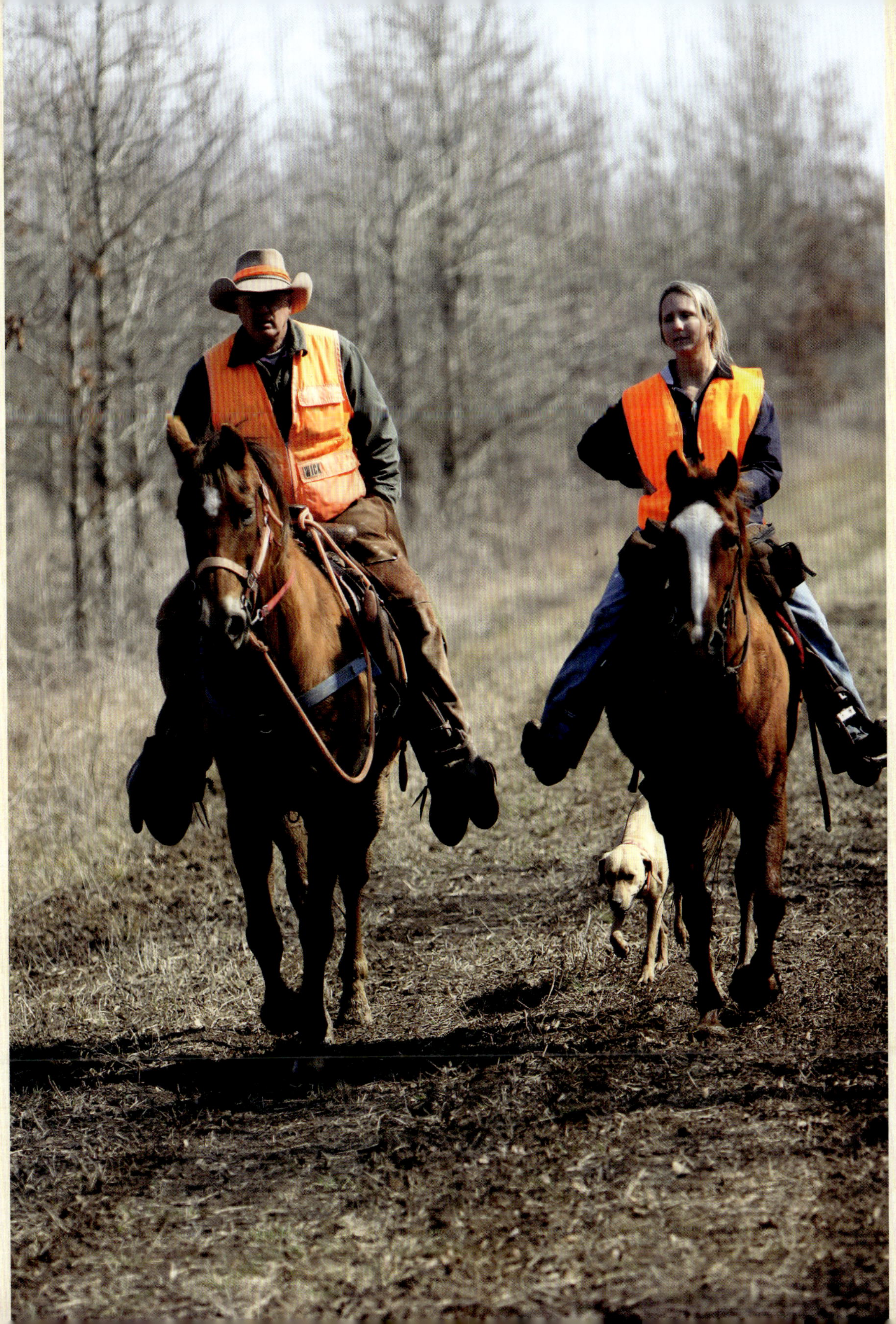
WICK

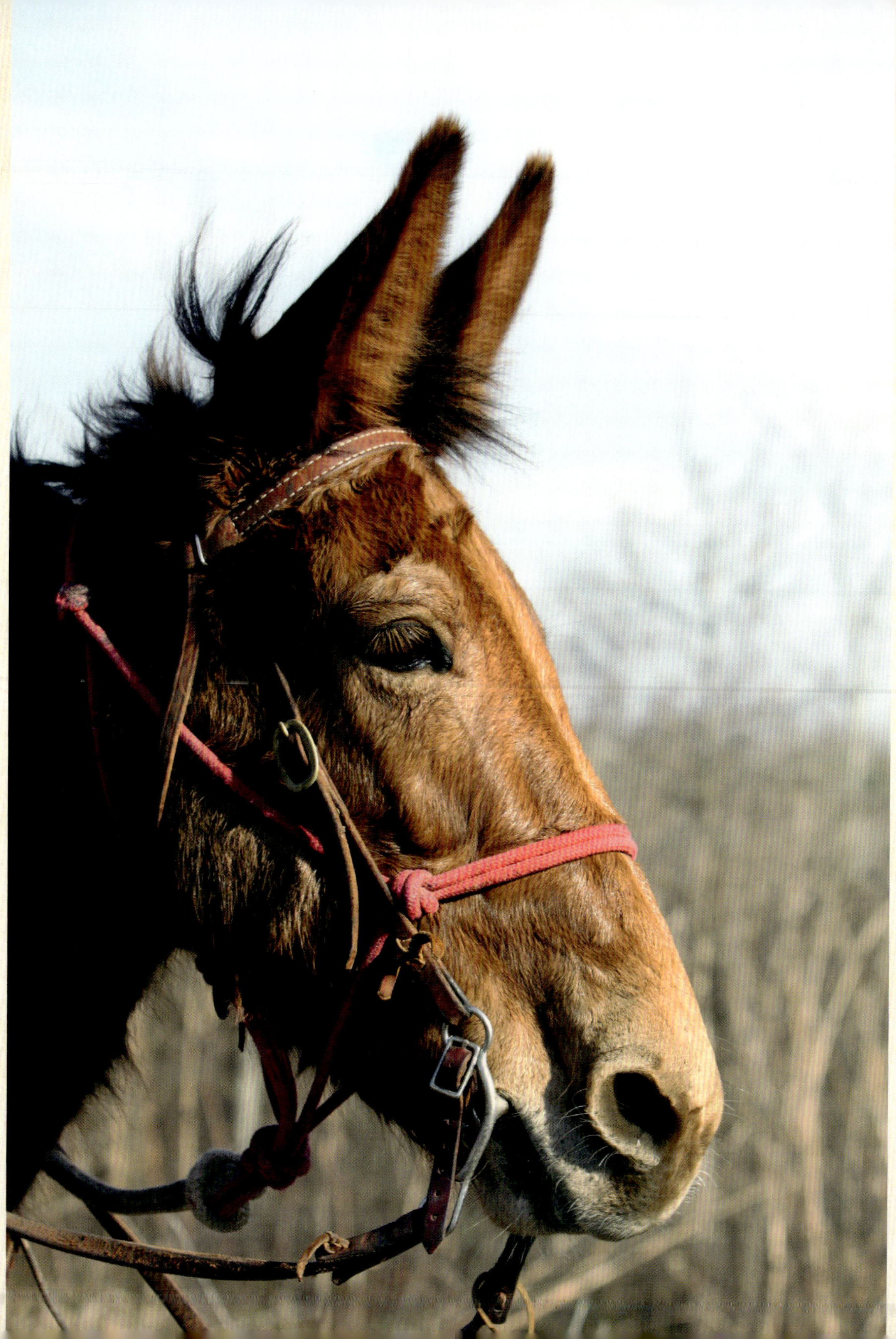

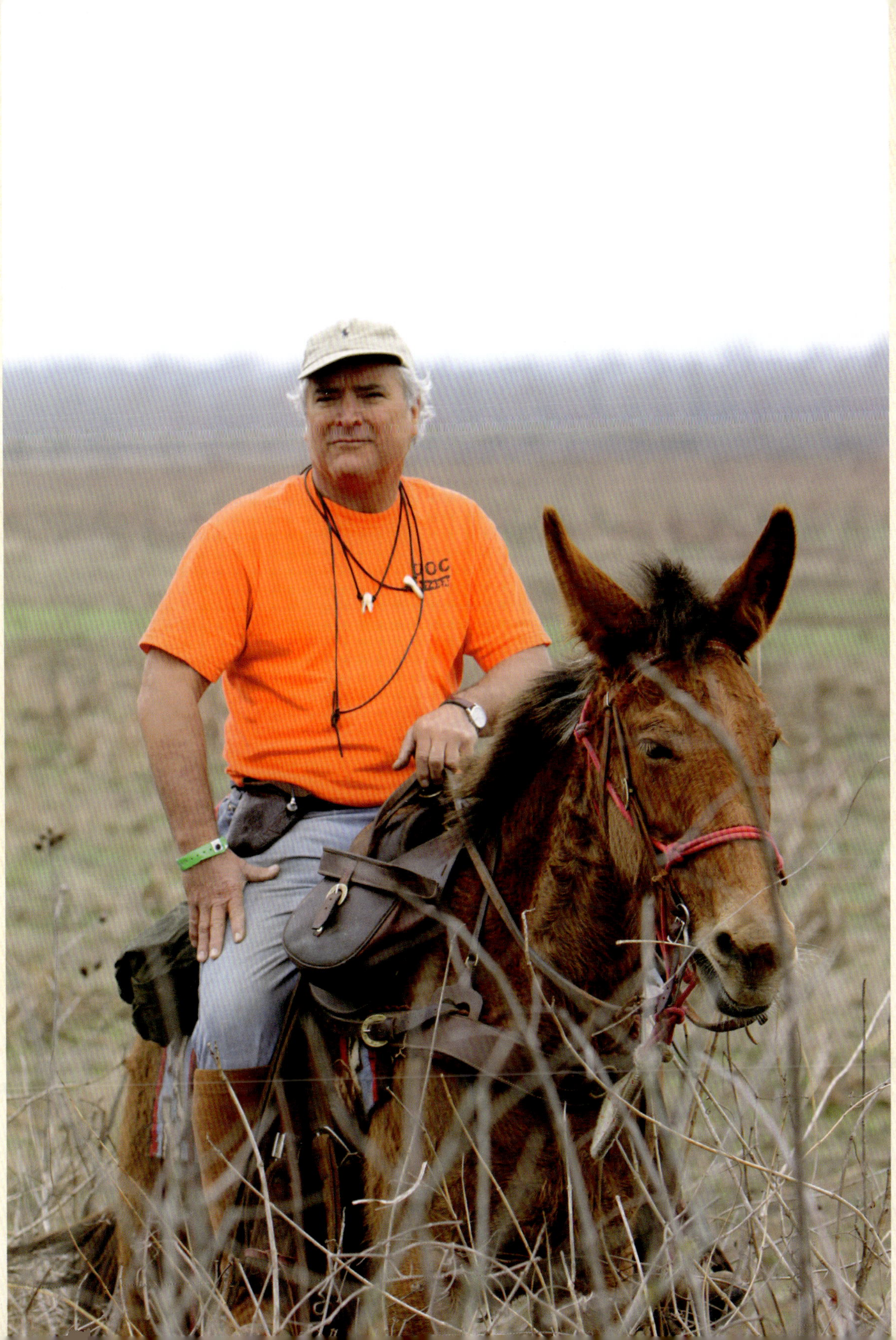

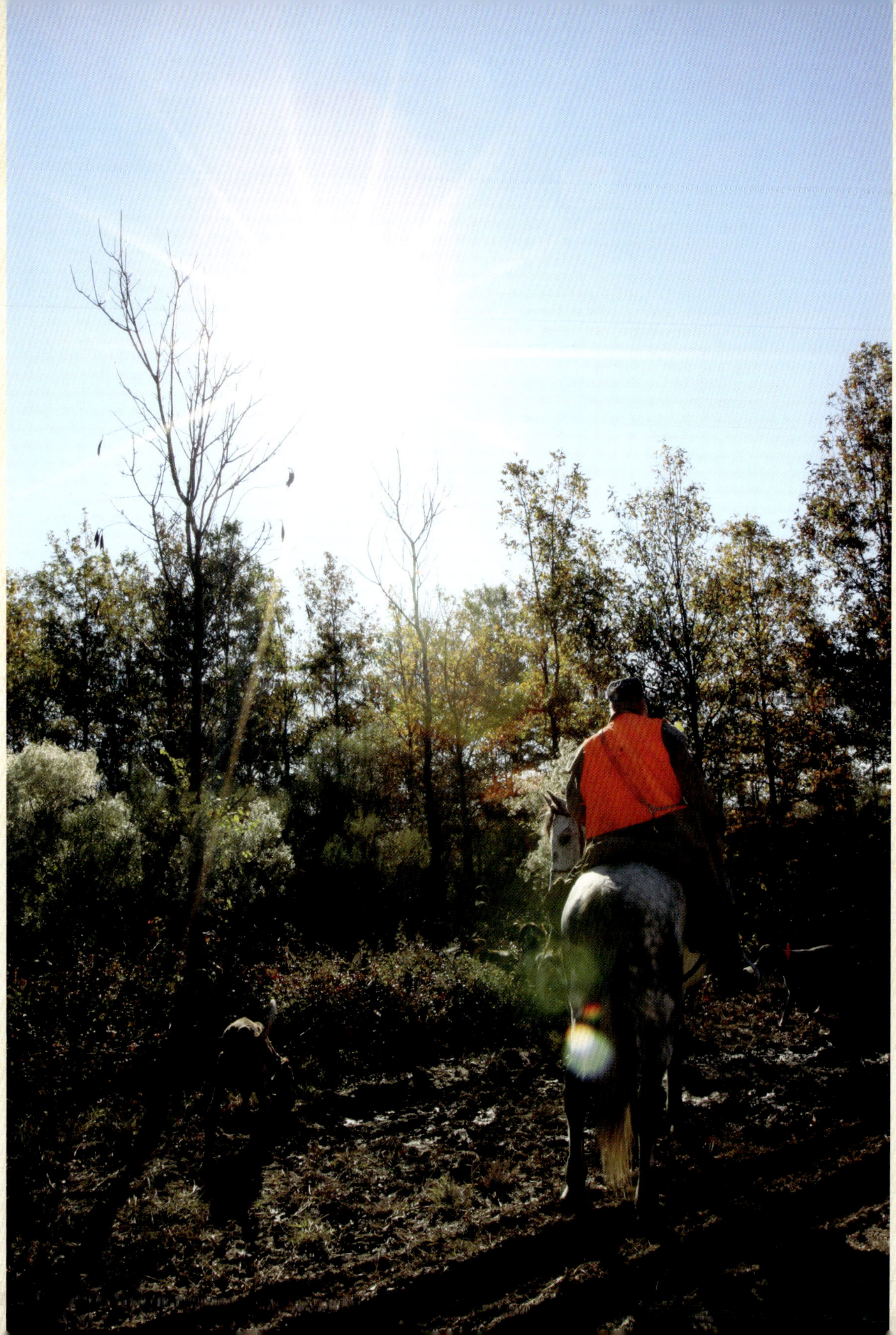

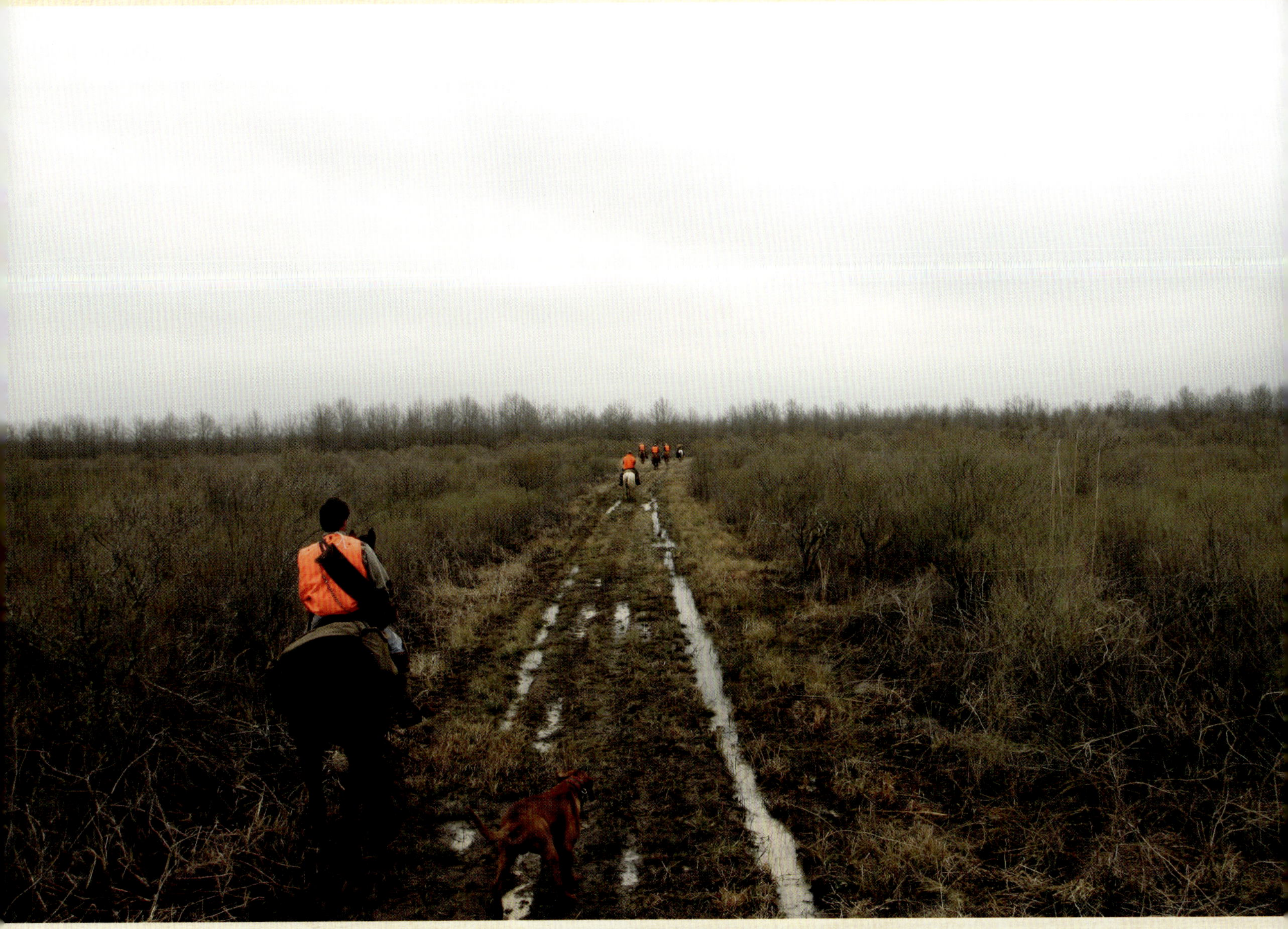

John Lowery, Harley, and Little Joe.

< Melvin (Bubba) Weeks, his horse Doug, and his red tick hound.

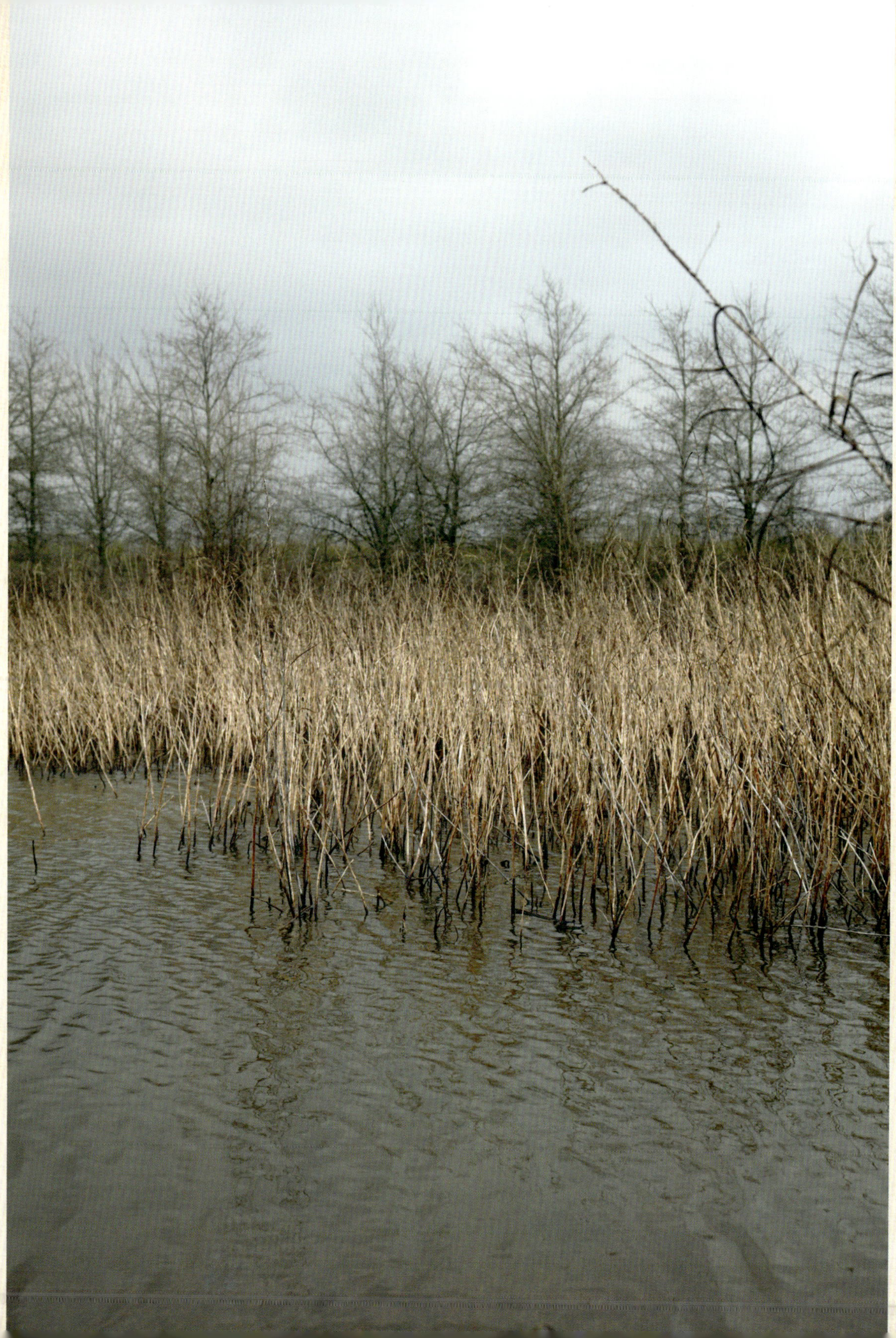

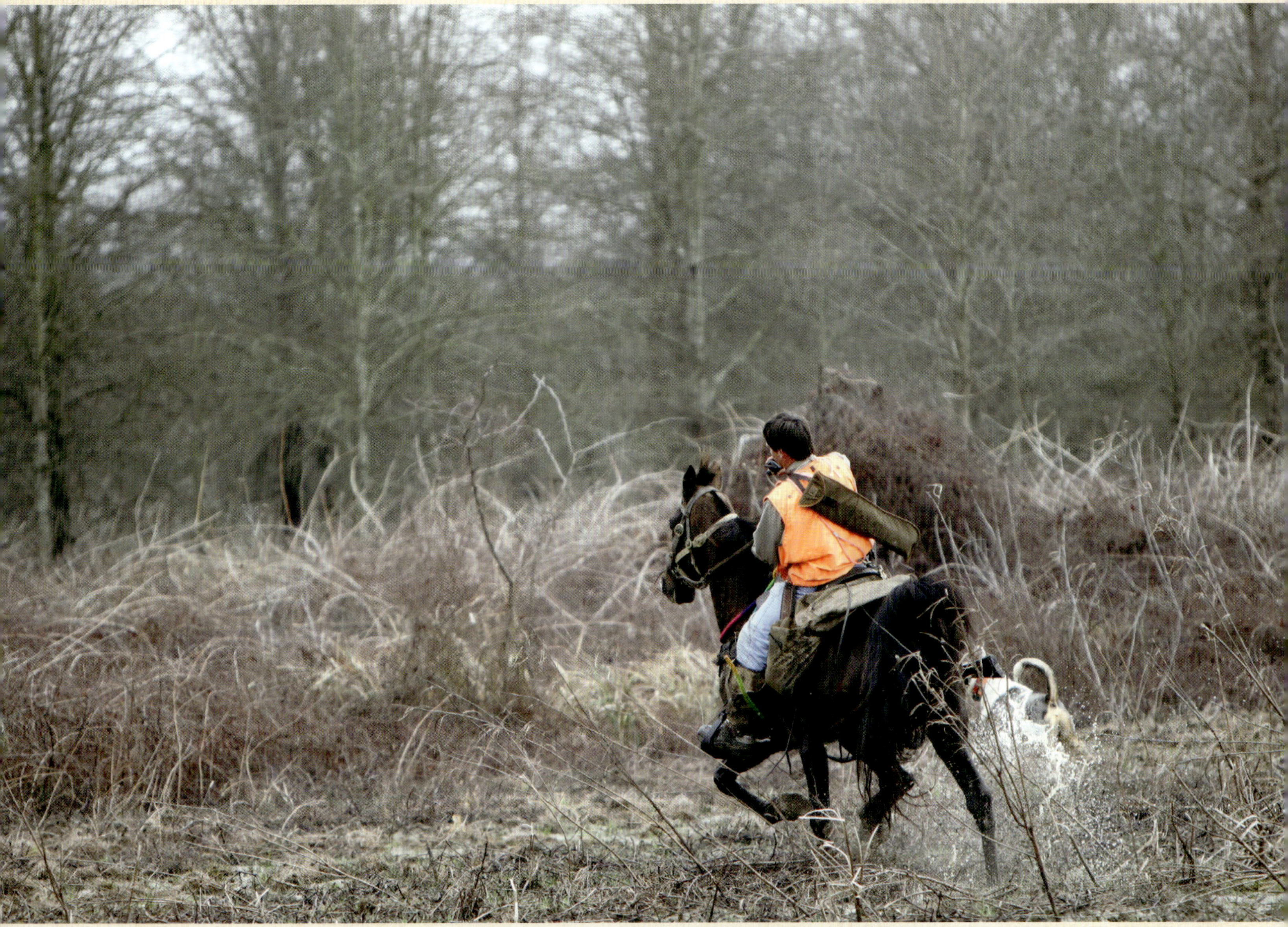

John Lowery, Harley, a Tennessee Walking Horse, and Jake, a Louisiana Catahoula hound.

< Grass in water.

Shootout at the OK Corral with the Dragon Slayer

MIKE LYNN

I was hunting wild Russian boar on the legendary Caulk Island Hunting Plantation as a guest of Alex Grisanti with another guest, Doc Sloas. We were driving the lower Mississippi River bottom Jeep trails, when suddenly, at the same time, we heard hogs squealing and Alex's dog went to baying. Doc and I rolled out of the Bronco after the dog toward the hogs. We busted through a cane thicket and there they were—eight or ten big hogs running everywhere. I had Alex's Marlin .45-70 and Doc had his HK91 .308 and we went to killing pigs like it was the shootout at the OK Corral. I rolled three big pigs and was outta ammo. Doc knocked down another couple, when I ("we") saw a huge boar barreling down (charging) me and here I am outta ammo. Doc was behind me to my left and I hear "step right!" So, in my mind I have a three-hundred-pound pig coming straight towards me and my very excited doctor friend holding the world's baddest .308 battle rifle with fifteen rounds still left in a three-fourths-full magazine fixing to go to town on this pig, and I'm not sure at that moment who I'm more afraid of! Well, Doc unloaded on that pig, who was a mere couple of feet away, and so was the muzzle of Doc's dragon slayer—smoke, dust, blood, mud, leaves, and everything else went to hell. Squealing, the pig veered off my path. I was not sure if he hit the pig or just scared it (and me) to death! But, in the end, it was a good rush, plus five dead pigs, a live dog, and a member, Alex who wasn't sure that he was going to bring these two guest/friends back at the same time again.

Hog Hunting in a Dry County

HOWARD BRENT

In May 1981, my two nephews, a couple of friends, and I drove to the Mississippi hill country for a hog hunt at a private camp. The Belmont Stakes was that weekend, and we listened to the race on the radio while we drove. We all listened as Summing took first place. Along with my hunting gear, I had packed a little whiskey but just enough for myself, not realizing how much of my whiskey I would have to share. Needless to say, we ran out of whiskey on that first evening. Then we found out it was a dry county—"Beer Only," the signs read.

I'm not from the hills. I'm from Washington County, and I'm proud to admit that Washington County held the first license issued by the state Alcohol Beverage Commission, the license numbered one. Sheriff Harvey Tackett, of that same Washington County, had presented me with a deputy badge in the late seventies, a badge I proudly displayed just inside my billfold. When we stopped and asked where the county-line liquor store was, we were told that it was a sixty-mile drive. "But," the gentleman said, "the man who runs the beer store just a few miles up the road sells 'laughing half pints' from the back of his store on the sly."

When we got to that store, my nephew got a couple of cases of beer and set them down on the counter in front of the owner. I pulled out my billfold to pay for the beer and whispered to the owner, "I'd like to buy six half-pints of whiskey you've got there in the back."

The owner jumped back and in a loud voice hollered, "You can't get no whiskey around here! This here county is dry, fella!" At that moment I realized he was looking at my open billfold with that deputy badge shining. I went ahead and paid for the beer, went outside, and sat in the car waiting. After about fifteen minutes, I stopped a regular customer before he walked into the store, gave him some money, and asked him to add my six half-pints of whiskey to his shopping list. He did.

When we got back to camp with the whiskey that night, I noticed that my hunting buddy, Charlie King, had brought his bow and three arrows with him. After polishing off my second laughing half-pint, I decided that I should try bow hunting for the first time during the next day's hunt. I pulled out the 8mm movie camera I'd brought with me and said, "Charlie, you ever saw one of these before?"

Charlie looked at the camera and asked, "What is it?"

"It's a Super 8 movie camera. You make movies with it," I answered. "You want to try it?"

Charlie's eyes lit up at the thought. "Sure," he said.

"I tell you what. I'll show you how it works if you let me use your bow tomorrow." Charlie nodded, and I handed him the camera. I showed him how to look though the viewfinder and how to start and stop filming. Then we went to bed. I drifted off to sleep, imagining myself in the movie, Charlie shooting film for the first time and me shooting a hog with a bow for the first time.

As we loaded into the guide's truck the next morning, he looked at the bow I was carrying. He raised an eyebrow and asked, "How long you been shooting that weapon?" Since it was just my first time, I answered, "Oh, about ten years now, I guess."

"Looka here, mister," he said, pointing to the dogs he was loading up in the bed of the truck. "I hunt bear with these dogs in Alaska. I've paid six to eight thousand dollars for some of these dogs, and I can't let you come over here and shoot one of 'em with a bow and arrow. When these dogs bay that hog they'll be moving in and out on him."

"Man, you ain't gotta worry about me shooting one of your dogs!" I said. Sure enough, he let the dogs loose, and in a little while, they bayed a hog. I reached for one of the three arrows but was real careful to raise the bow to the sky as I pulled back the string. The release on the bow was one of those old-style ones, the kind you punch with your thumb, and as I steadily pulled the string back, the release caught my chin. Off into the wild blue yonder the arrow flew! Swish!

The guide looked at me kind of mad and asked, "I thought you said you've been bow hunting for ten years?"

"Yeah," I said, "but not with that damn release." More determined, I reviewed my inventory of only two more arrows.

Just then, one of the dogs leaped head over heels out of the kudzu vines and the hog was gone with the rest of the dogs in swift pursuit. When they were almost out of hearing range, their barking voices changed tune. "C'mon! Hurry!" our guide shouted. "They've got him bayed!"

We were all gasping for breath when we reached the dogs high on a hill. Charlie was behind me with the Super 8 camera, and I drew back the bow, knowing that this would be a great shot for the movie. All of a sudden, the hog burst down the hill and headed right for us with the dogs nipping at his heels. I felt pretty safe until the hog was only ten yards away, headed straight for me. I decided we had enough good film. When I turned and shouted to Charlie, "Let's run!," I could see that Charlie was already fifty yards away, and if someone had checked him with a stopwatch, he would have broken the record hundred-yard dash.

The dogs bayed again in a creek bottom, and I saw I had a clean shot of this big, boar hog. I pulled the bow back smooth and steady and let the arrow fly. Swoosh! It caught the hog right through the neck. Perfect shot.

I was smiling all the way home, imagining my three minutes of Super 8 fame with the wild boar running crazed right for us and me shooting it clean through the neck. The day I got the film back from the developer, I was grinning, too, but when I put the film in the projector, that is not what I saw. All the film showed was one second of the hog running down the hill toward us and then sky and grass, sky and grass, sky and grass. It was Charlie running the hundred-yard dash with camera in hand. That's the last time Charlie let me use his bow and arrow, and that's the last time I let him use my movie camera.

The Hogbusters

DOC MERIDETH

This is a true hog huntin' story.

We all met at 7:00 a.m. on the first Saturday in February. The year was 1965. I was ten years old. James Crossing, Mississippi, was a big part of my life. Situated about thirteen miles south of Greenville, Mississippi, on Highway 1, James is where the infamous Jesse James and his gang purportedly crossed the Mississippi River. My grandfather, "Red" Merideth, owned and operated the country store at James Crossing. The levee, bordering the banks of the muddy Mississippi, was visible from the front steps of Paw Paw's store, to the west about a mile. That was where the hogs rooted around, over the levee, up and down the river.

The weather was cold and gray that Saturday morning, twenty-two degrees at 7:00 a.m. Snow flurries trickled down in the wind. The old wood-burning stove in the center of the store glowed with warmth. We all stood around it, warming our bones, waiting on the "Dog Man" to show up. The hunt party included my daddy, Sonny, our friend Hubert Hodges, Hubert's brothers, Herman and Herschel, Hubert's father-in-law, Mr. Buck Cornelius, and Malcolm Lowe. The "Dog Man," yet to arrive at the store, was Daddy's uncle, Uncle Pee Wee Horton. Uncle Pee Wee hunted 365 days each year, and he ran his Darlove cotton farm via radiophone from his car at hunting camp. He became a legendary Mississippi sportsman, hunter, and fisherman. Usually sporting a burned-out old cigar stub in the corner of his mouth, Uncle Pee Wee always had a dog for this and a dog for that—squirrel dogs, rabbit dogs, duck dogs, bird dogs, deer dogs, coon dogs, and, today, hog dogs.

We chewed on sausage and biscuits, and the men drank hot coffee and hunkered around the wood stove, waiting on Pee Wee. I was only ten, and did not like coffee. My "usual" down at the store was a "short" Coke with a bag of Lance peanuts dumped into the bottle. Later on in my life, having traveled the country and the world a good bit, I realized that it was only in the Mississippi Delta that I ever saw anyone at all pour a bag of peanuts into a Coca-Cola bottle and drink and eat the contents. My uncle, Champ Terney, had showed me this enjoyment about the time I turned six. I have savored it most of my life, usually shaking my peanuts down into my Coca-Cola bottle as I pump gas into my truck at local convenience stores.

Uncle Pee Wee arrived at Paw Paw's James Crossing store at about 7:45 a.m. "Hey, baby," he shouted, entering the store. "Ready to go chase some hogs?" The stuffed two-headed calf Paw Paw had mounted up in the corner of the store seemed to me to come alive with a smirk and a chuckle as Uncle Pee Wee unleashed his dynamic personality on his fellow hunters. Several farmhands, playing the pinball machine over by the tires for sale, acknowledged Pee Wee. "Morning, Pee Wee! Hello, Mister Pee Wee!" Everybody knew everybody down through that stretch of cotton fields on Highway 1 from Greenville to Swiftwater to Wayside to Avon to James and on down to Lake Washington. Mr. Mabry Wigley farmed a lot of the cotton. Mr. Murry Alexander farmed a good bit too. Murry donated the land so that the local families could build their church there at James Crossing.

Hog dogs were barking and "raising sand" out in the used car sales lot of Paw Paw's store where Pee Wee parked. The cold weather had them charged up, ready to push some hogs through briars and sloughs. Paw Paw "Red" Merideth made us a bunch of bologna and cheese sandwiches, placed them in a big old brown paper sack with some soda pops, and said, "You boys bring me back a couple of nice hogs to sell as sausage and 'poke chops.'" He was serious. Stuff like that sold good in country stores around Mississippi in 1965.

We loaded up in the trucks and departed west toward the levee. Uncle Pee Wee always wanted to be the life of the party. He was gregarious, boisterous, a braggadocio, loud, and fun. Pee Wee led the group, driving his Jeep truck the short distance over the levee. We were supported by the eight Catahoula cur hog dogs Pee Wee sported in the bed of his Jeep. They were howling continuously before we even topped the levee. Fiery steam emerged from the dog kennel as all hog dogs hyperventilated, anxious to start the chase. The pace of the snow flurries hastened. It began to snow more, quite a rare treat in the Mississippi Delta. The temperature remained very cold, rising only to twenty-five degrees at 8:00 a.m. The sky filled more with puffy white clouds. 'Twas the perfect hog huntin' day! The hog dogs run better and longer in cold weather. If any hogs are taken, the meat stays better preserved in the cold. Taken hogs are usually immediately gutted and left trailside to be picked up later in the morning. If there is not the luxury of cold weather, then the gutted hog is best left in a cold body of water, a pond, a lake.

In 1965, there were numerous old-fashioned "redneck deer clubs" scattered around the Delta, especially along the river. Some of these clubs, like Catfish Point, north of Greenville, along about the 1980s were developed into high-dollar "blue blood" clubs. Fine hunting lodges and homes replaced the old trailers, mobile homes, and school buses that hunters slept in, froze to death in, and called home during deer season. But this day of this hog hunt in 1965, there were no such elite deer clubs or "timber clubs."

Uncle Pee Wee, because he was a "Dog Man," could fairly easily gain permission to hunt, and especially hog hunt, any of the clubs in the Delta and along the river. February, after deer season, was down time for the clubs. The deer hunters had cleaned up and shelved their guns 'til next season. Most clubs too wanted to rid themselves of the darn hogs, who root and destruct forage and food plots prepared for deer. So we were sort of VIPs for the day, not "ghostbusters" but "hogbusters," on the three or so clubs that the hog dogs would run this glorious day.

Topping the levee, we proceeded on it, passing oxbow lakes until we came to our destination. We parked vehicles at a cleared landing. The river looked frigid with whitecaps. Arkansas was visible in the west. The hog dogs sensed that their hunt was imminent. Their barking intensified. I was a young boy. I was excited. In my mind we were aristocrats, akin to English fox hunters, that frosty morning. I expected a trumpet to sound off, signaling that the hunt was on. Uncle Pee Wee blew his cow horn. Wow! A youngster could not have more fun than this! Six of the eight hounds were released. Experienced hog hunters always hold back a few dogs for fresh replacements later in the day.

There were hog tracks at the landing where we were parked. Daddy and Hubert, Malcolm Lowe, and Buck stood over the tracks, analyzing them, deciphering their meaning. The committee concluded that, yes, hogs made these tracks only a few hours earlier. They were fresh. Oh boy! My adrenaline was going as I grabbed my old used single-shot .410-gauge shotgun. Daddy got it from the legendary Frank Carlton of Greenville. Everyone in The Delta came to know Frank—lawyer, legislator, district attorney, practical joker, humorist, and the host of Christmas in July. Christmas in July was Frank's BBQ social event feeding hundreds of Deltans on a July weekend each year, mostly on his own nickel. Frank was particularly well known for having parachuted at night in National Guard training drills, breaking both legs on landing. He frequently collected good used shotguns and rifles as a hobby. My .410 shotgun was special, as the stock had been sawed off to make a youth's gun.

The dogs hit the scent right out of the Jeep. It sounded like a locomotive going down through the woods. I was hooked. This was my first hog hunt. Before, we had usually only rabbit hunted in February. Now life had more variety—hog huntin'! My Daddy, Sonny, and I had a barrel of fun just watching this cast of colorful Delta characters and hog dogs get in gear. We learned that the joy of hog huntin' is in the people you meet and the hog huntin' stories and other tales that you hear told. The friends that you make are special.

The dogs ran straight south, zigzagging between the levee and the banks of Ole Man River. Whatever it was, running out in front of these dogs, it was putting on a show. It ran five miles south, through thickets, swimming water holes, occasionally losing the dogs. We pursued in trucks and in Pee Wee's Jeep, on muddy camp roads partially laced with ice. Whenever it sounded as if the dogs were bayed up on the hog, we'd set out on foot. There were no four-wheelers in those days. For some reason, that day no one was riding horseback, though we'd hunted deer and rabbits by horse quite routinely.

As the hog made a U-turn heading back north, Mr. Buck spotted him. "It's a big ole black boar hog," Mr. Buck hollered. He called the dogs up to the fresh track. "Mickey, Sally, Bad Eye, Judy, Mamie, Rosie, get in here, get him up, get him up!" The dogs got after him again, hot as ever. They ran out of hearing. We knew they went north. We drove on back up to where we'd started the morning. There was no hint, no sign or sound of dogs running a hog. Just as we were about to proceed further north, Mr. Mabry Wigley, gentleman farmer, topped the levee in his old International Harvester Scout. "Your hog went east toward James Crossing, and so did the dogs," Mr. Wigley chuckled. He laughed some more.

We sped past what is now John Thomas Greenlee's

farm headquarters. Sure enough, as we approached Highway 1 at James, we could see that the ole boar hog had run right up to the Pure gasoline pumps at Paw Paw "Red" Merideth's store, as if to refuel. Paw Paw lived in a big white house behind the store. The boar had run up under the house. The dogs were up under there too.

Uncle Pee Wee nervously pleaded, "Someone get under there and get him! He's gonna kill my dogs!" I had some bottle rockets and cherry bombs left over from the holidays up in Paw Paw's kitchen. Also, I knew there was a hole in the pantry floor that opened directly to the house's crawl space. I put a match to the fireworks and began dropping them in.

It worked. Hog and dogs took off from under Paw Paw's house. Paw Paw had perched on the back steps of the store with his shotgun. On seeing the boar, he saw dollar signs, as sausage and "poke chops" sold well at the meat counter. Paw Paw fired three shots at the sprinting boar. Must have missed. Hog and dogs ran straight east of James to the yard of my great-aunt and great-uncle, Ellene and Mike Tavenner. We all drove up as Aunt Ellene stepped out of their house into her yard. The boar had backed up into her nice landscaping to defend itself. The dogs were in a standoff with the big, black beast. Aunt Ellene did not care much for my beloved Uncle Pee Wee, who was once married to Ellene's sister. "Pee Wee," she said, "you load up those damn dogs of yours, leave that poor hog alone, and go to church tomorrow with your lonely wife!" We never saw that big hog again.

This is a true hog huntin' story.

The Great Mississippi Boar Hunt

PARKER HALL

When we were in India in the 1970s I remember seeing old paintings of the Moghul rulers, and later the English colonialists, spearing wild hogs from horseback. In the book *Raj, a Scrapbook of British India, 1877–1947*, Charles Allen notes, "Pig-sticking—more properly the spearing of the male wild boar with 'hog spears'—was a peculiarly Anglo-Indian sport that flourished in the plains of upper India. It was a risky business, demanding cool nerves and a high degree of horsemanship."

Now we fast-forward twenty years to a presentation on Tuscany by Patsy Ricks, the noted Jackson teacher/guide and expert on all things Italian, and here again were photos of wild boar heads mounted on the walls of butcher shops in San Gimignano and Lucca. It turns out that there is a thriving business in that area for cinghiale, or wild boar meat. It amazed me that there are enough wild animals in Italy to make a steady market. But like here, they are a real nuisance and the farmers welcome the hunters to get rid of them. We were told in Sicily by Turi (Salvatore) Siligato that they use dogs and "beaters" or "drivers" to move the boars, and they station the shooters in an area where they are likely to run. (Turi, by the way, has a wonderful restaurant in Taormina called Osteria Nero d'Avola, at Vico Spuches, 8, just a few steps down from the main road. He is an avid hunter and wonderful storyteller.)

While everything I had read and heard of these boars was interesting, I never thought that I would be personally involved with them. In the spring of 2007, however, I returned home to our house overlooking the Mississippi River one night from a five-day trip to Toronto. The next day I noticed a disturbed area twenty feet by thirty feet down the hill, in the field toward the river. It looked as if someone had dug it up with a rototiller. I naturally assumed that my brother, Tully, had discovered a newfound interest in gardening and was preparing for some planting. It turned out that there was no interest in planting. The damage had been done by three wild hogs that had apparently swum across the Mississippi River during the high water. We later were told by the construction superintendent at the nearby Riverwalk Casino-Hotel that their crew had managed to get a rope around one of the big males who had fallen into a hole on the casino property. They finally were able to pull it out of the hole. What looked like a thoroughly beaten pig and a sure candidate for a luncheon BBQ turned into a mad, raging 250-pound wild animal. With only one rope and no gun, the crew thought better of the adventure and managed to avoid any damage. The hog scampered off into the woods.

The nightly digging continued and the hogs seemed perfectly happy with the cool muddy place to dig, with ample water and plenty of dessert provided by the pear trees that had been planted in the field years ago by Rodney McCann. The yard really had become a mess and Tully could stand it no longer. I advised on the use of a large-caliber deer rifle with a scope so that the hunt would be short and deadly. Tully, unfortunately, had lent the gun to son Ben, who had taken it with him to North Carolina. "No problem," Tully opined. "The trusty 12-gauge shotgun will do!" So one Sunday night, with a failing flashlight, one slug in the shotgun, and a .22 caliber rifle, we mobilized for the hunt. All three hogs were just down the hill, no more than one hundred feet away. Tully took careful aim and shot the small one with the single slug. Big Blackie and the other one charged into the woods and away. It was time for us to complete the job. The hog was badly wounded but still alive, so we shot it five or six times with the .22. Surely it wouldn't last until morning. At first light I proceeded down the hill armed with my trusty Nikon camera. The hog was lying with his back to me, apparently dead. I saw what I thought was a bird moving on the top of his head. Oops, it was not a bird, but an ear. It had heard me coming, and jumped up ready for a fight. Fortunately, it was so badly injured that it could not catch me as I retreated up the hill to find a weapon a little more lethal than my Nikon camera. Six shots with a .38 pistol sadly were not enough and it took a single shot with a .44 magnum to finally do it.

It was sad for me to see a badly wounded animal like that, but fortunately no human was hurt. It proved to be a good lesson in how tough those wild hogs really are. The larger brown one was later caught by a crew from Mississippi State and weighed two hundred pounds. The big black male left the area when the construction crew began a big dirt work project in the field. Perhaps he will return when the work is completed. You can bet that if I am involved in any more hog hunts, I will have the biggest gun that I can find.

Hogs and Bears

WES LAWRENCE

On February 20, 2010, we awoke to a pleasant morning about forty-two degrees. We were on a hunting club off of the Mississippi River. Along with me were my hunting buddies: Billy and Charles Stone, Billy Mitchell, David Ubanks, Justin Davis, Tres Hienz (twelve years) , and Jeff Loveless. After breakfast, we suited up the dogs with tracking collars, cut vests, and cut collars, loaded the Rangers, and were off to a day of hog hunting. On our way to our hunting spot we found a good hog wallow with fresh tracks on the edge of a cane thicket. So, we decided we would start there. We turned three or four bay dogs loose on the track and immediately started baying hogs. Our first hog was about 140 pounds, a boar that we caught and killed.

Minutes after that we bayed again. On the way to the bay I saw three hogs between me and the bay. One was a white spotted hog and I knew I wanted that one. We caught the second hog and the dogs immediately left, pursuing those that I spotted on the way to that bay. The third race lasted longer and ended up on the edge of some water with the dogs bayed solid. This time we had a pretty good boar hog bayed in a very dense thicket. We turned both bulldogs loose and they went in there and caught him. That is when Billy Stone and I went in and dispatched the hog. We gathered up all the dogs and went looking for more hog signs. It was not long before we found another good track. We turned the dogs loose and had another race. We bayed a small group of shoats with several other big hogs in one bay. We eased in with the bulldogs, Blue and Boss. This time each bulldog caught a hog and one was the white spotted hog from earlier. Billy S., Billy M., Tres, Justin, and I went in and caught the hogs. Billy S. and I legged each hog. Then Tres went in for the "stick." After those hogs went down, Charles's little white cur dog, Jackie, and my Plott/black mouth cross, Gretta, were five hundred yards away, according to the GPS system, by themselves. So, Justin, Tres, and I started tracking those two dogs. We got within a hundred yards of them on the Ranger and we turned the Ranger off. Immediately we heard them baying. But both bulldogs were still at the other bay. So, we ran to the bay. As we got closer, we saw that they had a two-hundred-plus-pound boar hog bayed. So, Justin eased in and shot one with his new Glock .357 Sig. He was definitely excited that he broke his new pistol in right. We loaded the hog and the dogs up on the Ranger to go meet back with our group.

It was almost dinnertime so we stopped for a light lunch break with our group. After enjoying a little conversation, fellowship, and snack, Billy Mitchell decided he was going to walk the dogs into a thicket right beside us. After a few minutes, Hoover, an English coonhound of mine, struck a good track and away they went! They followed the track across a deep bottom of water and we quickly made it across the water on the road with the Rangers and got on the same ridge with them and stopped to listen. Billy Mitchell had made it across the water in his hip boots and radioed to us that they were bayed. We immediately drove to the end of the road, which put us within a hundred yards of the bay in a cane thicket. When we turned the Ranger off, the dogs were baying like crazy. I jumped on top of the Ranger and turned both of my bulldogs loose, Blue and Boss. I took off running with them towards the bay. Then Justin, Billy M., and I got to the treetop that they were bayed in. Billy M. turned and hollered, "It's a *bear*!" As I turned to look, I was looking at the bear eye-to-eye level. He stood as tall as I am and weighed around three hundred pounds. From my right, Boss dived over the log in the treetop and caught the bear on the neck with Blue hitting him in the side. We looked at each other and said, "What now?" All we knew was that we had to separate this somehow or the bear was going to kill the dogs, but killing the bear was out of the question. The bay dogs were trying to get in there but the bear kept swatting them back out. We had to do something quick because the bear was now biting the bulldogs. Our lives were in danger and our dogs' lives were in danger so we just reacted. The rest of our group was quickly approaching this bay. I told the kids to stay back, and several others had their guns drawn for backup because we had no idea what was going to happen. I immediately fired four shots in the air trying to break the bay up. Ubanks hollered, "He is killing the dogs!," so our adrenaline kicked in and Jeff and I jumped into the treetop with the bear. We were poking and swinging pecan limbs at the bear while he was chewing on Boss, the red bulldog, like corn on the cob. After several good hits with limbs the bear decided he was ready to go. (Thank God!) He let go of the red bulldog and turned to leave. I dived over a log and grabbed both bulldogs and curled up with them on the ground. I was about two to three feet from the bear at this point. The bear exited the treetop straight toward our entire group that was watching, which sent them scattering like ants. Immediately, I started checking dogs for damage. A few bay dogs had small superficial wounds but nothing compared to the bulldogs. I attribute the reason that our bay dogs were not hurt to the style of dogs that we hunt. We hunt a strictly "bay only" dog, not a rough catching bay dog. This is the reason we use bulldogs to catch. Both bulldogs had canine wounds on the face and head areas. But both will live to hunt again. Who ever would have thought we would have bayed a bear? This incident killed the momentum of our group for the next hour, that is, until we caught the next hog.

Mississippi Huntin'

CLARK GORDIN

Wild hogs are hunted throughout the United States from Georgia to California, with a special love of the sport existing in the Deep South. My love affair with this species began in a swamp just north of the Ross Barnett Reservoir on public hunting land known as the Pearl River State Wildlife Management Area. This area of pine flats and flooded backwater from the Pearl River is open to the public for hunting and is managed by the Mississippi Department of Wildlife, Fisheries and Parks.

While disco tunes were topping the charts on the radio in 1978, bow hunting for wild hogs was at the top of my chart on the fun-things-to-do-outdoors list. Having read every magazine article I could find about how to approach this sport, I was ready to dive in and concentrate on sticking a pig. My arsenal at the time included a four-wheel compound bow, some heavy aluminum arrows with stainless steel broadhead tips, and a large bottle of strong insect repellant. You have to play the wind to hunt pigs anyway, and the repellant was a necessity for swamp hunting. Mosquitoes in the Pearl River valley are measured as clouds, not as individuals. You have to be able to stay to play, and without protection the mosquitoes will send you packing to your truck in short order. The mid-October temperature was perfect for major insect activity but the slight breeze moving up from the river helped just enough to make my wait possible.

I had picked a tree (in the dark) and was perched on a small climbing tree stand about fifteen feet up. One of the oak limbs prevented further ascent; however, it did provide some support to lean on and get the weight off of my hips from time to time. This was one of those early model stands with a hand climber to assist someone in going up the tree. You sat flat on a small plywood platform and your legs would go to sleep in minutes. I carried a length of three-quarter-inch nylon rope which I tied in a square knot around the tree, then under my arms at chest height. My thought was if I slipped off the stand it would be better to die from the rope squeezing the life out of me than to break my neck hitting the ground. At least I could have a few minutes to enjoy the scenery while hung up there and ponder my sanity for being up there in the first place. The foolhardy side of me wasn't interested in ending my days with a dull thud. In my more recent hunting years my friends have quizzed me regarding the danger of going to Africa and other exotic locales. My standard reply has been that I would rather them discuss my demise from a mamba bite than talk about some idiot who ran a red light on me.

My tree was only about ten yards from shallow backwater that led through the switch cane, palmettos, and mixed forest to the river. Some whitetail deer had moved through at first light and they had been quite a sight. The early rays of the sun filtered through the trees and I could just make out the glint of reflection on the horns of a young buck that was trailing two does. Any of the three would have been legal game, and I was glad they were out of range. I would have been tempted to try a shot, but that would probably have ruined my chances at pork.

I heard the splashing first, then the low grunting noises as the hogs moved through the water directly toward my tree. There were three of them but unlike the nursery rhyme they were far from little. Two would have gone well over a hundred pounds and the largest was well over twice that size. They were picking up floating acorns but moving rather steadily and making plenty of noise. At about the thirty-yard mark I drew my bow and held on the largest of the bunch and followed it with my sight pin on its chest.

For some unknown reason and much to my good fortune the hog stopped dead still under my tree and at that instant I released the arrow. With a loud squeal the hog bolted and ran through the shallow water. The other two members of the trio scattered in other directions as I kept an eye on my pig until it disappeared from sight and the splashing faded into the distance.

This was my first hog to take with a bow and arrow (or any other means) and I would have jumped from the tree stand had I been a little closer to the ground. I walked back through the timber to my truck and told my hunting partner that there was possibly a hog down. As we walked together back to the flooded area we met another hunter who joined us and offered to help with the search and recovery. My prize had only traveled about seventy-five yards from my tree and was very much unalive. We all breathed a sigh of relief. The prospect of tracking wounded game is never appealing, especially a hog that is traveling through muck and mud and leaving a weak blood trail. This particular specimen had really nice jaw ivory and when the conservation officer weighed her the scales showed 285 pounds. All hunters want to take boars and I suppose it does boost the bragging rights a bit; however, this sow is a trophy in my book and I have always been proud of it. The meat was extremely lean, and the sausage was a welcome gift to many friends. If you enjoy hunting and have not tried the pigs, I highly recommend the sport. You are sure to create memories such as this one, which took place thirty years ago.

There have been many opportunities to chase pigs across the United States. One location, however, always stands out and I relish each invitation. Mr. Bobby Dunlap of Oxford, Mississippi, has a little patch known as Caulk Island and it is widely known as pristine pork habitat.

At whatever time the good Lord and Ole Man River had decided, the path of the Mississippi changed many eons ago and Caulk Island was formed in Desha County, Arkansas. To be more specific, this property lies south of Christmas, west of Benoit, and northeast of Eutaw. That should clear the location up a bit if you know anything about the delta region of north Mississippi and Arkansas. Not unlike most of the lands inside the levee system, this ground is a sportsman's and naturalist's dream. The diversity and density of the flora and fauna are just amazing and I never miss an opportunity or an invitation to visit.

On this particular occasion in mid-February we were invited to pursue "hawgs" on a post–deer season weekend. As most people will attest, if you ever have wild pigs visit your property they are not soon to leave. By nature they really don't have a home like the whitetail deer, but tend to roam according to the availability of food. And in this pursuit, using their amazing senses of smell and their flexible snouts as tillers, they can really wreck a pasture or painstakingly prepared food plot. If you can't identify "hog sign" on a property then you are probably as "blind as a hog." The eyesight of pigs is poor and this allows for some really exciting opportunities to stalk them on foot. Firearms are fine; however, with archery equipment this type of hunting can be really interesting. The fact that hunting pigs can be such an exciting sport and affords the use of a variety of methods has led to their being distributed over much of the United States. Many times this anticipation of having a fun new hunting resource has led to dismay as the hogs begin to do permanent damage to native plants and destroy farm crops. Now comes the task of trying to eliminate what you have started by introducing a nonnative species such as a wild pig. Well, you say, we will just hunt them out. The problem is . . . can you keep up! Talk about prolific! A mature female, and by maturity I mean six months of age, is capable of producing three litters annually with as many as fourteen in a litter. After three months, the piglets are weaned and although they will still travel with the female, they are on their own as far as foraging for food. Read back over the above statement, and you do the math.

It's no wonder that gaining access to hunt wild hogs is not too difficult if you know someone who has them on their property. The rules and regulations in most states are very liberal when it comes to taking pigs under most hunting conditions. This particular pig tale (sorry, couldn't help it) involves a trip to Caulk Island to do some "spot and stalk" and fill the cooler with pork for sausage and also to help lessen the area population.

February weather in Mississippi can be a fickle friend. A wet bone chiller in the morning may quickly turn to T-shirt time by midmorning. First light found Jim and me parking our ATV and entering a hardwood bottom that ran parallel between a mile-long field planted with wheat and ryegrass and a low man-made levee on the other side. You will find these low levees throughout the Mississippi River system, constructed many years ago to help control flooding. It's difficult to imagine the amount of labor needed to yield such a structure made well before the days of backhoes and bulldozers.

My hunting partner, Jim Polles, has a dental practice in Jackson, Mississippi. Since he is a professional when it comes to teeth, his admiration of the ivory that wild hogs can sport seems easy to understand. This day, however, we were not trophy hunting but trying to take enough pork to the skinning rack to make a trip to the local processor for a batch of sausage. We began to move quietly through the trees, keeping about thirty to forty yards apart and well in sight of each other. Hunter orange is a necessity for this type of stalk and it was easy to identify the other guy as we moved forward. The idea is to spot hogs before they see us and make a stalk to within reasonable shooting range. We were both carrying high-power centerfire rifles—no prisoners on this hunt. I have done this type of hunting with archery gear and with some success. It's great sport and a whole mess of fun but the odds stack heavy with the pigs. We only had a couple of days to hunt so we chose the rifles this time to try to insure some success with the harvest if we had an opportunity. As mentioned, the bottom we were moving through is about a mile long and it should take at least an hour to travel this or you are moving too fast. Slow and easy is the way so as not to miss any telltale signs of hog activity. The idea is to hear them rooting in the leaves and brush or grunting and squealing to each other. If you can accomplish this there may be a chance to make a stalk. We had been moving along for about twenty minutes and the woods were active with woodpeckers rattling on the timber and squirrels barking at us as we moved through. Jim was to my right and I had been looking to my left and forward along the low-running levee top. A shot rang out abruptly and immediately I saw a whole crowd of pigs running about forty yards ahead of me, moving from Jim's direction and crossing in front of me. A very large three-hundred pounder with long curly tan hair and a very prominent tail was in the lead, followed immediately by a two-hundred-pound class solid black one. I got hair in my scope and touched off a round. The woods got quiet again very quickly and then about a hundred yards ahead there was a lot of loud squealing by a very upset hog. I saw Jim moving up to my right and I walked over to sort the scenario out with him. Sure enough he had already followed the short blood trail to a very nice two-hundred-pound sow. Jim had fired a second round for insurance at exactly the same time that I had fired and neither of us heard the other shoot. We marked the spot and began to search for any evidence of my pig. After a lengthy attempt to solve the puzzle we saw no sign that my bullet had found its mark. We had hoped that the squealing we had heard was my hog but we could not find anything except the spent shell case from my gun.

It's always a pleasant walk back to the vehicle after a successful hunt. We were making plans for sausage, gumbo, appetizers, and the like as we walked the green field toward the ATV. Jim's prize had a really interesting copper tan and black coat and he was thinking about a pigskin rug in addition to the butcher shop goods. The end of deer and duck season doesn't necessarily mean the end of hunting opportunities. On property like Caulk Island it's easy to, as Jim says, "make some memories."

My First Wild Boar Hunt

WALKER SWANEY

For the last twenty years, I have been very fortunate to be a member of the Caulk Island Hunting Club located between Lake Whittington and the Mississippi River north of Greenville, Mississippi.

We really did not know what "hog hunting" was all about, however, until about ten years ago. This land, about eleven thousand acres, is probably the most "hog infiltrated" land along the Mississippi River from Memphis to Vicksburg! In the dark days we would sit on a deer stand and count twenty, thirty, forty hogs each morning and in the afternoon in the large fields we would see over a hundred!

No one told us about the thrill of dog hunting for hogs until 1998. A good friend of mine, Bob Whitwell, former U.S. attorney in Oxford, Mississippi, said we should invite his buddies from Arkansas over because they had really good dogs and knew what hog hunting was all about. So we said, "Come on!"

The river was high and we could not drive in. We had to push a barge across Lake Whittington three times almost one mile each way back and forth to unload four-wheelers, a truck, eight dogs, crates, and supplies. The guys did not know us and we did not know them but pushing a twenty-foot barge in the middle of the old river we found out quickly that we could get along.

The wind was up blowing waves over the bow of the boat and we were taking on lots of water, cold as it could be, and it was quite dumb I guess, but exciting! We obviously made it.

We quickly unloaded, regrouped, got the dogs out and within one hour we were on our way. Within twenty minutes we were on our first hog.

Now, the truth of what we didn't know. The boys from Arkansas brought with them a secret weapon—Hammer! We really were too much in a hurry on the barge trip to notice the eighty-pound pit bull, but, my goodness, what an animal! They held him in a crate separate from all the other dogs. Instructions were "don't try to pet Hammer." We pretty much figured out why.

The first hog was a three-hundred-pound boar. The dogs ran for only ten minutes before the boar turned to fight and faced us. We came up on eight dogs barking, delirious, excited, and all in a huge ditch going against a vicious angry wild animal. Well, then, you have never seen such excitement!

The Arkansas boys, Larry and John, were old pros at this and had hunted all their lives. They were both about thirty years old, and John's dad, Dick, was the old pro. He was about fifty-five and he had been on six hundred hog hunts. Paul rounded out the group. Paul owned Hammer.

As the dogs closed in on the boar, Paul yelled, "I'm getting out Hammer! Don't get in his way!" With that, Hammer was released.

We had pulled in within a hundred yards on our four-wheelers—ran about fifty yards more, then watched Hammer being released. Probably no more than four seconds later Hammer hit the boar with the momentum of a freight train! Hammer knew what to do. He bit down on the boar after a minute. You have never witnessed anything like the way that Hammer bit into the back of the hog's head, somewhere behind the ears and that's exactly where he was attached. The fight with this boar was on! The dog boys quickly moved in, with Hammer attached the whole time to the hog's head. They spread the back legs of the hog and flipped him on his back.

The other dogs also immediately moved in to get at their piece of the hog and for thirty seconds it really was a mass of chaos!

With a mad-as-hell boar, attached pit bull, eight or nine dogs biting and barking, this is the extreme excitement for dog hunting! Now, our next job was to get our hands on the dogs and not get accidentally bit. One dog at a time we chained them and pulled them off the hog.

Needless to say, this hog was very upset. With eight or nine dogs biting at him, an eighty-pound pit bull attached to his neck at the back of his head, and twelve guys watching the controlled chaos, this hog was in a bad, mad mood.

With a hickory mallet, the dog owners slowly attempted to get Hammer's mouth open and release him from the hog. We could not open Hammer's mouth—he would not let go! So we had to cut the hide so we could pull Hammer off. All the dogs were still trying to bite the hog and we were pulling them apart. Paul said, "Don't get too close to Hammer's face! He is pissed!" We let Paul put the muzzle back on him with the hog's skin still in his mouth.

Now, the dogs were off the boar and the dog guys said, "What are we going to do with the hog today?" Hell, what does that mean? I thought we would shoot him and we would move on. No, that was not how it worked with these guys. Larry said that this wasn't a good eating hog so we should turn him into a "barr."

You know what that means? Yes, you're right, that means cutting off his testicles. Let him grow up to be a big fat lazy barr hog to be caught at a later date.

We agreed that this was a good idea. Except who does the cutting? Well, I got selected to castrate the boar hog and they showed me the technique, saying not to throw these nuts away, that we were going to eat them for breakfast. We promptly put them in a Ziploc bag for the next morning's breakfast.

And that is what we did!

Back to the dogs. The handlers had the dogs chained back. The testicles were gone and the boar was really, really mad. The hog's tusks are about four to five inches long and razor sharp. He would kill you if he could. So the handler said, "Boys, we are letting the hog go! And it's everyone on his own when we release him!"

With that I looked for a tree, except no tree was around that I could climb. In that moment Larry let go of the boar and jumped back. The hog jumped up and lunged at him, then turned and ran right toward us! We scattered and this hog came right at us, with tusks going right and left. He chomped my buddy's hunting coveralls and ripped them about eight inches. No blood though—lucky him.

Never in your life do you anticipate the adrenaline high that this gives you when you witness the action of man and beast. You simply cannot believe the excitement, the chaos, the frenzy, the savagery of these animals involved in a fight for life, especially the catch dog, Hammer.

That weekend started with baying hog after hog. Hunting in the night we bayed nineteen hogs! Having a big boar in the absolute still of night, hearing the dogs barking, watching the dogs fight the hog is something everyone should experience. Coming up on a hog at night with nothing but a small light and a knife is pretty mind-boggling. You just hope in a dense cane thicket you don't step right into the middle of the fight.

Later that weekend, my son, Walker J., was on his four-wheeler when just he, Paul, and Hammer were on a big boar. Paul let Hammer out and he bit down on the hog and missed. The boar turned towards Walker J. just feet away. He turned to run (or climb a tree) and fell. The boar was right on him. Paul immediately pulled his gun, shot once at ten feet, and hit the boar in the head, a very lucky shot. Immediately the hog was dead. He was four feet from Walker J.'s head. Walker J. was very fortunate. Lying on the ground, you are no match for an angry boar. Paul saved him from a very serious injury or worse.

Later, neither Walker J. nor Paul could speak. It scared them both that much. Walker J. went back to camp to regroup. This first weekend will be remembered as long as I live!

In most of these hog bayings we do kill the hogs. The method of taking the hog is to use a knife with a long blade and it better be sharp. After the hog is flipped over on his back and he's pinned down by your knee, you sever the vein under his right front leg and he will bleed out in about three minutes. Very rarely are guns used. In Walker J.'s case the gun probably saved his life.

On another hunt Dick had his favorite dog, Buck, with him. This dog has more scars on him than Frankenstein, all from being cut by hogs. The dog is stapled back again for another hunt. After two weeks, old Buck is ready to hunt again.

On this hunt Buck was punctured and cut deep into the chest fighting a big boar, exposing his lungs and diaphragm. Poor ole Buck was not going to make it! Then our camp co-owner came to the rescue. He is an Oxford surgeon and he got Buck back to camp and operated on the picnic table. We cleaned ole Buck up and disinfected him with alcohol (we had some left over). Dr. Lovelace completely sutured inside and out and Buck lived. Probably Dick gave him two weeks off. Later on Buck got run over by a damn gas truck in front of the house after three hundred hog hunts. Dick was ill.

We have been going back each February for the last ten years with these dog handlers. Each year the hunts are complete with unbelievable excitement and every year something new happens that we've never seen before.

Hog hunting is not for the faint of heart because the dogs pay a price—sometimes with their lives.

However, if someone asks you, "Would you like to go hog hunting?," just answer them with "if you have a Hammer and a big sharp knife" (sometimes you need a gun), "I'll go."

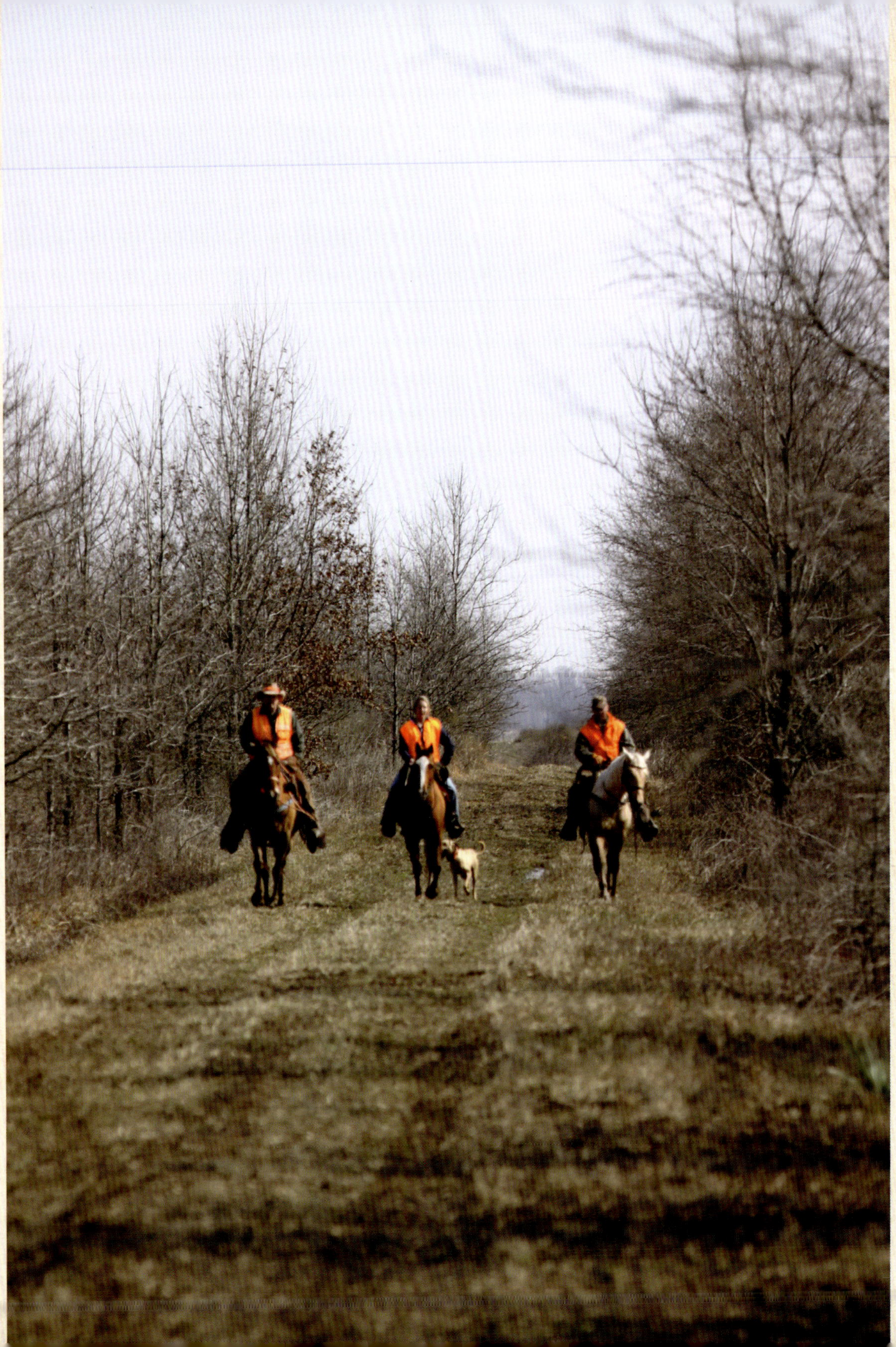

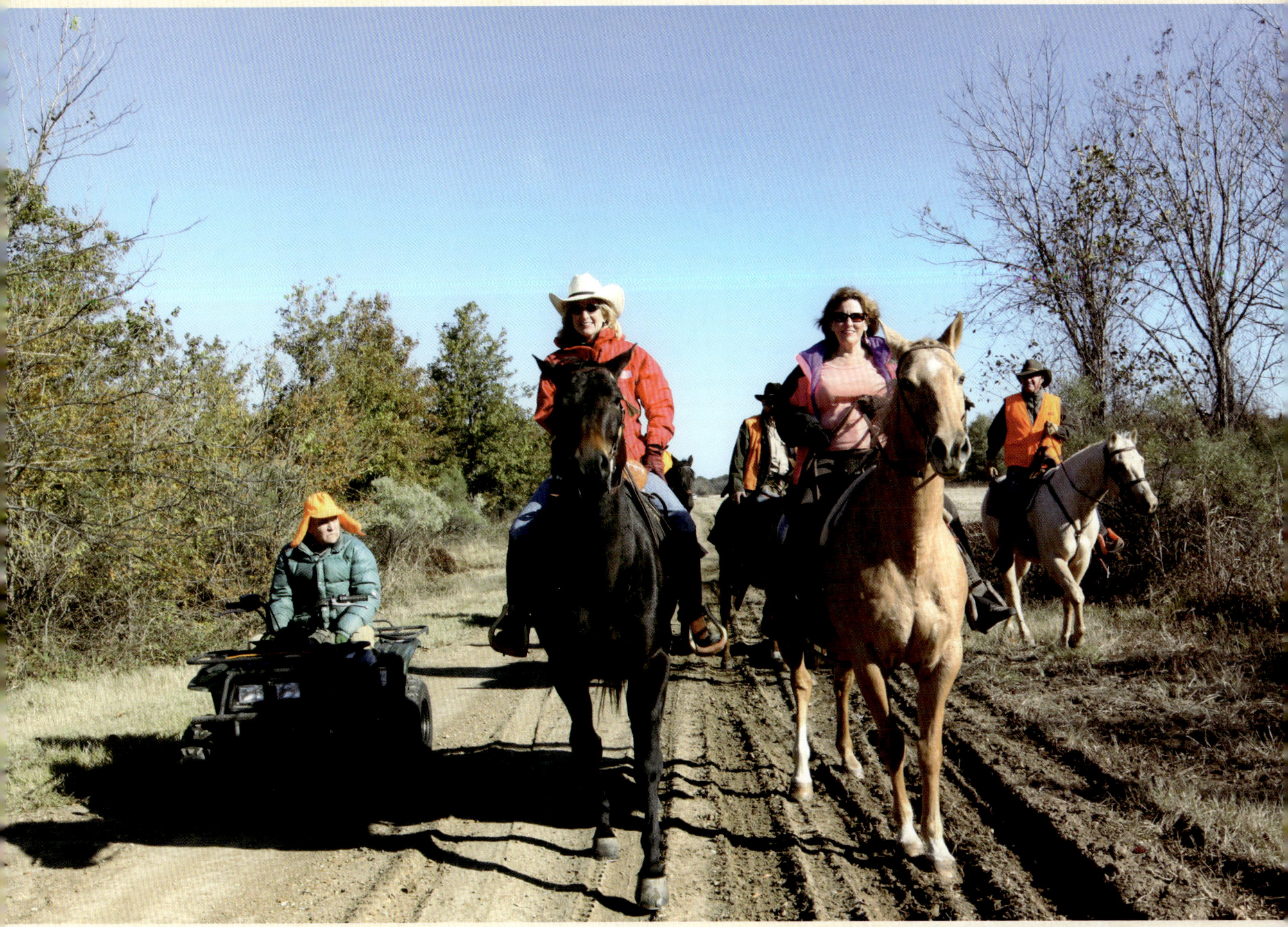

Sonny Merideth, Mary Hardwick, Kay Shipp, and Steve Golding.

< Wick Eatherly, Stacey Waites, and Andy Wilkinson.

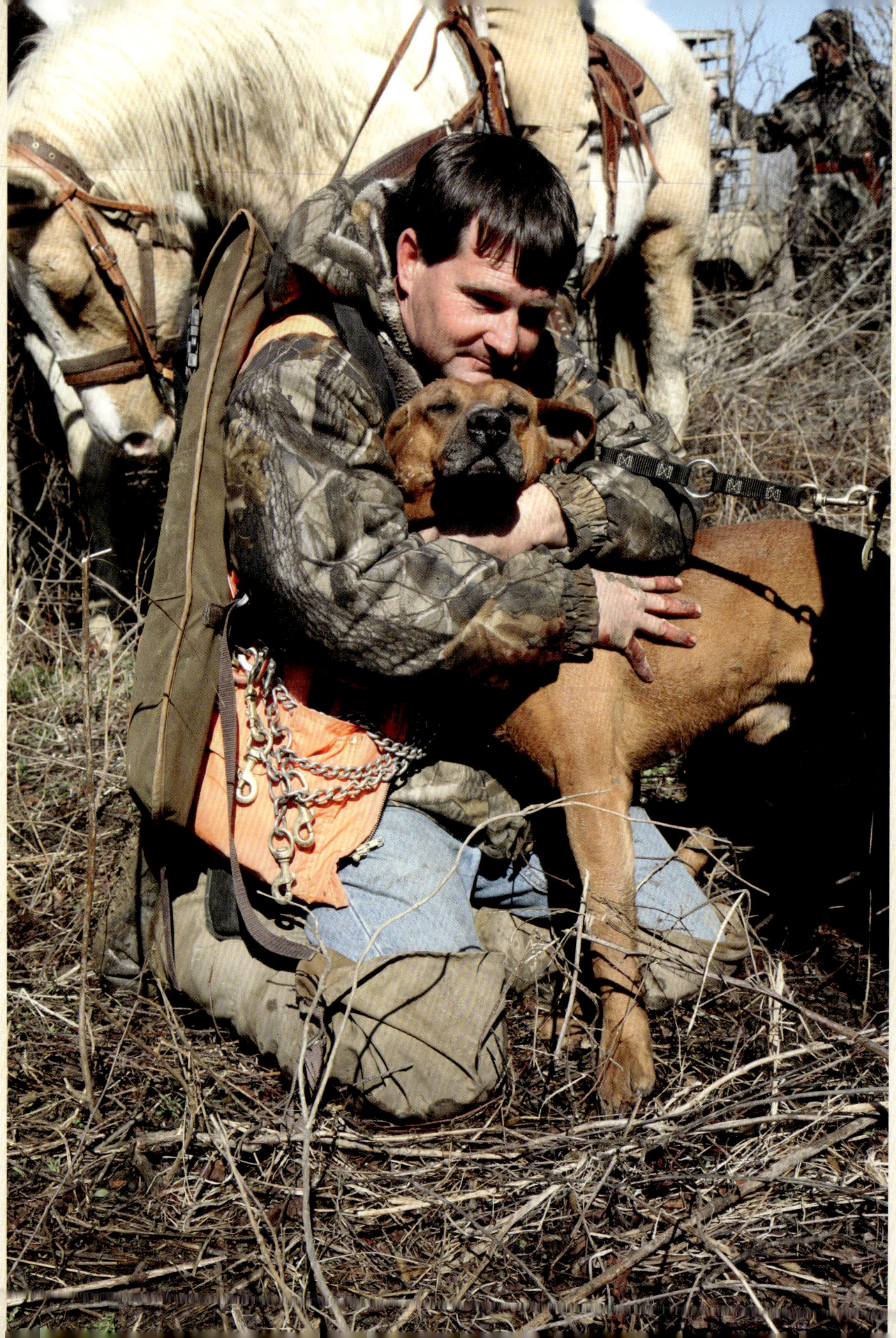

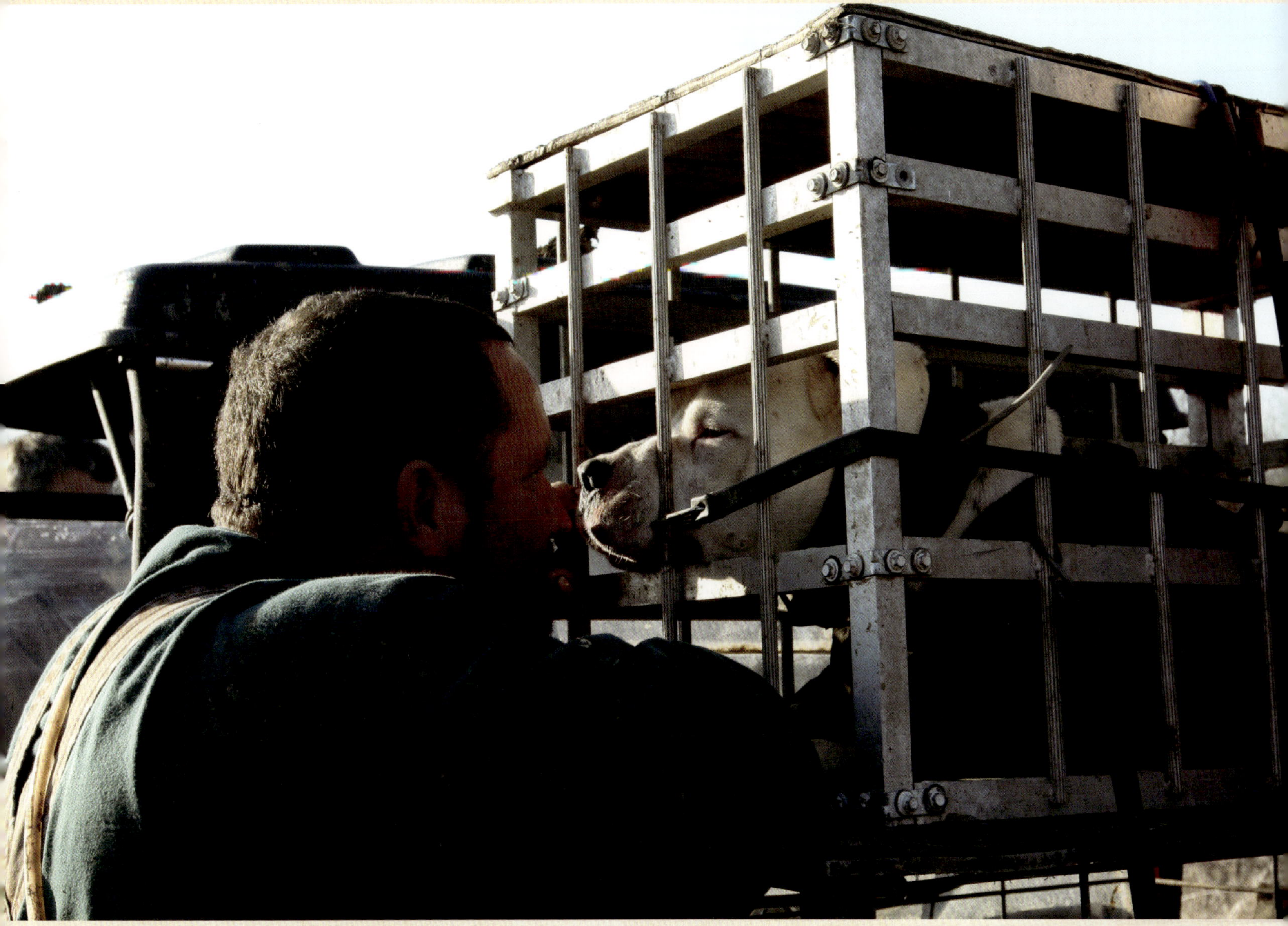

Jeff Cremeen and his bulldog, Frog.

< John Lowery and Bo, redbone and black mouth cur, and Sid "Bo Weevil" Law (in back).

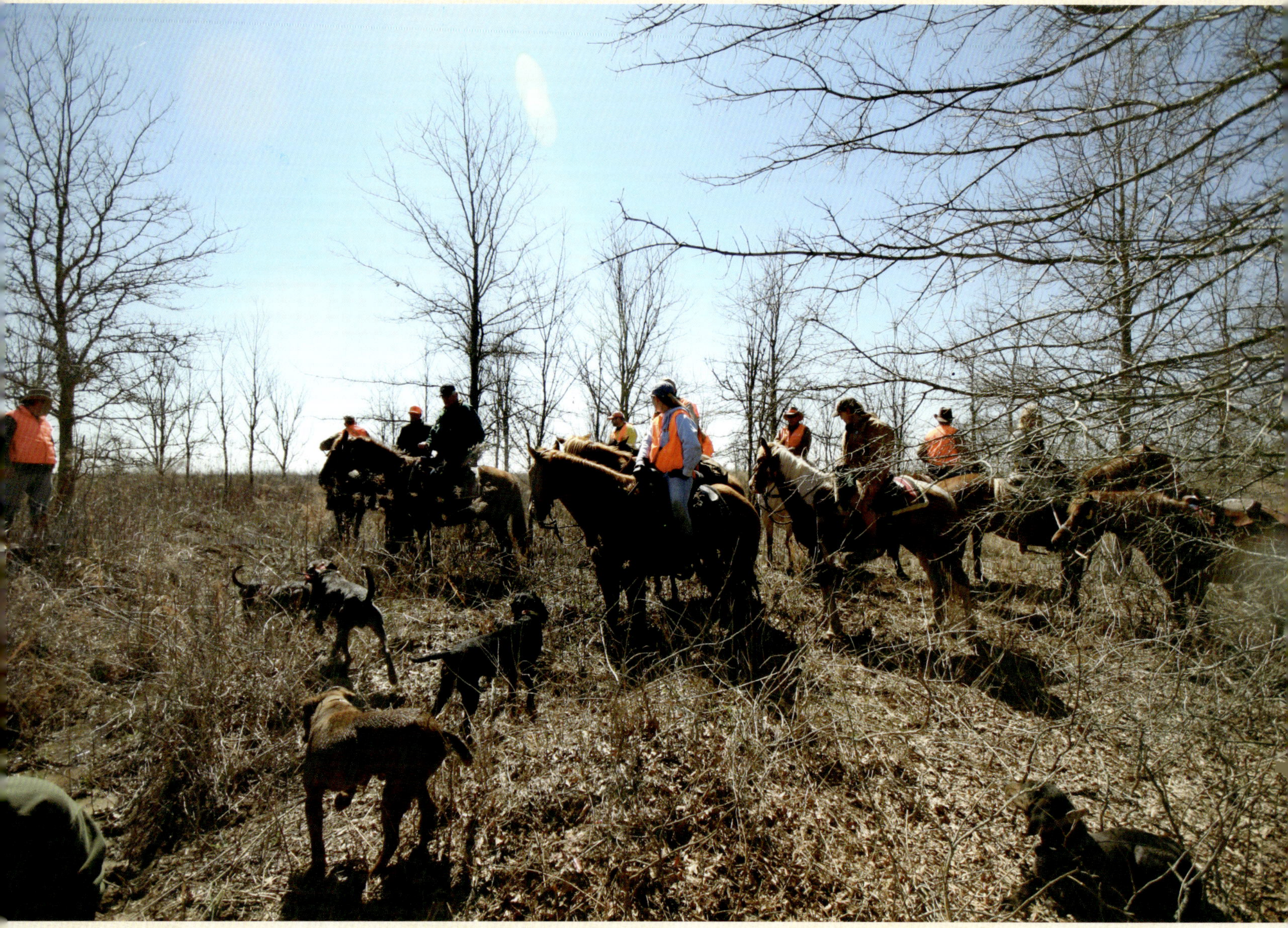

At the hunt.

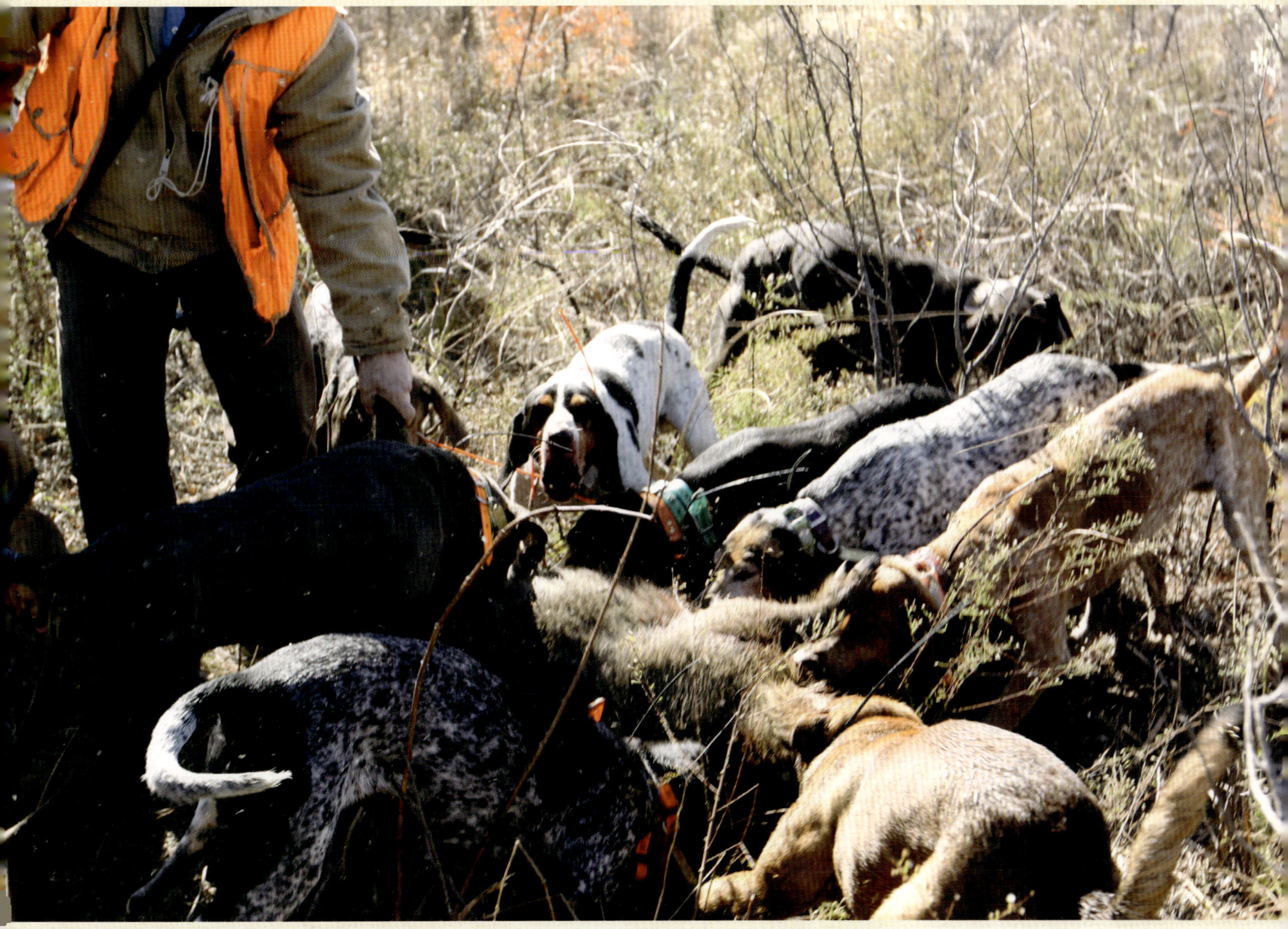

Hounds at the kill.

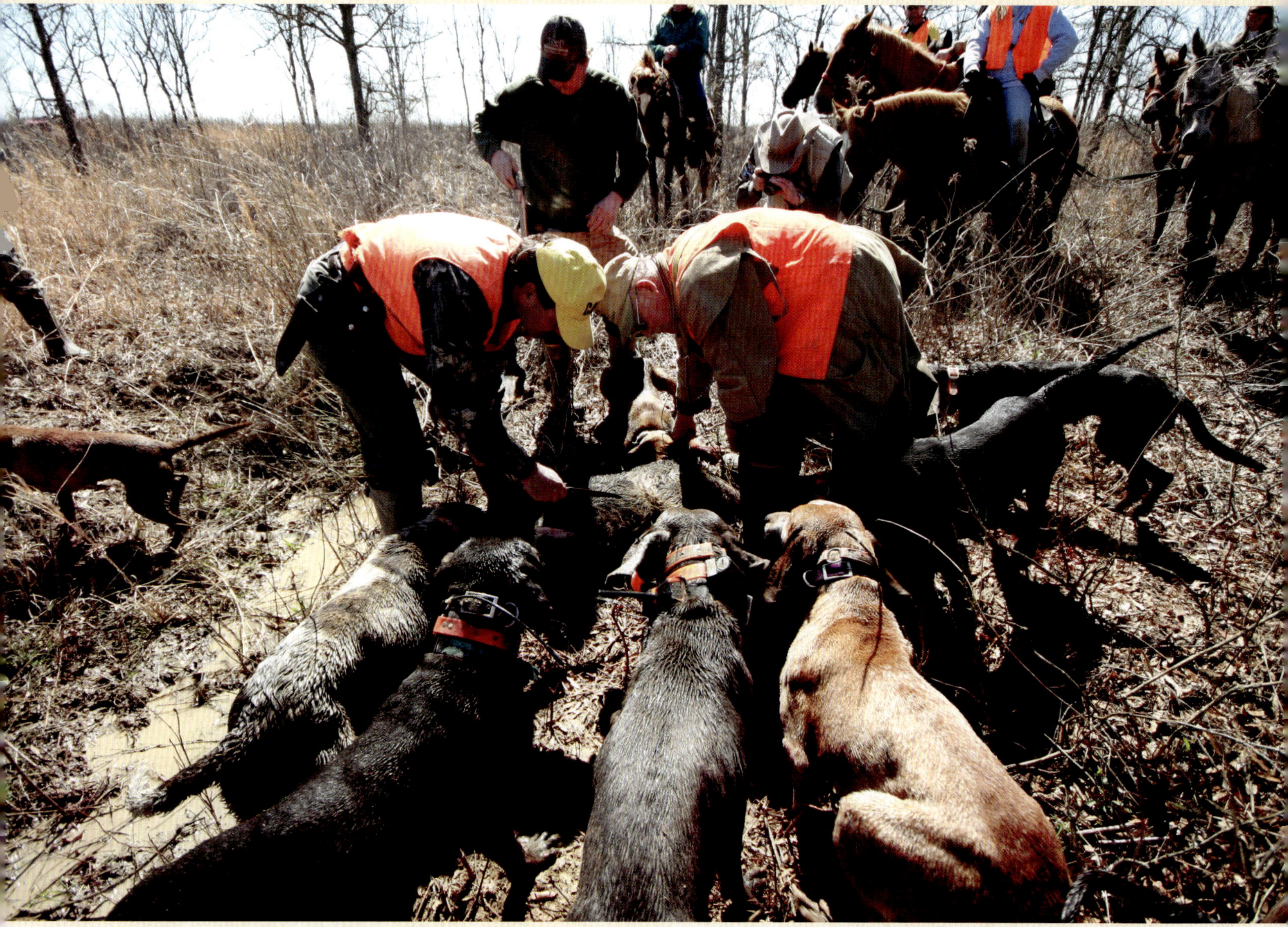

Wild boar with tusks.

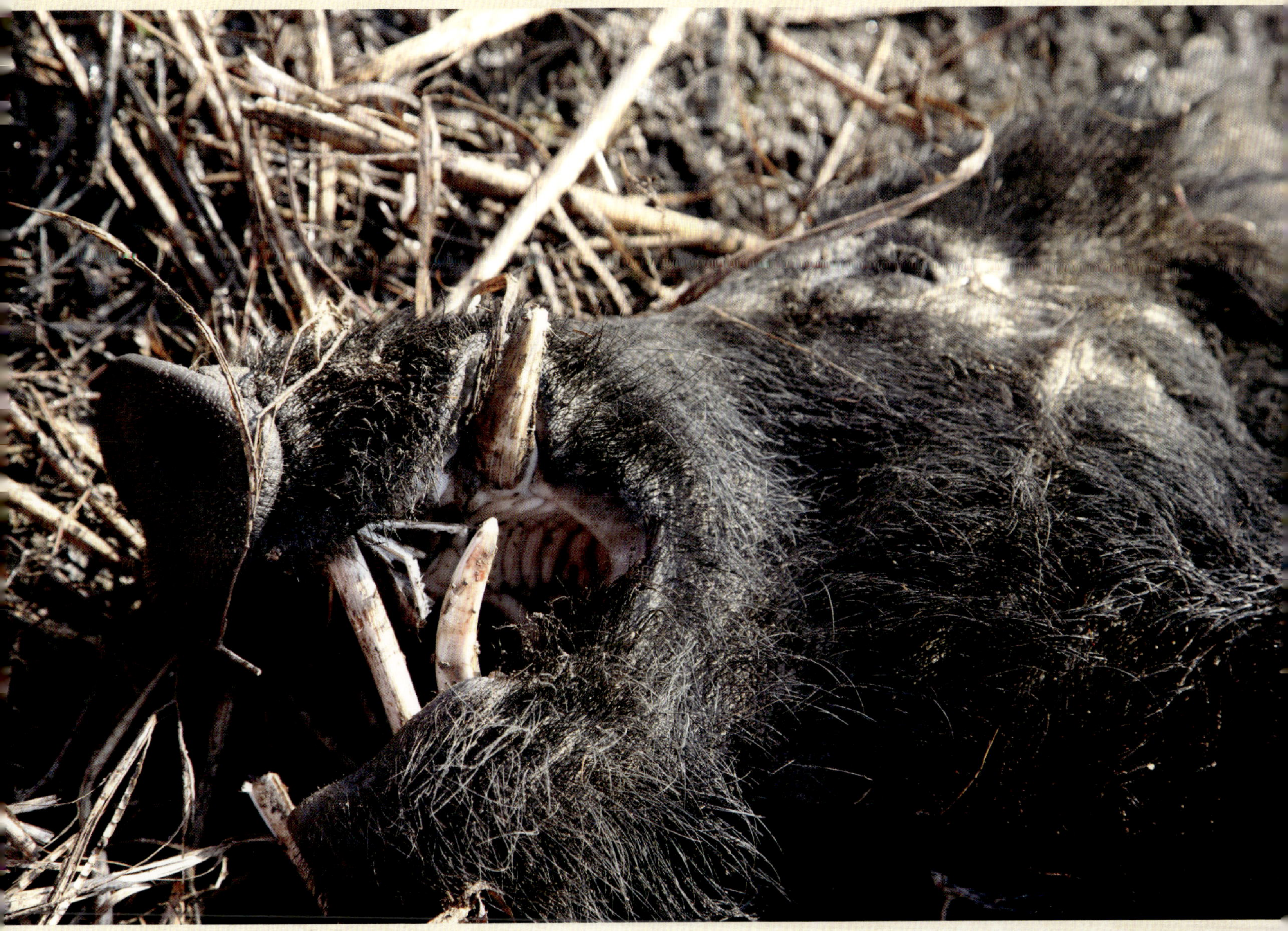

Pork chops.

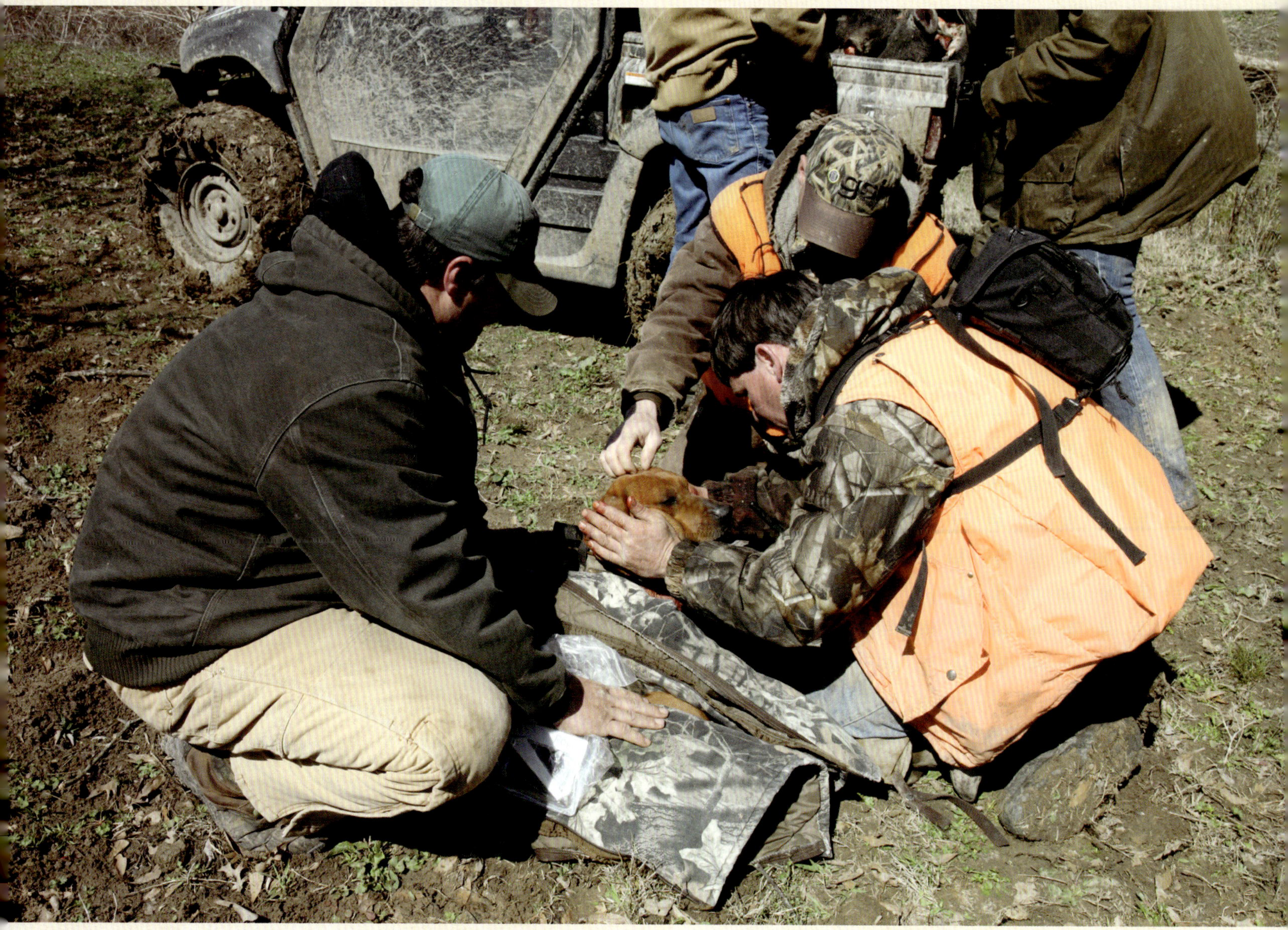

Stephen Davis, Little Joe, a black mouth redbone cur, John Lowery, and Andy Wilkinson.

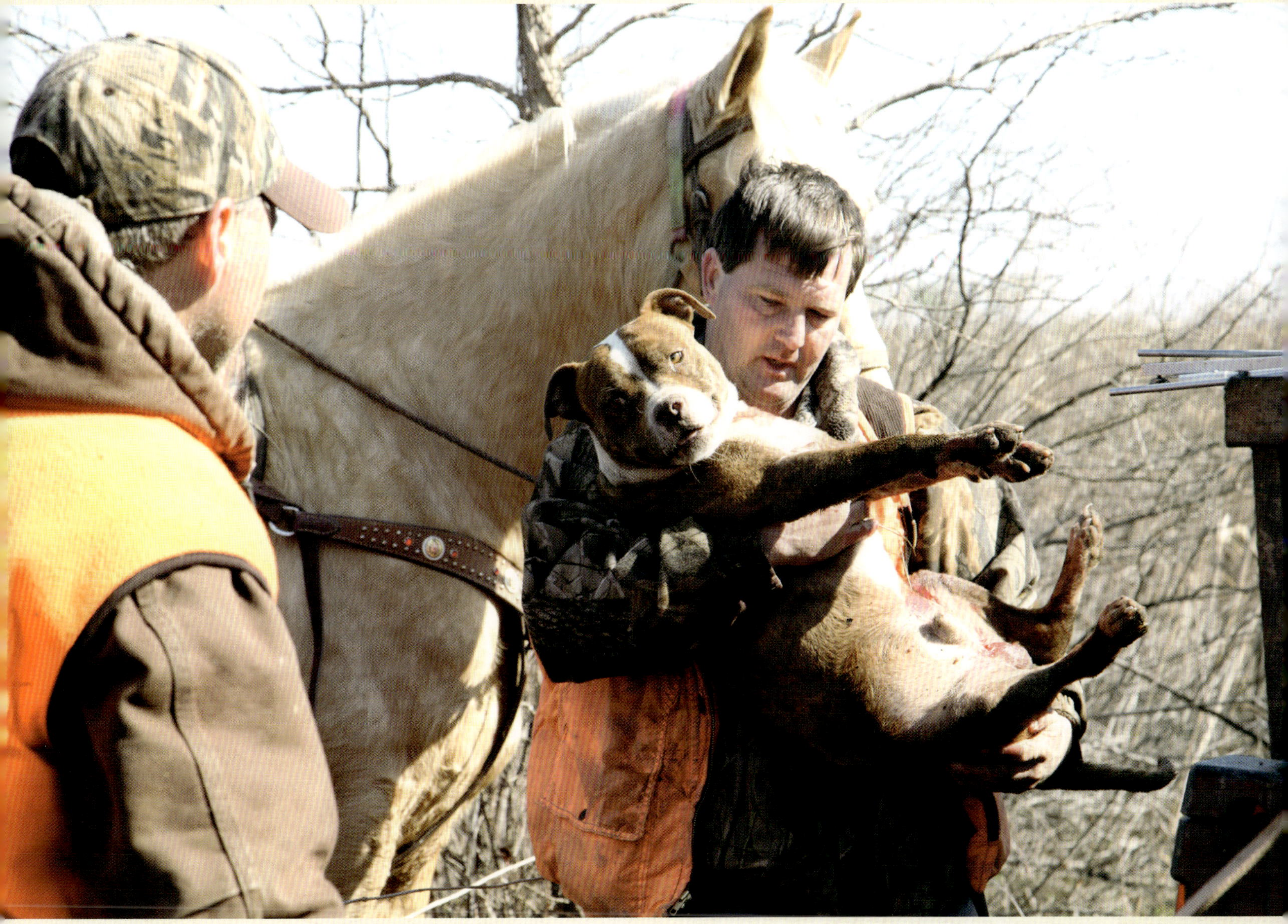

Andy Wilkinson, John Lowery, and Homicide, a rednose pit bull, and Wyatt, a Tennessee Walking Horse.

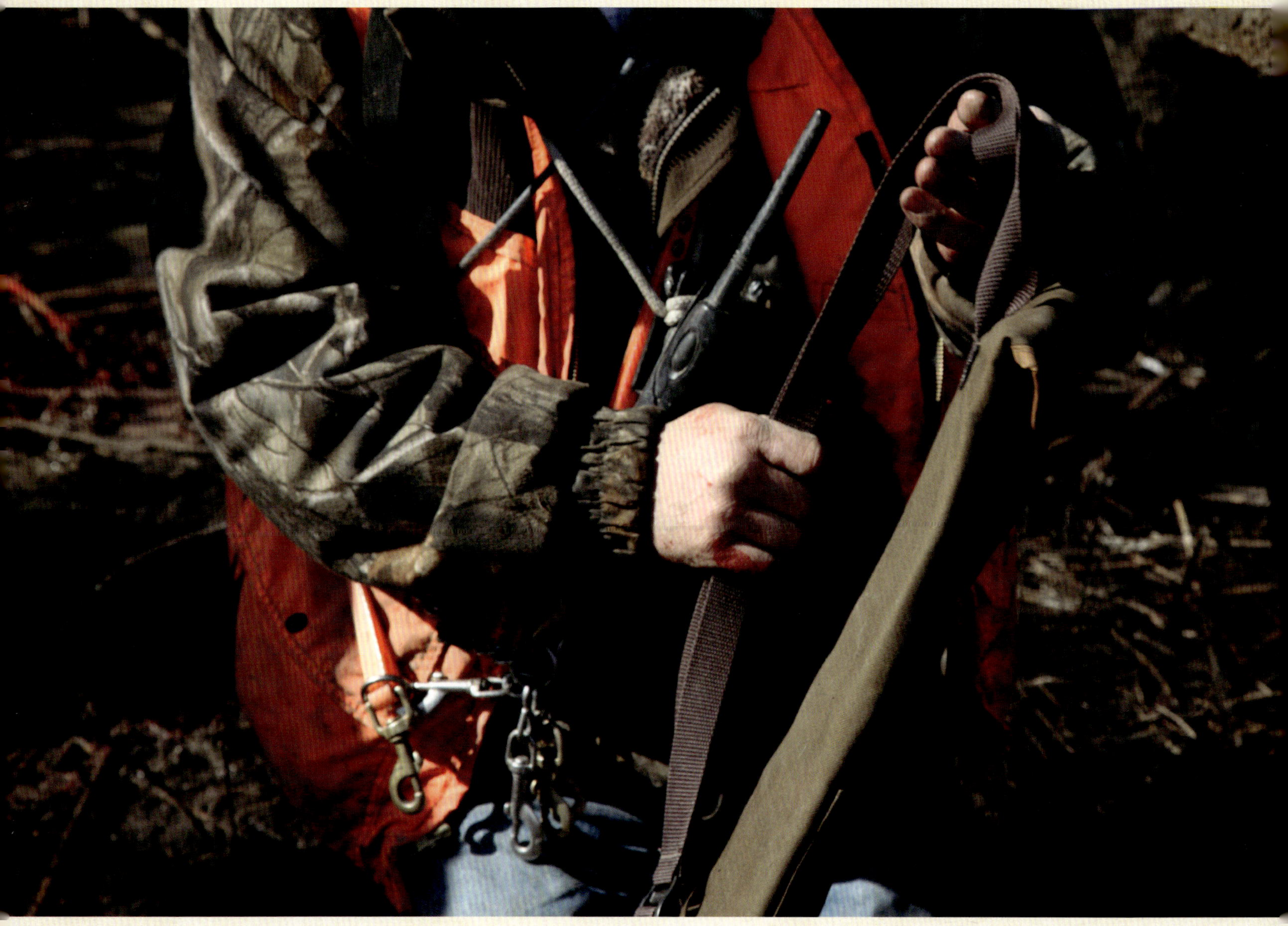

Holding a tracking system.

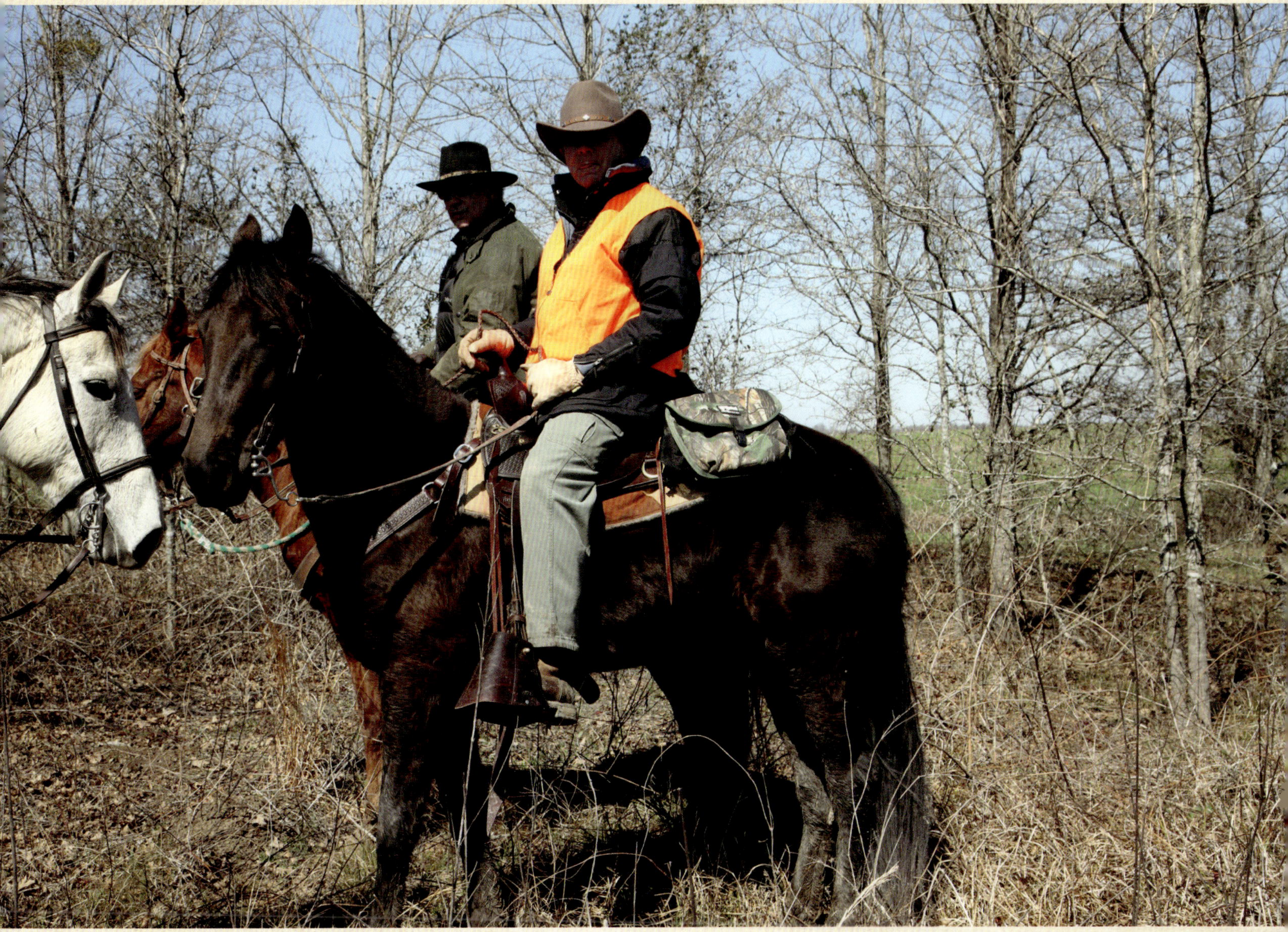

Steve Golding and Frank Dantone, Clue and Red.

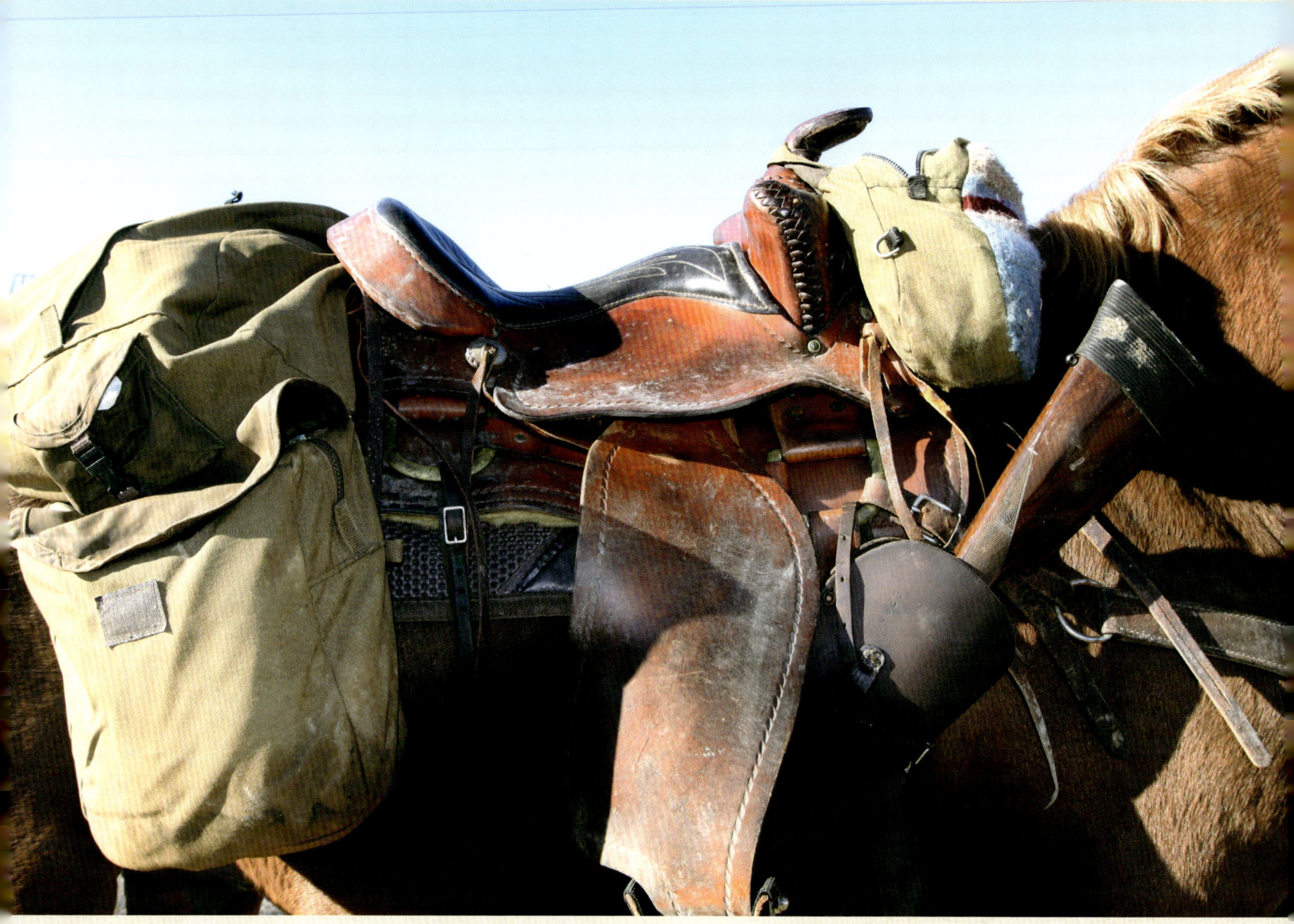

Horse hunting attire.

Frank Dantone and Howard Brent. >

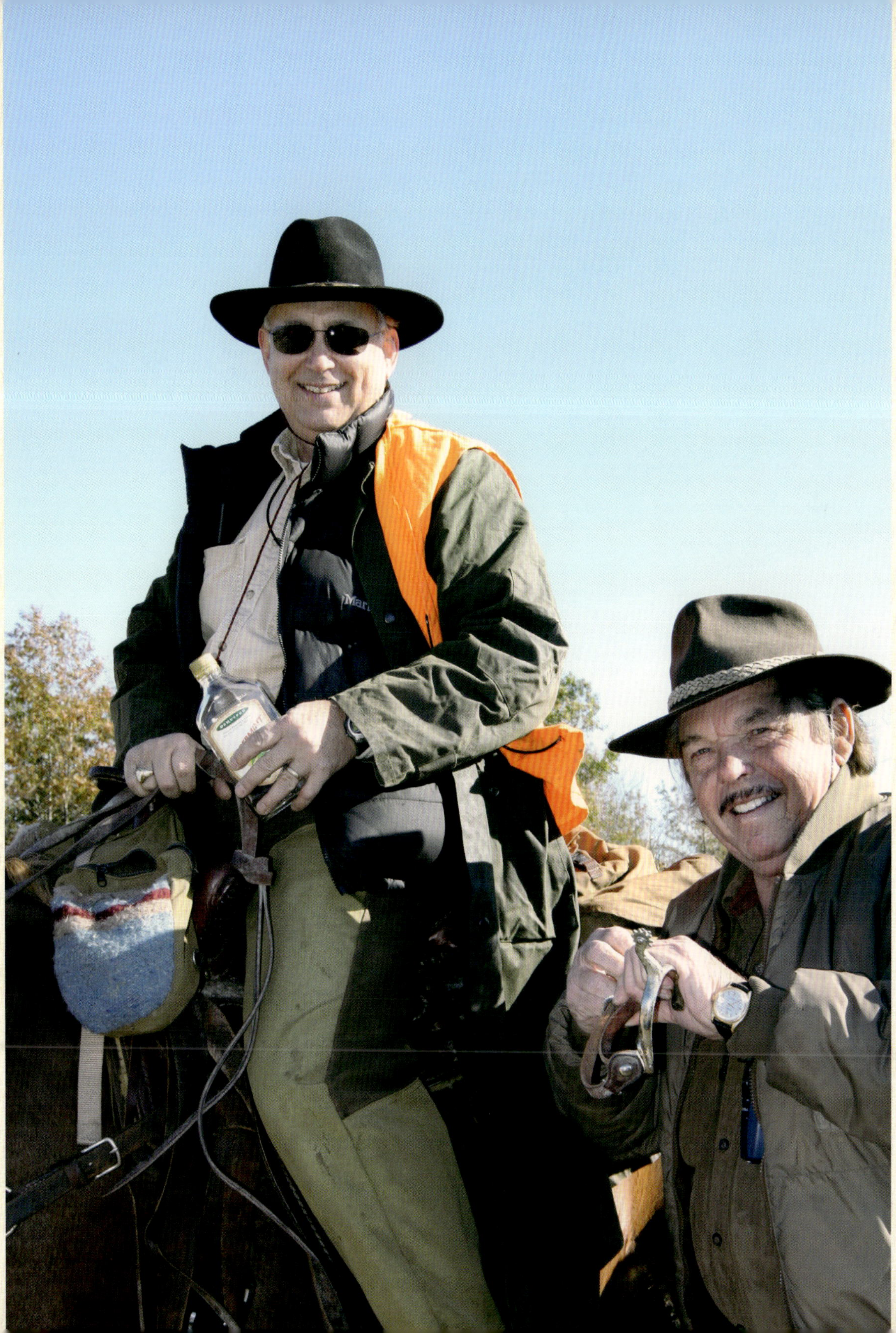

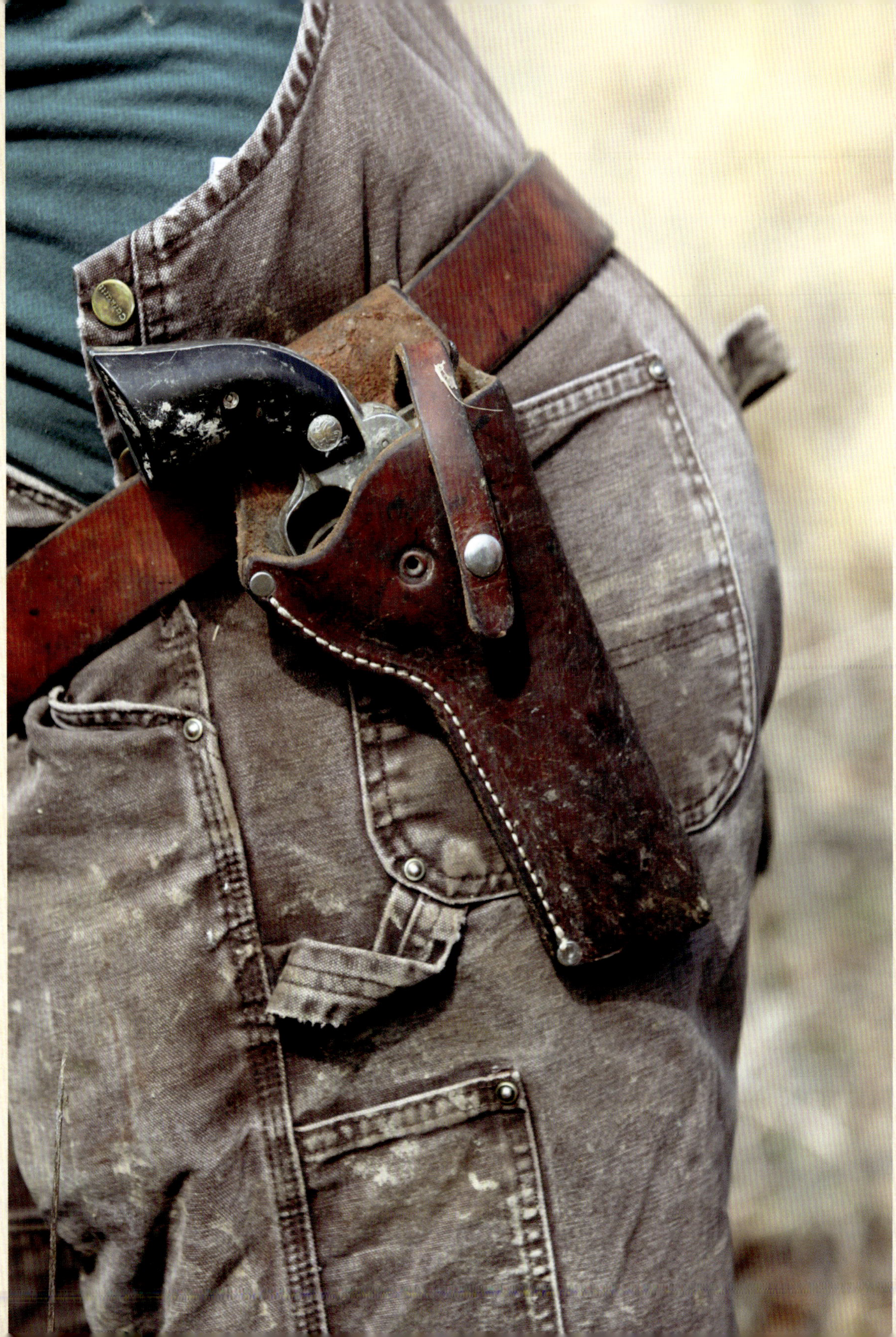

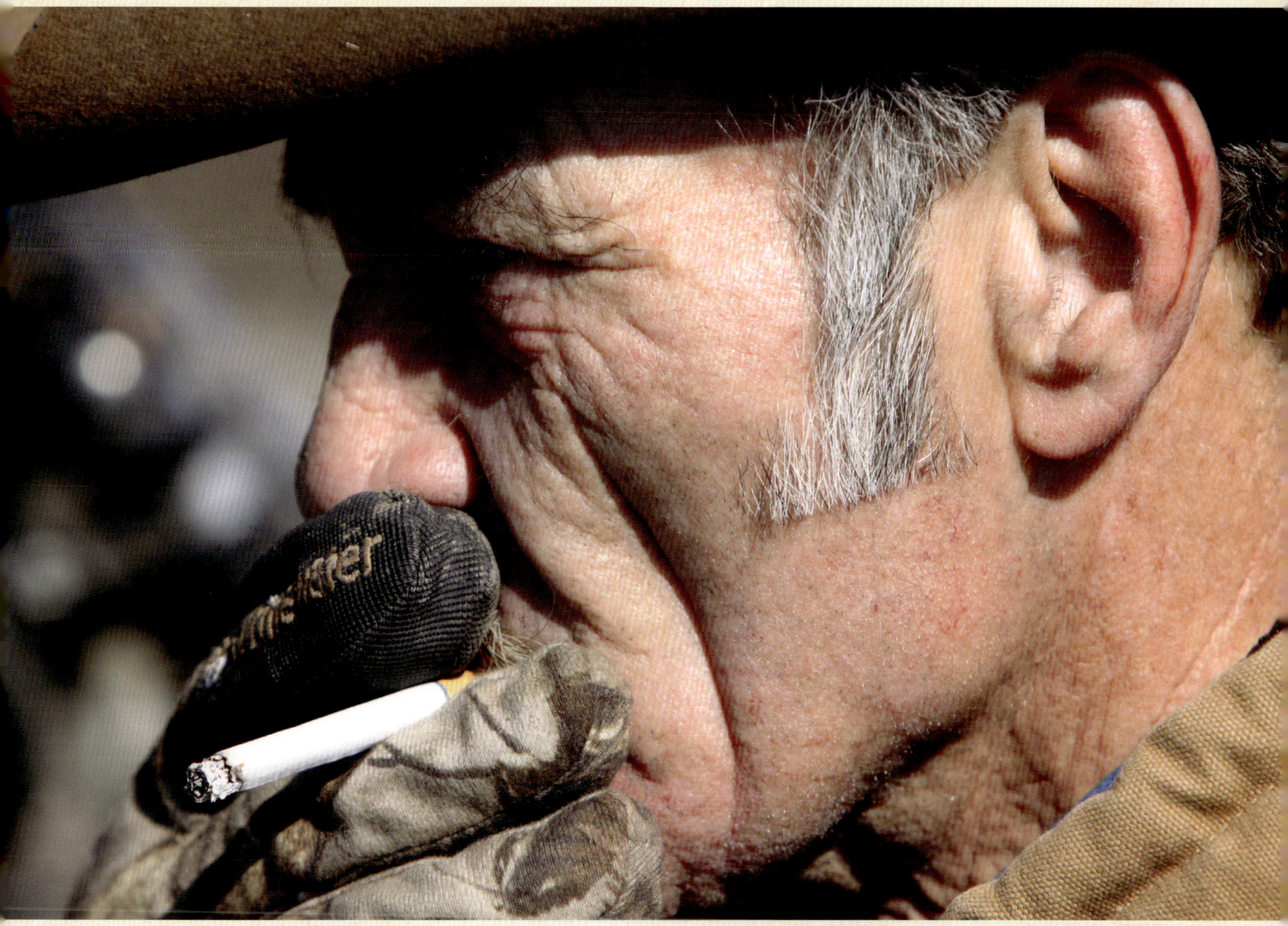

John "Wildcat" Stephens.

< 22 magnum pistol, Jeff Cremeen.

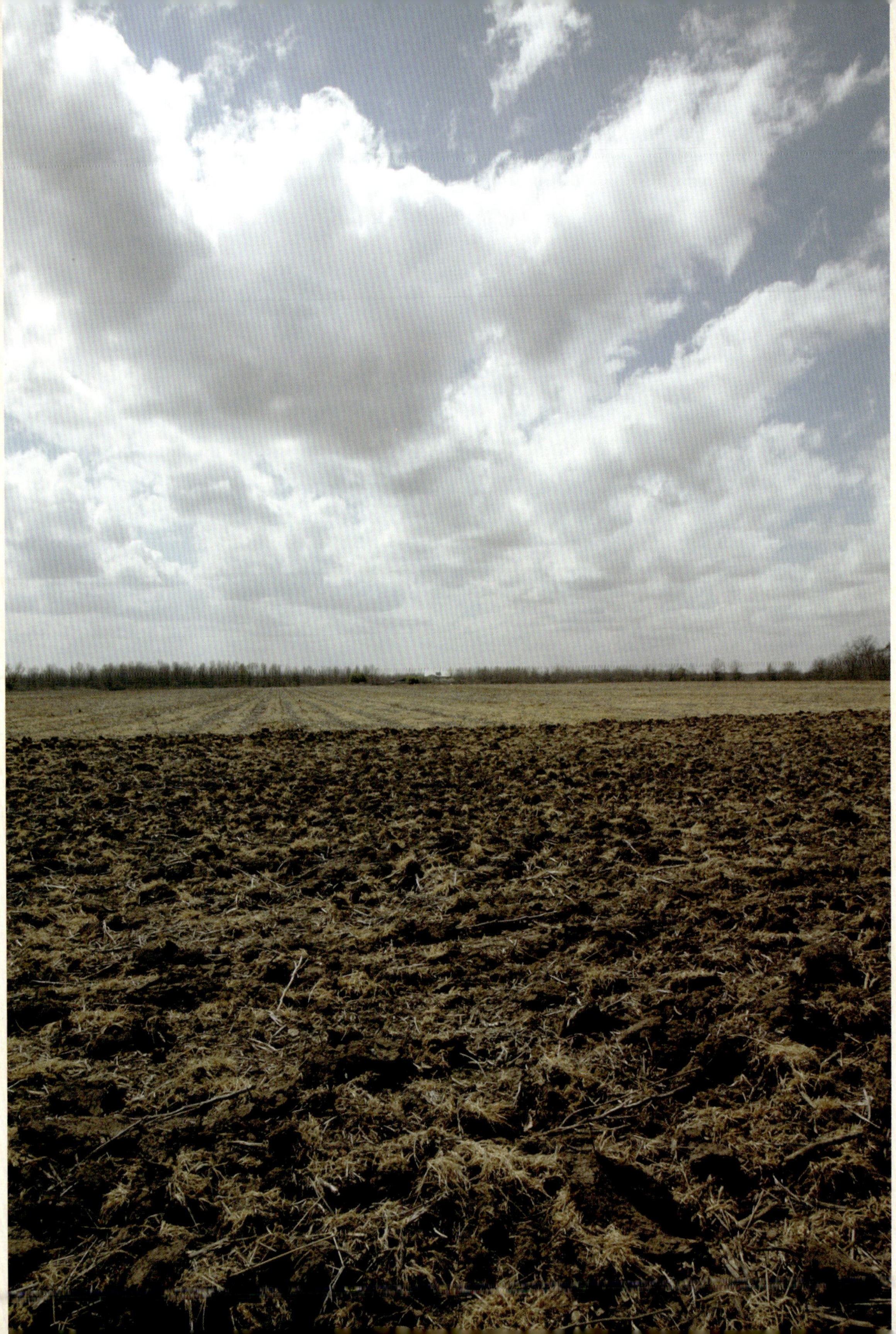

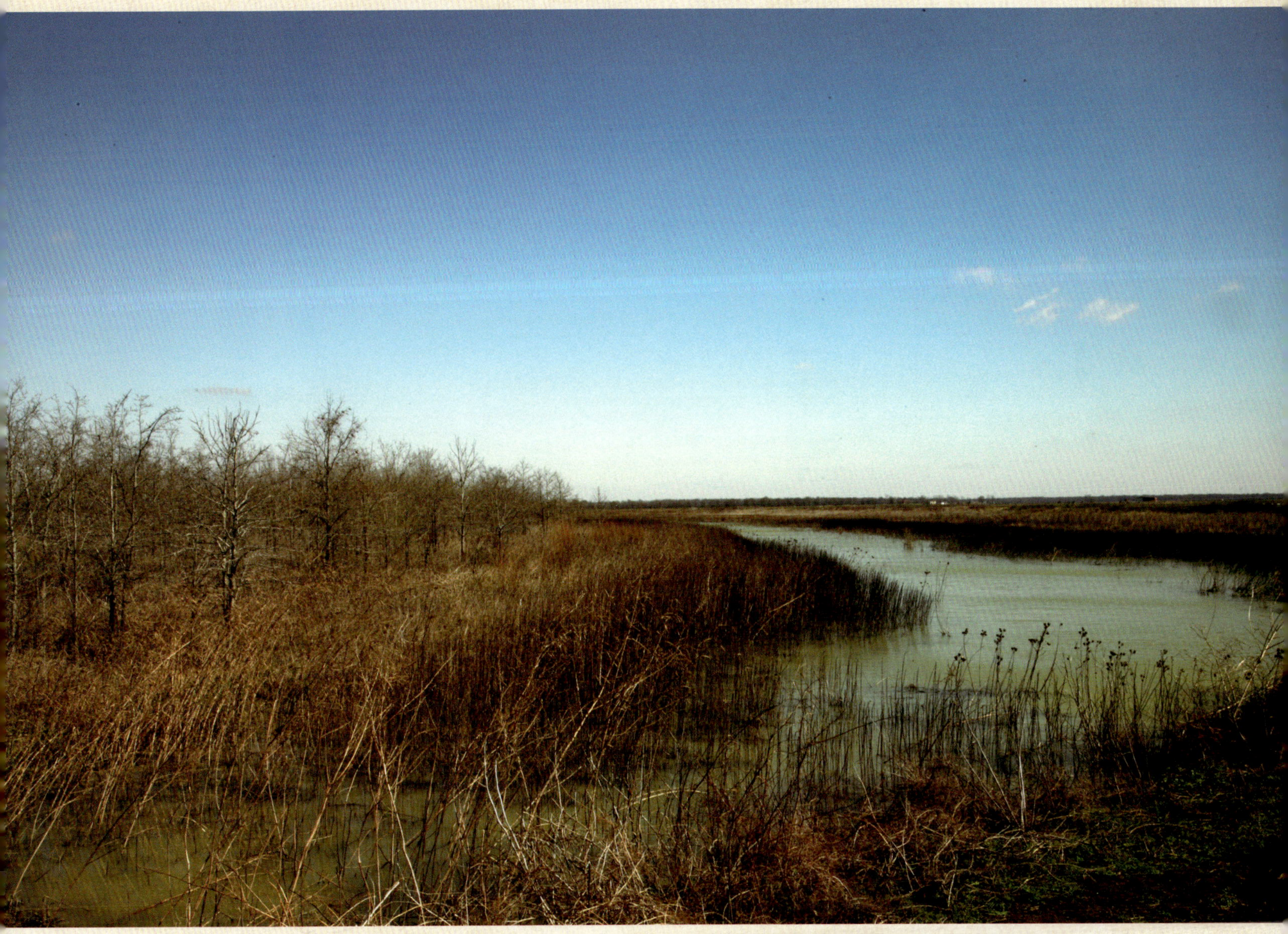

Upper Lake Discovery.

< Field.

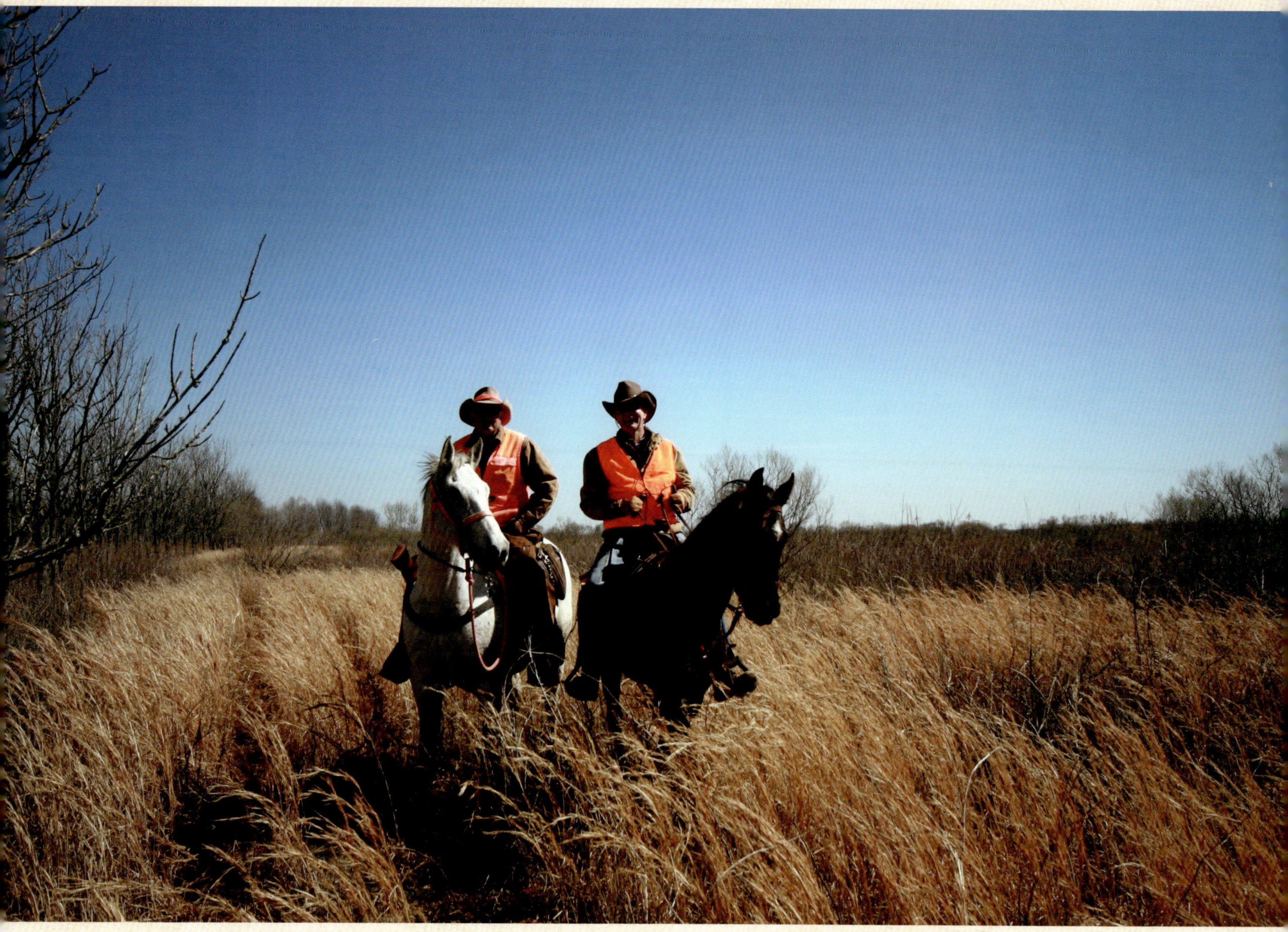

Wick Eatherly and Steve Golding.

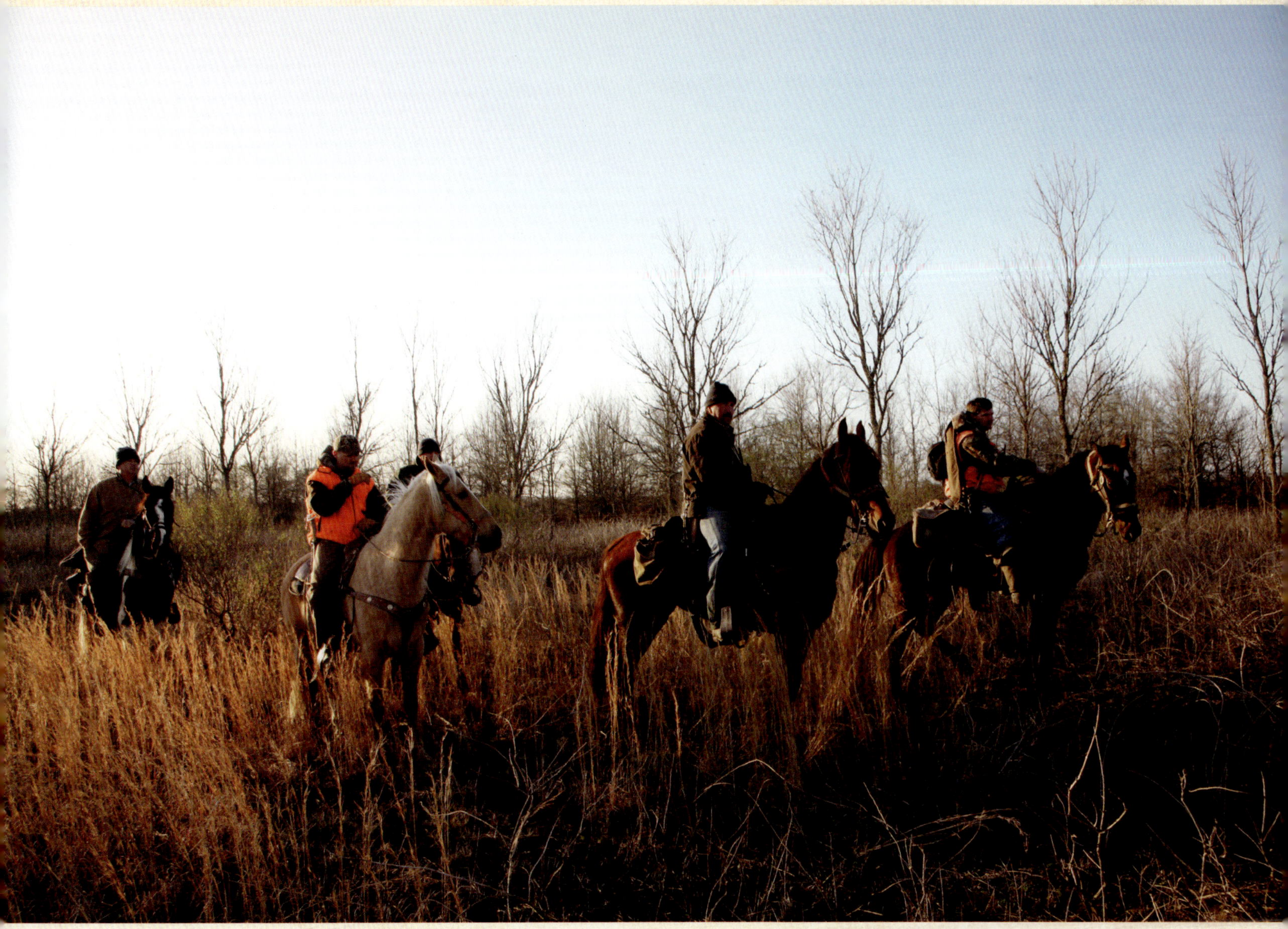

Late afternoon hunt.

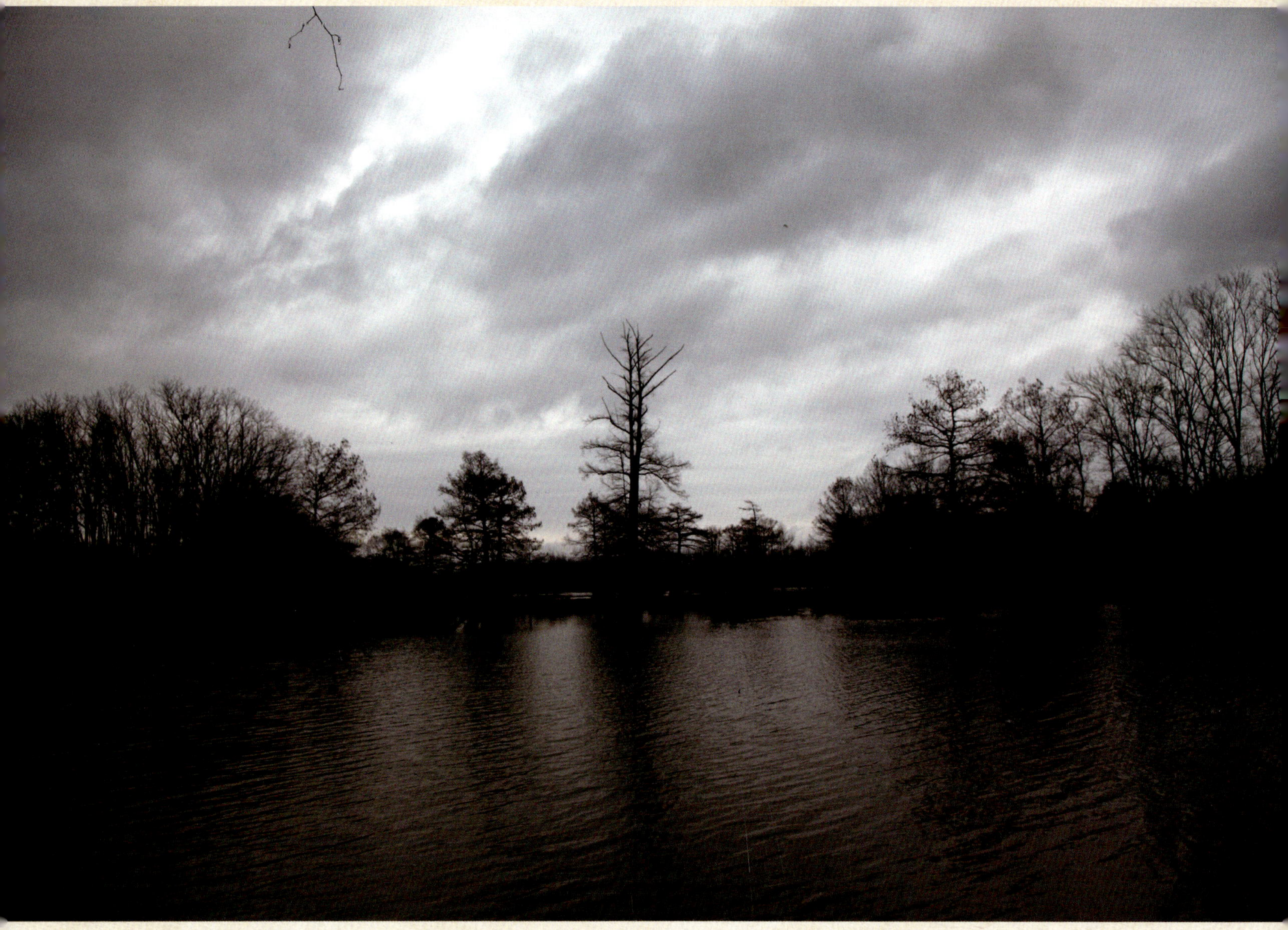

Boobie Lake.

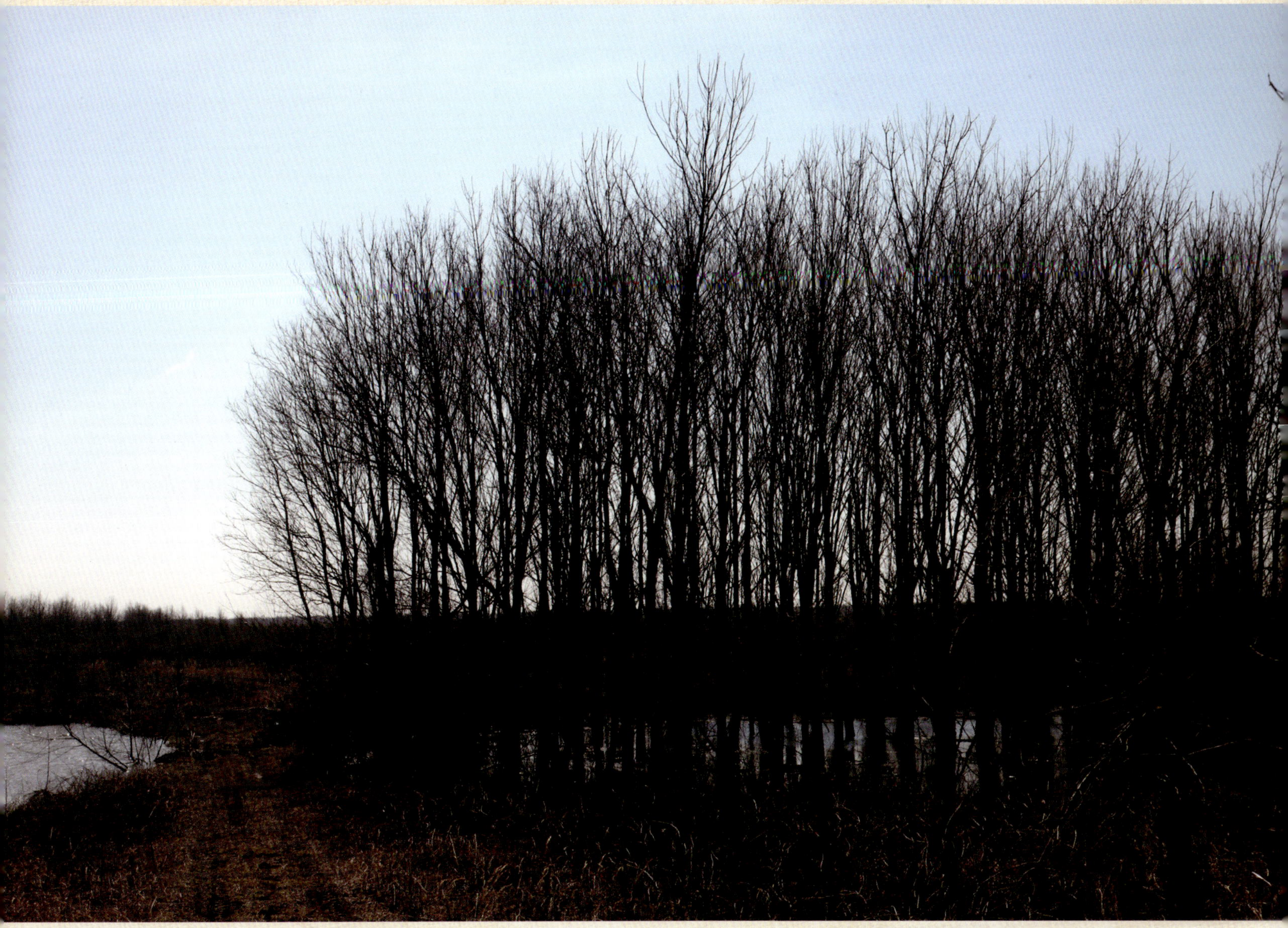

Upper Lake Discovery.

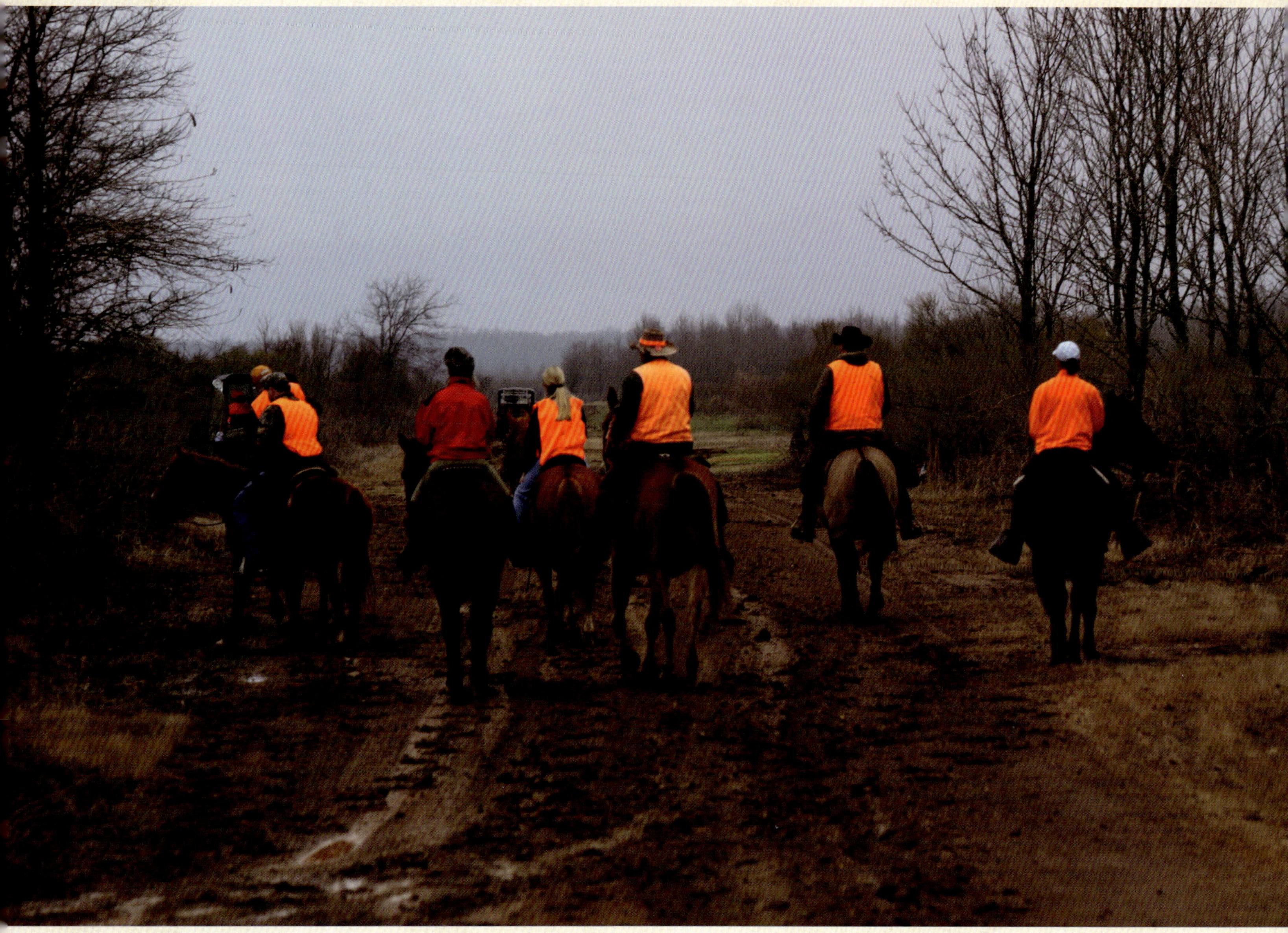

Hunters on horseback.

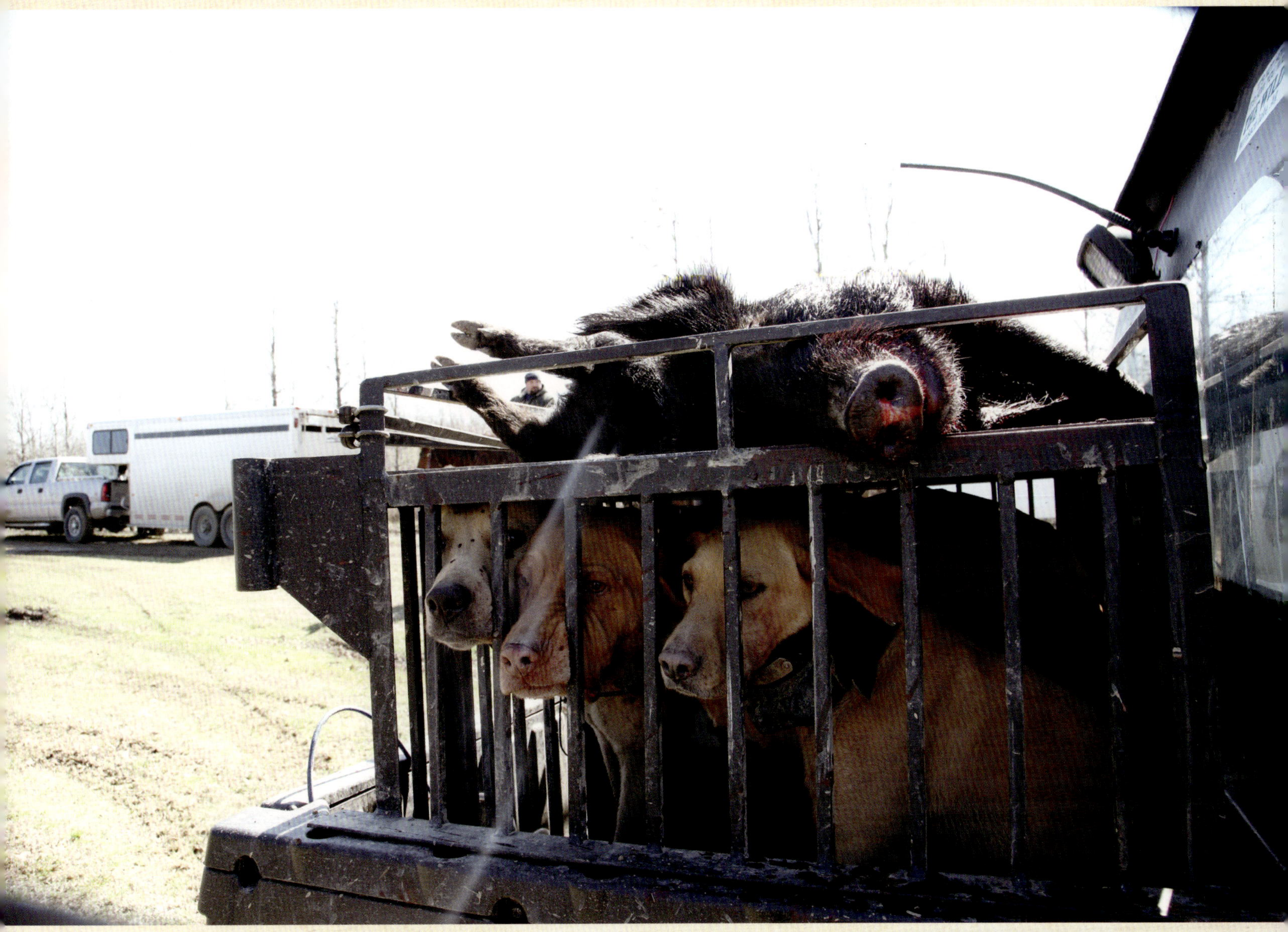

Hogs and dogs, Molly and Junebug, mountain curs, and Rooster, black mouth cur.

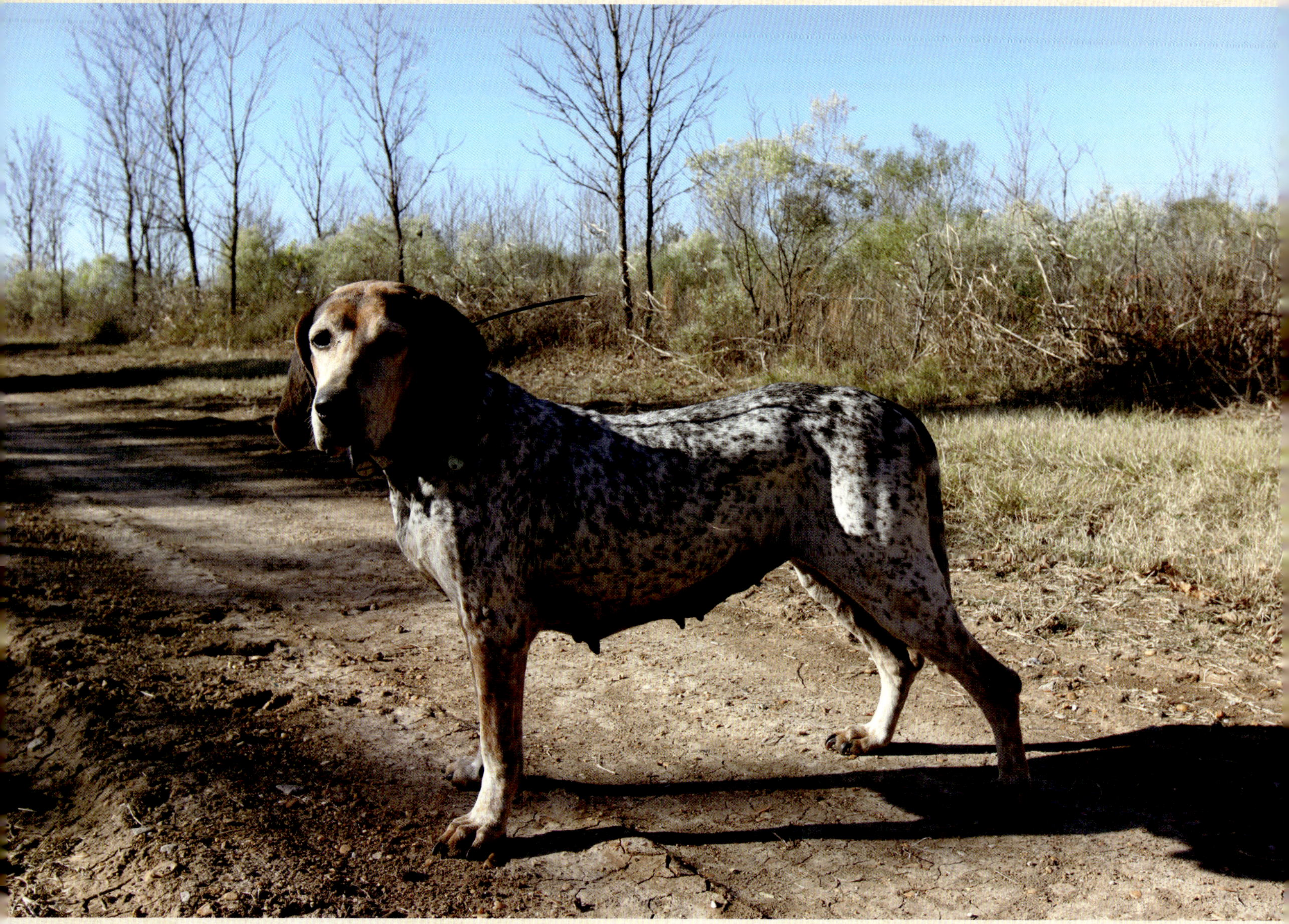

Mabel, blue Plott hound.

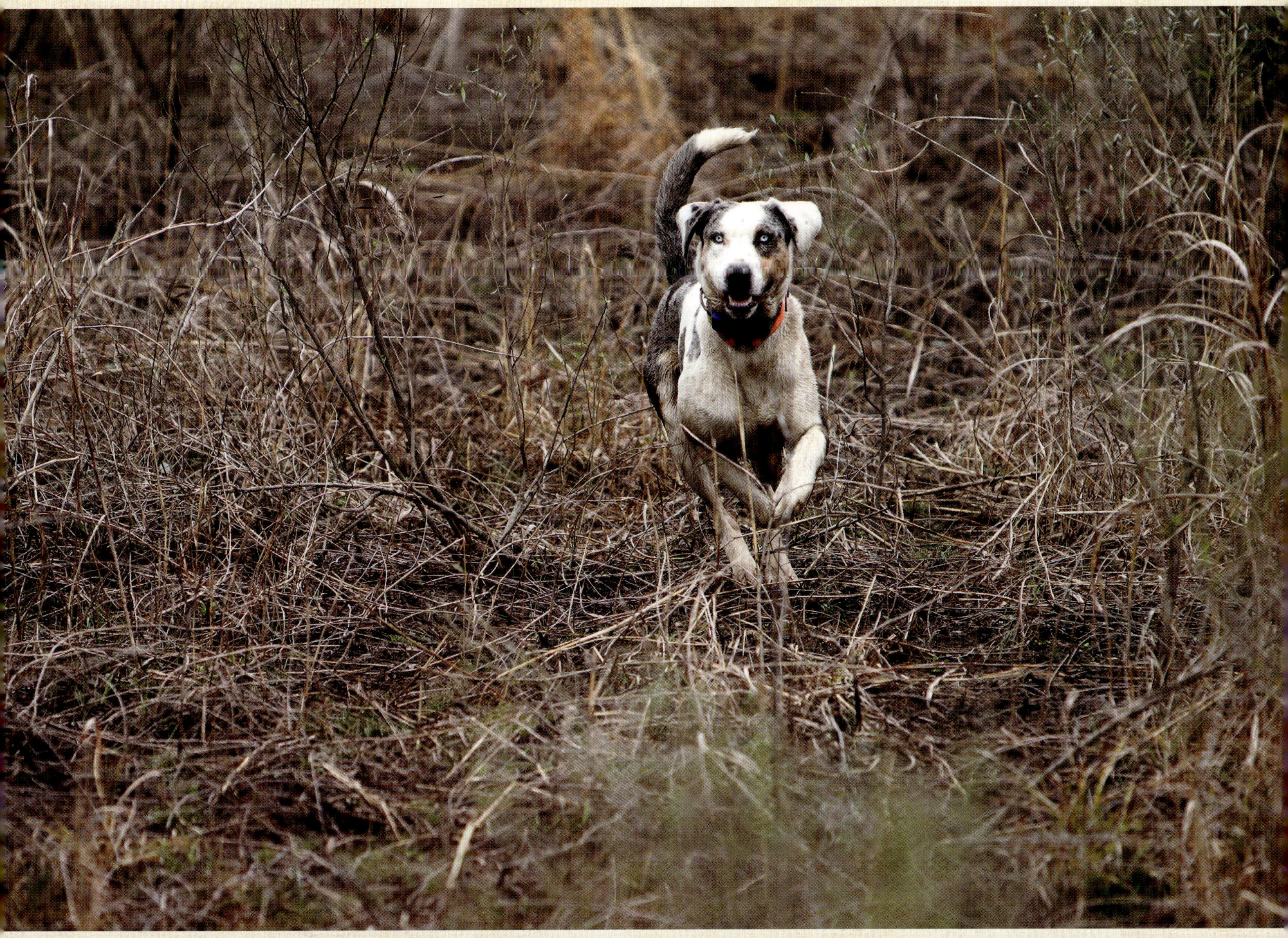

Jake, Louisiana Catahoula cur.

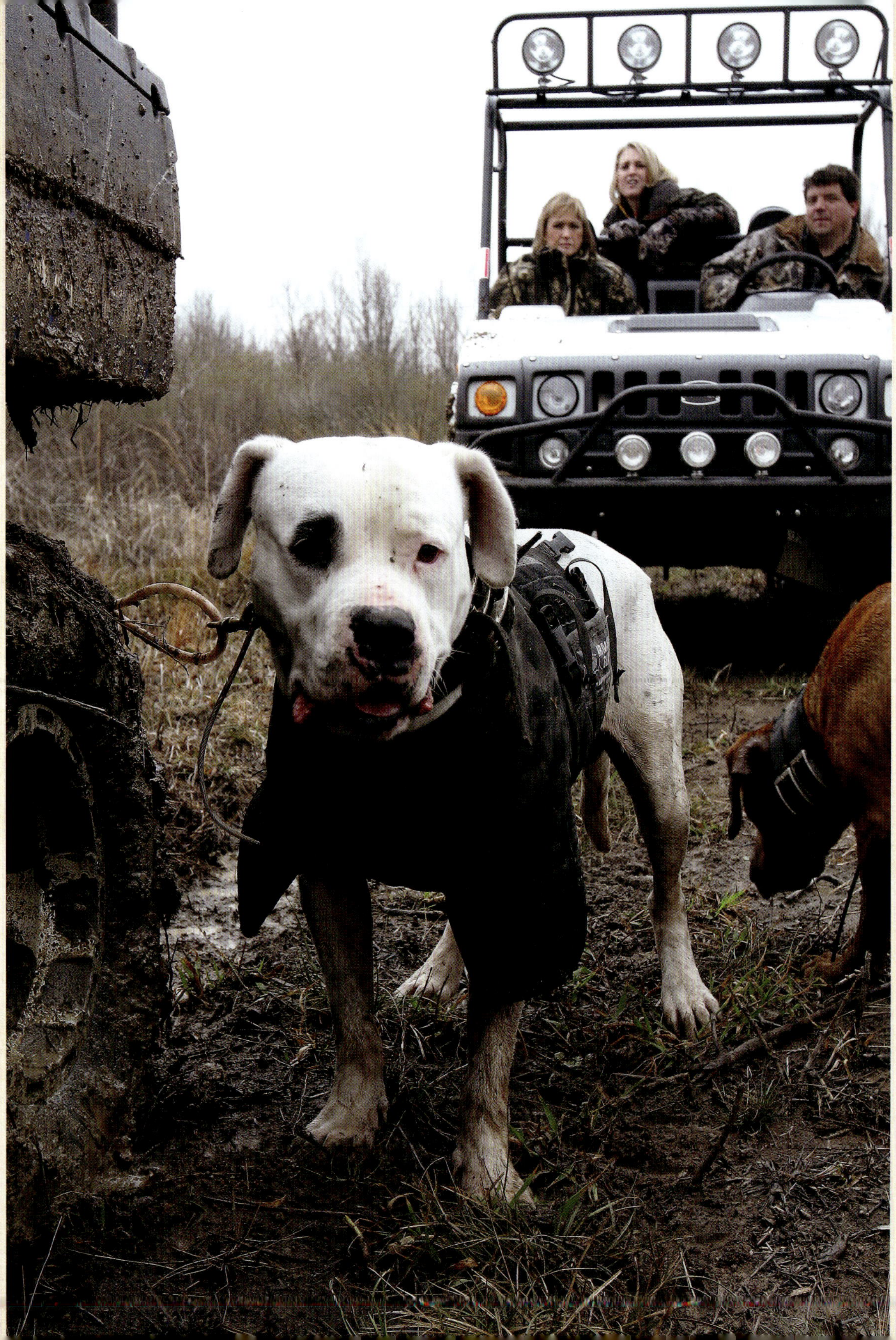

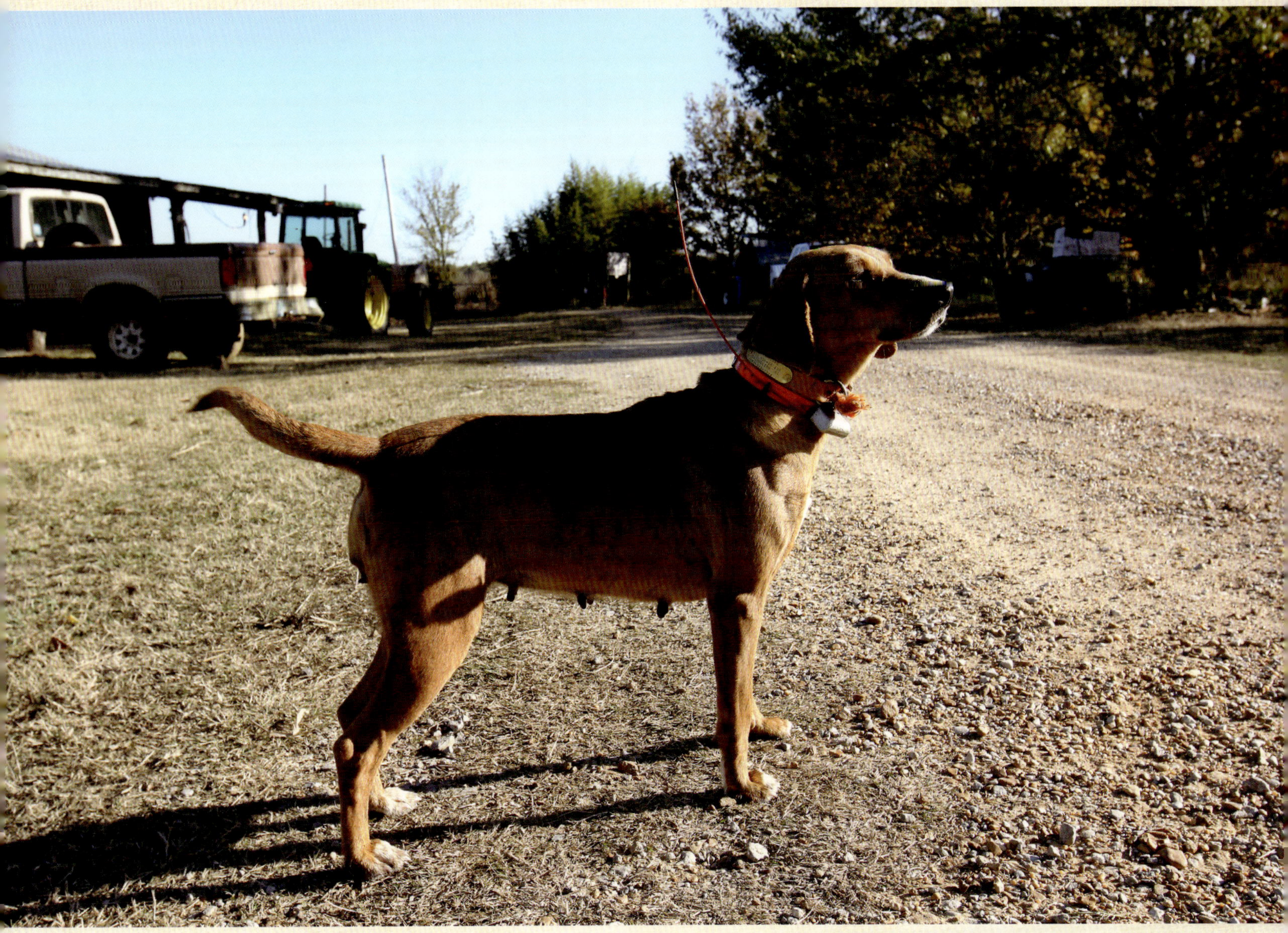

Plott hound.

< Frog, bulldog, Mr. and Mrs. Mark Tate, and daughter, Ashley.

A Louisiana Visitor

NANCY BANCROFT

I received an invitation for my first hog hunt. It seems these hogs are descendants of European boar that escaped from the Spanish conquistadors and now are a nuisance to farmers. I had no idea what to expect or what was expected of me and waited with great anticipation for the big event. I woke up early while it was still dark and quite chilly, loaded the walking horses, quarter horses, and even a thoroughbred, and drove off to a farm of several thousand acres that butts up to a state forest. The prerequisites for the ride are a pair of comfortable pants, boots, layers of jackets, and preferably a flask of something with alcohol. Any combination of English or western tack and clothes is acceptable. We had it all. I then was introduced to the specially trained dogs sporting radio collars and even some wearing Kevlar jackets for this hunt. These dogs, mainly beagles, bird dogs, and pit bulls, can run for hours and are the most essential ingredient of the hunt.

Our mixed group of male and female riders separated into teams and the dogs took off. Riders must follow these dogs through ditches, ravines, woods, and open pastures. Horses need to be able to swim too. When the dogs catch the scent of a hog, they take off and so do the horses. It's hang-on time! Quick ducking and riding low can often prevent a limb or two from swinging back in your face.

We ran all over the countryside that day and never did see a hog. Of course, a few hunters swore they saw one. The lack of hog sightings did not dampen the excitement of the day for me. I was hooked. The country-style barbeque and party afterwards was extra icing.

Thankfully, I received a second invitation. The next hunt was to take place on a sprawling farm of several thousand acres owned by Howard Brent outside of Yazoo City, Mississippi. This time I loaded up my thoroughbred horse, Baron, picked up good friend and avid rider Mary Hardwick and her horse, and off we went. We had heard the famous Weeks brothers were going to guide the hunt and we couldn't wait to meet them. Stories of these four fearless brothers and their incredible hunting skills are recounted all over our area. Known to grab a four-hundred-pound boar and stab him with a Bowie knife, the Weeks brothers were sure to show us a special hunt.

The weather had been especially wicked all night, with lightning and thunder continuing until 6 a.m. We showed up anyway, knowing we might just be having a nice lunch. The weather, incredibly, let up but the fields were soggy and ditches full. Our horses and dogs trudged through miles of gumbo mud fields and ditches. At one particularly steep ravine that was filled with waist-high water, the horse and rider in front of me fell over backwards trying to climb the steep slope to get out. I came off my twirling horse and waded to the rider, Pete, who had fallen from his big 16.2-hand grey quarter horse and was being repeatedly kicked in the leg as his horse tried unsuccessfully to get up. Several times he would almost make it up and then fall over on Pete again. I was trying to hold Pete's head above water. After what seemed thirty minutes but was actually no more than five, both my horse and Pete's managed to climb out of the ravine without us! We were wet and cold. Luckily, a kind gentleman on a four-wheeler arrived and helped up locate our horses. I had broken reins and a wet saddle. That was good news!

We rode back to the trailers where Wick, a fellow hog hunter, loaned me dry socks, warm pants, size twelve boots, and a large jacket. Inside his trailer he had ten of every clothing article one could need, in his size, of course. That was a lesson: take spare clothes.

Live music and a huge barbeque with venison, pork ribs, chicken, beans, and salad followed the hunt. Our host, Howard, sang a few tunes in Spanish and English for us and even played the harmonica along with the guitar player.

The hunts just get better and better!

Horses and Hunting

STEPHEN SMITH

I have been hunting for about ten years. I'm usually right by Justin Braswell when I hunt, and I started hunting at Delta Wildlife. One time, when I was about twenty years old, I fell off my horse. They say that I fell asleep because I was nodding my head and that's why. I was young and inexperienced and always hunted with Mr. Wick and Mr. Braswell because nobody else would take me and I wanted to ride horses. And nobody else would go with them but me!

Once, we were going to get a hog the dogs had bayed up and I got off my horse to find the hog because the hog was in a big briar patch and it was really thick. The hog was only about ten feet away and started to charge at us, and I should have shot it then, but didn't. I was trying to back up but Mr. Billy George pushed me forward and my spur got caught in the briars. About that time I fell back and I pulled out my gun and shot off to the side of the hog's head. It was enough to scare him off!

Another time I fell off my horse in about chest-deep water off Nixon Dump road here in Delta Wildlife, crossing a slough, about two or three miles from the camp house. I was trying to see how deep the water was and Thomas Eatherly was behind me. I had my feet up in the saddle so they wouldn't get wet. My horse started lunging because his feet got caught up in the grass and since my feet were up in the saddle I just rolled right off the back of my horse in about thirty-degree weather. It was really, really cold!

Lionel—A Child's Story

ANDREW WESTERFIELD

Some boars become so big that they become larger than life and achieve mystical proportions. Such a boar lived and still exists in Yankum. Actually it's deep in the heart, the interior of Yankum. Now two things: Yankum does exist at the Merigold Hunting Club and this boar has a name, Lionel. He was named potentially for the "king of the jungle," the lions of Africa, or maybe the old Lionel train sets we used to have as children because trains are big. This is about Lionel the Great Boar Bush Hog that exists inside the forest of Yankum.

Although roads surround the forest, none venture through its inner reaches. Although partially traversed by hunters, Yankum has never been fully explored because of the impenetrable, vast expanses of cane breaks interwoven with saw briar and fallen trees which form huge go-arounds. These areas are only penetrated by game trails and in the thickest parts with "holes" which man can only attempt to crawl through. Hence, it is without doubt that parts of Yankum still remain unexplored and thereby hold secret places where mysterious and large boar live to escape the arrows and guns of hunters.

Attempts have been made to send dogs into these parts of Yankum in an effort to dislodge the Great Boar that lives within. Large packs of Catahoula curs, "mean dogs" with different-colored eyes that you could see through, and pit bulls with spike collars that, left unrestrained, would finish off hogs before you get there, have been sent into Yankum's interior. Although the sounds of terrible combat could be heard coming from within, silence would always return to the distant battlefield, and no dogs ever returned because they were eaten by the Great Boar as the hunters waited anxiously and futilely for any survivors.

Now, how do we know that Lionel exists? Like all kings of their domain, such figures seek to extend their range—venture outside their protected area—and, just like the Loch Ness Monster, surface from time to time. These sightings, though rare, give proof and testament to the existence of Lionel.

Here is one such sighting.

It was a cool fall day with wet ground and leaves, where when a branch breaks you know that something is there. I had been easing into a light wind through the edges of Yankum since daylight on a morning when the moon was still in full view to the west. There was the sharp sound of a branch breaking behind me and I knew I had been winded. Freezing, I took the Voss Drilling 7.57Rx16GA from under my shoulder and without stepping merely turned in my tracks to my rear. I was in one of those immense cane breaks which run into and out of Yankum. My view was limited to twenty-five feet and I was on a game trail which on either side could not be seen through, never mind trying to move through silently or otherwise. But from time to time these "holes," which only game could pass through, appeared in the sides of the trail and I had just passed one on either side of me. As in all such hunting moments time stands down; then a large boar crosses the trail within view and is gone just as quickly. He is a definite shooter, without a doubt. Gone . . . not even a chance of a shot . . . wait . . . another hog, bigger than the first, enters the trail. I see tusk—I say to myself, "Raise your gun" and I set the trigger on the rifle as I do. It comes to my shoulder but the boar is gone behind the first one into the "hole" and I wait, really hoping that I haven't missed these chances—wondering how I let two of the biggest hogs I had ever seen get away *without a shot*—when the trail is filled suddenly with what must be a *huge bear*. It can't be a hog but the head is huge, long and pointed flat with tusks—they can't be tusks, they are too large, like white bananas sticking out of its mouth. "What am I doing looking, just shoot," and I touch the hair trigger of the drilling and the sound of the firing pin hitting the primer is almost as loud as the shot would have been that never occurred! What? A *misfire*? But now that huge head turns towards me with eyes six to eight inches apart and coal black and the shoulder turns with the head and now the back end turns and enters the trail and now he is facing me . . . where to shoot is a flash of thought . . . just shoot comes next and I pull the back trigger on the now "little" 16-gauge Breinke slug that I load in the drilling confidently just for such occasions . . . another almost ear-splitting *snap* and now a step towards me he comes. His jaws snap on those huge tusks like a butcher knife scraped over steel. I break the barrels down . . . *yes, it was loaded* . . . two imprints on the primers. All this occurs instantly in my eyes together with an almost pitiful glance down the trail to see this creature boar coming towards me but no . . . he is still standing there—eyes glued to mine or mine really to his . . . a thought . . . they don't see so well—I heard that somewhere . . . a step forward by him and a real shock to my system occurs—I am just frozen—I can't move . . . a long time ensues that probably lasts just seconds, then another movement towards not me but the other side of the trail, then a thought—*he is wider than the trail*—as he slides into the thickness and is gone into the "hole" . . . did this really happen—I think—only silence—no sound—what's that smell? The wind has changed into my face suddenly and I realize I haven't even closed the gun and there I see again the two indentions on the primers . . . more silence. A thought comes to mind—I hope he's gone—no—did the great hunter really think that?

When I leave the cane break (after reloading), not trying very hard to be quiet or slow about it, and get back to the YoYo (a 1970s Toyota Land Cruiser), I take out the two "good" shells and reload the rifle barrel with the "bad" one from in the cane break on the trail. I take really no aim at a tree thirty yards away and jerk the trigger, maybe thinking to help the hammer within when the concussion and jolt of the shot rocks my shoulder since I didn't have it on there tight. *It shot*! After being loaded so did the "misfired" Breinke slug! Another jolt to the shoulder but an even more severe jolt to the mind. I peer over the hood of the YoYo into the sky which looks down into the interior of Yankum wondering what really happened and what mysterious powers this Great Boar must possess.

(One of the Lionel stories told to Jack and Jon Westerfield when they were probably six and four years old and at ages when their young hunting minds were susceptible to visions of Great Boar in the fall woods.)

Hog Hunting

CROSS CROWE (AGE 8)

I have been going hunting with my dad since I was really little, like around eighteen months old. I shot my first deer before school one morning, when I was in kindergarten. I started hunting hogs with my dad a few years back, mostly stalk and shoot in the woods behind our house.

Last winter, when I was seven, I started hunting hogs with dogs at Caulk Island. My pappaw and his buddies had been doing this for years, so had my dad, but this was the beginning of my adventure.

Together, we had assembled some pretty good dogs. My favorite was Snoopy. We loaded them up every chance we got and headed to the thickest woods anyone has ever seen. Sometimes when they would bay a hog, they would be in cane so thick we could hardly move through it. We always left the woods all scratched up from thorns and briars, and the dogs left in worse shape than that because they had been nose to nose with the baddest hogs in the world!

Early in the season, we let our dogs out on what we thought was a pretty good hog. They split up in groups and bayed more than one. We later found Snoopy dead in the canebrake with his throat cut real bad. We never saw the hog that got him, but his tracks were there and they were as big as my dad's fist and both of mine together. My dad said Snoopy died because he was brave and determined to do his job. He would not back down or give up. I want to be like that too. We brought him home and buried him in the woods behind my house. That was a sad day.

We rallied the troops and kept on hunting. About three or four weeks into the season, someone spotted a track in the bean field that was a pretty close match to the "big one" that got Snoopy. We were the ones that were feeling determined this time. We were hunting the bean field. Everyone came that weekend and everyone was excited!

My pappaw had loaned me his very own .30-30. They were all hoping I would be the one to shoot, and I was pretty nervous.

We set out for the bean field and not long after we turned the dogs loose, they had bayed up a big hog. The dogs were ahead of us and we could hear something awful going on. When we got there, he was fighting three of our dogs: Big Head, Jim, and Rob. Big Head had him by the ear and the hog had cut him twice and then threw him through the air! I was nervous and could not get a clear shot. Dogs and people were everywhere and I did not want to shoot any of them. Finally I shot, and missed. The next shot I fired went right through his heart and he ran about fifteen yards before he fell. He was bleeding good but it took him a little while to die. I remember watching and waiting. My heart was beating so fast. He was the biggest hog I had ever seen.

We all celebrated and took lots of pictures. There I was posing for a proud picture holding my pappaw's gun and our cousin Clint's dog Jim when Jim decided to run off, dragging me behind. I held on tight until my dad caught up with us. Kinda funny now, not so much then! Just one more story to tell.

Of course, we had to get him mounted, all four

hundred–plus pounds of him, full body mount in my dad's trophy room, and I got the bragging rights! But not without thanks to all of the men who helped make that hog season one for the record books. Thanks to my pappaw, Don Blackwood; my dad, David Crowe; my buddy for life, Mr. Bill Mann; my uncle, Jim Blackwood; our cousin, Clint Mabus; my pappaw's two best friends, Billy Marlow and Bobby Maxwell; my good friend, Mr. Larry Rial; and all of our dogs, brave and determined.

Springtime brought new puppies, and thanks to Bill Mann I have one of my very own. Her name is Dotty and she is a Catahoula hound with double-glass eyes. She is trouble! We brought home a baby pig from the woods last year that we bottle-fed until she was too big to stay in our house. She has adapted to life on our river and loves wallowing in the mud. Her name is Buttermilk. She is white with black spots. She can often be seen running across the pasture with Dotty close behind. Come January 1, Dotty will join the big dogs for her first of many trips to Caulk Island.

Caulk Island Hogs!

LARSON FREY

Growing up hunting on an island in the Mississippi River, I heard many stories about hunting wild hogs and the difficulties in pursuing them. Hunting hogs on the island always seemed to elude me, with Christmas holidays ending and new semesters of school beginning. I always resolved that if given an opportunity to hunt with my father and his friends, I would grasp it.

The invitation for my first island hog hunt came in my freshman year at Ole Miss. My father called and explained that it would be a great deal of hard work, but the reward was worth the trouble. As the proverbial rookie of the group, I was assigned the task of videographer for the trip.

I arrived at our point of demarcation with quite a bit of apprehension about the hunt, but also excitement that I was about to fully experience hog hunting firsthand. Preparations for the hunt began well before we arrived at the boat ramp. Hunting dogs had to be gathered up, along with the food and medicines needed to sustain them. Horses, hay, oats, and tack were also assembled for the trip to the island.

Provisions and clothing were boated across the half-mile-wide oxbow to camp first. A barge was brought across to the ramp to load the horses, dogs, and the rest of our party. It must have been quite a sight to the fisherman motoring up and down the lake as we made our way across, the horses looking over the side of the barge in trepidation and the dogs howling and glaring at the men who put them in this predicament!

Once at camp, the animals were unloaded and put in temporary stables and kennels, given a ration of feed and water, and allowed to rest for the night. We continued to unload our personal belongings and check equipment before heartily eating supper and retiring for the morning hunt. Although the hunt had not yet begun, I was already physically exhausted from the energy expended preparing for the battles to come and quickly drifted into a deep sleep.

The next morning started well before sunrise with a quick breakfast, followed by saddling the horses and placing radio tracking collars on the lead dogs, lest they get out of earshot. It was a beautifully clear January morning and cold! I vividly remember the iciness of the saddle as I eased my horse onto the road leading out of camp. There was a heavy frost on the ground as the hunters began to fan out into the thick grass and briars of the cottonwoods a few hundred yards from camp. We immediately jumped several deer from their beds and they bounded out in front of us in herds, making their way to the safety of thicker cover. The dogs skirted in and out of openings in the grass before us, trying to pick up a fresh trail.

I was excited at seeing the whole scene in front of me, and the open woods made for easy riding. My excitement quickly faded, however, as we left the cottonwoods and entered into the native woods of pecan, hackberry, cane, and vines. Walking through these same woods a month earlier, I had regarded them as somewhat open. The view from atop a horse, however, was much different. Every vine became an obstruction, and every branch a potential whip on my face and hands.

Just when I thought things couldn't get any worse, the dogs bayed, and the other hunters quickly spurred their horses in pursuit. Quickly left alone, and novice horseman at best, I spurred my own mount to catch up. We had traveled about fifty yards before a low-hanging hackberry limb caught me in the midsection and toppled me backwards from my saddle to the ground. Stunned, I checked to make sure nothing other than my pride was broken, and sheepishly mounted and rode toward the sound of the dogs and gunfire.

Upon my arrival a minute later, the deed had been done. My father and the rest of the group stood above a large black and white sow that had been dispatched only seconds earlier. I was completely deflated. I had failed in my responsibility to capture the hunt on video! I was quickly reminded that there would be more opportunities, so I settled for recording the men as they tied the sow to the saddle horn of my father's horse and dragged her to the road nearby for cleaning.

We rested for a minute and gathered up all the dogs before heading down the road again in search of new quarry. The group eased down the road for a while, the dogs crossing in and out of the woods on either side of us in anticipation of a chase. Just around a bend in the road, the lead dog entered a patch of cane the size of a small house and bayed. We quickly galloped to the edge of the thicket just as a large boar crashed out of the side of the green and brown thicket in full run toward the slough below us. He had sounded like a bulldozer as he made his way through the cane, and I marveled at how fast he was in the open, the dogs trailing farther and farther behind.

Two of us immediately gave chase, skirting around the canebrake and onto an old roadbed that led to the slough. Unlike the earlier episode, we found ourselves ahead of the other hunters as we made our way to the edge of the ironwood thicket and the squeal and barking it encapsulated. As we dismounted and tied the horses off, my partner suggested that I leave the video camera behind and grab my gun. We crept silently toward the battle and hunkered down twenty yards away behind a large mound of earth. My heart was racing as we peeked over the crest to glimpse what was unfolding behind it. The large boar had decided to make a stand at the water's edge. He lunged out at the half dozen dogs in his midst, grunting and squealing with each thrust. Dogs flew in and out of our view, grabbing at his hindquarters and ears.

We quickly formulated a simple plan. I would ease over the crest of the mound and carefully pick a kill shot without injuring any of the dogs in the process. The key was to get into position without the boar realizing my presence and bolting again. I eased over the mound, rifle up and aimed, and prepared to fire. The harassing dogs were making a shot almost impossible but I finally found an opening and squeezed the trigger, with no result! In my excitement I had forgotten to take my safety off! The boar quickly spotted me and broke from the dogs toward us before veering to my left. I quickly removed the safety and fired at him on the run. He hunched up at the shot, but was quickly out of sight with dogs in tow, racing back toward the canebrake.

I had made a critical error, and I realized it immediately. A wounded hog was a dangerous hog, and could put the dogs or my fellow hunters in danger. The prospect of entering the canebrake after the boar gave me chills.

Before we could mount up again to give chase, a shot rang out in the direction of the rest of our party. We quickly made our way to them, hoping for the best.

I was lucky. The boar's path led him directly to one of our riders, who finished him off from the saddle. In hog hunting terms, he was a "bad one," with large upper and lower tusks protruding from both sides of his mouth. His coat was mostly black, with red on the belly, and the longer coarse hair on his back was still bristled upward even in death. I recorded each of us posed behind the boar, a stick in his mouth showing off his impressive tusks as well as the story of my misfire and eventual bailout.

We hunted hard the rest of the morning into early afternoon until the dogs gave up in exhaustion. Our final tally: three large sows, two boars, and enough scratches, aches, and bruises to keep me sore for a week. As we made our way back toward camp, we talked about what had transpired, laughed at my inexperience in the saddle, and began to discuss plans for when we would return.

Three Hog Stories

LARRY MCALEXANDER

Three years ago we were hunting an extra-large boar that had long hair and weighed about 350 pounds. He was huge. We tracked him into some very thick bushes. There were six of us hunting together and three dogs. The dogs had the hog bayed up in the thicket. He was mad and fearless and ran out charging at us. He ran past the dogs and so I grabbed him by his back legs and flipped him over on his back. One dog grabbed him by the ear and one other guy ran in and slit his throat with a big long knife. This hog is the one that is hanging on my cabin wall.

Several years ago my brother-in-law, Walker, my friend, Foshee, and I were hunting another big three-hundred-pound boar hog. We had gotten him hemmed up in a thicket and all of a sudden he turned and charged right toward us! Walker and I jumped up on an old fallen tree but Foshee could not get up the tree—he was frozen in place! The boar hog charged under the tree towards Foshee. As he ran past Foshee's leg he rammed into it. The large and sharp tusk ripped through his new double-layered canvas Carhartts and sliced his leg wide open about six inches below his knee. As the big hog ran past we shot him and he dropped in his tracks.

Then, last year, I went hunting with two friends, Tyler and Earl. We had three dogs with us. We had tracked a three-hundred-pound boar into a large growth of cane. The boys had no gear so I gave my pistol to Tyler and my hunting knife to Earl. I wanted them to go in for the kill because they had never been on a hog hunt without others, so they also took the dogs. I decided to circle around behind the cane and cut the hog off so the guys and the dogs could trap him.

Well, when I went into the cane I was met head-on by the hog. He was mad because the dogs were right on him. There I was with no knife and no pistol. As he charged at me I picked up a big stick lying nearby and hit him, but he kept coming right at me. The dogs then rushed up behind him and one dog bit him in the back leg. The hog turned quickly toward him. All three dogs then jumped on the hog and backed him up to a tree. He was trapped. The hog became more fierce and aggressive and charged the dogs. He literally tore one of the dogs up. I kept hitting him again and again with the stick to get him off of the dog. This gave the other two dogs a chance to back him to the tree again. At this moment, Tyler and Earl finally made it into the thicket. I grabbed my pistol from Earl and shot the hog behind the ear and that ended it, except for the brave dog that had to be stitched up by Tyler's dad, the surgeon from Oxford.

A Wild Hog Hunt

GENE AND SARA JANE PARKER

It is hard to make this sound real, but it is, we promise and swear.

One fall day we were invited on a wild hog hunt with instructions to bring our horses and weapons and ride along with the professional hog hunters, or "do as you will," so to speak. This wild hog hunt was to take place on forty-five hundred acres of flat Delta land owned by Howard Brent west of Yazoo City, Mississippi. Since we have always believed in walking the walk and talking the talk, in that order, we slicked down our horses, washed our three-quarter-ton Chevy pickup truck and hooked up the gooseneck trailer, loaded the horses, and headed north on Highway 3 to Highway 49. A left turn led us to the unexpected, i.e., Howard Brent's hunting lodge, a three-story home moved to the site and reconstructed on the spot. It was so gorgeous; you expected to see Scarlett on the upstairs balcony and anticipated another unforgettable memory.

We had never had any reason to doubt our friends, especially those two with whom we have had many wonderful trips and rides over the years within and without the United States.

On the top level of the lodge was a wraparound porch and a hired three-man band. As our old and new friends arrived, we greeted them and their horses and went upstairs to the porch on the second floor where you could view the farm as far as you could see containing gorgeous fields, sloughs, and lakes in the Mississippi Delta. No one even suspiciously looked like a professional wild hog hunter though. As we were enjoying the food, music, and libations (the latter of which you have to have several prior to going on a wild hog hunt) and after scores of other people showed up with anticipation and horses, a small, covered, overly used truck, without any advertising on the side other than rust, pulled into the yard below, driven by, we later discovered, the hog hunter extraordinaire, his brothers, and their horses and dogs, who were all in the back of the van mingling with each other.

With our Bloody Marys in hand we had no idea of that which was about to occur below.

We are both from Mississippi so none of this should have bothered us, and it did not, but you just had to have been there to understand. The next sight was to behold. The driver got out of the cab of the van, went around behind, threw open the back door and out jumped, one by one, six horses, twenty-one hog dogs equipped with radio locating collars, and one pony.

The back of that van was probably two or three feet off the ground and it takes every effort to coach our three horses into a three-horse slant load which may be six inches off the ground. Never could we have imagined what we were witnessing with our own eyes. After the hunt and later that day the horses and dogs were taken to the back of the van and they all jumped in together, on command, as everyone watched in awe!

After saddling up, we launched upon one of the most amazing events we have ever experienced. The hog hunter, his brothers, sons, and cousins outrode Teddy Roosevelt and the Rough Riders, through sloughs, thickets, briars, and forests. They followed the dogs until a wild hog was cornered or caught, at which time the hog hunter, his brother, son, or cousin dove off the horse and onto the top of the wild hog, slit its throat, and then threw the hog on the back of a four-wheeler. One hog escaped the grasp of the hunter by diving into a deep slough in the cold of November, but no problem, the hunter shed his overcoat and boots, dove into the slough and retrieved the hog from the bottom, returning to the bank and shaking himself off like a dog, took a slug of whiskey, donned his coat and mounted his horse, and continued the hunt in pursuit of the next wild hog. Mind you, this hunt was on a weekday and according to our gracious host, Howard Brent, the hog hunters, sons, and cousins simply called in sick that day in order to lead this hunt.

After several hours we returned to the lodge to refresh our libations and rest. About two hours later the hog hunter and his brothers and company returned to the lodge with eight or so wild hogs in tow and with enough energy and excitement to hunt through the night and probably the next day but their dogs and horses were too fatigued to continue.

When we returned to the lodge we were greeted by a cadre of chefs, cooks, and bottle washers with an elaborate and tasty meal of coleslaw, baked beans, barbequed deer brisket, cookies, pies, jams, peanuts, and other hors d'oeuvres. This was certainly no fox hunt. Boots, camouflage, coats, gloves, and weapons are required.

We will never forget one of Howard's young friends showing off a new stud that he had just purchased and never ridden, though he had been assured by the prior owners that the horse was accustomed to hunting. The friend was probably six feet three inches and weighed at least 250 pounds. Sometime during the hunt he and his horse became separated while crossing a slough. The rider later returned, soaked to the bone, in a four-wheeler. Sara Jane was given to tell him that she had a horse for sale and his quick retort was "so do I." Sometimes horses are like dogs—don't brag on them until they are dead!

It will be difficult for Howard Brent to top this outing but we are staying tuned and ready!

The Hogs, Dogs, and Hunter Hit the Catfish Pond!

STEVE GOLDING

We were hunting on horseback at Panther Tract. It was getting late in the afternoon so three of us decided to head back toward the camp house. It had been a great day of hog hunting in the Mississippi Delta and we thought it was coming to an end. All of a sudden, we spotted a 250-to-300-pound wild hog crossing the road about three hundred yards in front of us.

Robert McConnell, who was riding with us, took a quick shot at the hog with his rifle but missed. We yelled for the guys to bring the dogs to where we had seen the hog, and then the last and most exciting chase was on! The dogs pushed the hog to the banks of the catfish pond and the big hog did not hesitate to hit the water and start swimming.

John Lowery's dogs also did not think twice about swimming after the fleeing hog. It was an amazing sight to see this huge hog swimming the length of the catfish pond with the dogs swimming and barking loudly behind him. We knew that we needed to get to the other side of the pond before the hog emerged from the water or it might get away. The problem was that there was a really deep ditch between us and the levee for the pond. The angle of the slope on the ditch was really severe, steep and deep for a horse to cross. There was also a fair amount of water in the ditch. This did not stop John Lowery from following his dogs nor did it deter Robert McConnell from asking his horse to make the crossing and continue to chase across the pond levee. When John got to the other side he shot the hog in the water with his rifle. The hog was hit but still not wanting to die. John then got off his horse and swam out into the pond to meet the half-dead three-hundred-pound thrashing boar hog. He stabbed the hog twice while in the water until the hog was motionless. He then swam back to the pond levee as he dragged the hog with him. A true Mississippi hog hunter will never give up on his dogs' pursuit. He expects his dogs to give him everything they have in their pursuit of the wild hog and in return he will give them everything he has to support them in successfully getting the hog.

I have been hunting all types of game in the Mississippi Delta for forty-five years and I can honestly say that I have never seen the level of passion that exists in hog hunting in any other type of hunting. The dog owner and the hog hunter will not hesitate to swim a river, slough, or even a catfish pond in freezing water to help his dog who has a hog bayed up on the other side.

It is truly an amazing experience to be on horseback and to be able to be a part of the chase and to see firsthand the bond of love and respect that exists between a Mississippi hog hunter and his dogs.

Tusks and Teeth

HANK BURDINE

"The planters of the South, more than the citizens of any other section of the Union, indulge in the manly excitements of the chase; they are, without exception, excellent horsemen and have a thorough knowledge of woodcraft." So stated *Harper's New Monthly Magazine* in October of 1885 regarding the men that rode to the hounds in pursuit of the black bear in the thick and very dense canebrakes of the Mississippi Delta. According to Paul Schullery in his book *The Bear Hunter's Century*, the aristocratic flair of the southern gentlemen enabled them to hunt with "the British formality of the fox hunt and the aboriginal wildness of the bayou alligator hunt. Bear hunting for those ardent sportsmen was almost a religious activity."

There is a very intense correlation between the bear hunters of yesteryear and wild hog hunters, the catch and the kill being very similar. Danger abounds at every juncture. A bear will stand to fight once cornered, as will a wild hog; however, a hog will rush and attack with a viciousness that will place the seasoned sportsman in awe. The excitement of the chase is an adrenaline rush not experienced by commonplace deer hunters or turkey hunters, for at the end of the chase is a life or death situation for either man or beast. If you are not mentally prepared for this, don't go hog hunting.

Robert Eager Bobo stated in an interview in an 1887 issue of *Forest and Stream*, "One year we rented a farm and spent the entire time out in the swamp. We didn't come out for three months at a time. We killed 304 bear, 57 deer, 47 wildcats and 9 panthers." According to Mrs. Fincher Bobo, daughter-in-law of Robert Bobo, "There was a festive mood in their setting out for the wild country, with the string of four-mule wagons, the dozens of dogs racing here and there, and the hunters themselves, mounted on their fine spirited horses. The men were gone for weeks and lived on bear steaks and stew. The bear meat was quite tough and coarse, but good."

In an attempt to not spook his dogs once they had bayed the bear, Captain Bobo carried a Colt pistol instead of a rifle. He found that in order to be effective, the pistol had to be fired from close range, often spooking the dogs or, worse yet, killing one. The dogs would often back off when they saw him coming or run away in fear of the gun. Heavy hunting knives were used as the hunter rushed into the marauding dogs and the bear fighting for his life. Captain Bobo later claimed, "A bear not distracted by dogs is a hard bear to kill with a knife." The same stands true today with the wild hog hunt.

A modern-day hog hunt is enjoyed by young and old, rich and poor, with an intensity that is not seen by the common everyday sportsman. In the days of the infamous Teddy Roosevelt bear hunts, a sitting president leapt off into the woods with governors, planters, business owners, common sharecroppers, and an ex-slave named Holt Collier from Greenville, Mississippi. Collier had ridden with the Texas Cavalry during the Civil War and now he was serving as guide for the president of the United States. According to Holt, "Money don't buy nothin' in the canebrake, and a man's dog don't care whether he's rich or po'." It is the camaraderie of the hunt, the interdependency of the men and their expertly trained horses and dogs, and the respect for the danger of the beast that makes a modern day hog hunter a very unique sportsman.

To sit in a deer stand and wait patiently for a big buck to come walking by is a great way to experience the outdoors and to learn about woodcraft and the ways of the whitetail. To come to the hounds on the back of a well-trained horse, through swamp and brambles, briars and thick woods, rushing into a marauding pack of dogs in a life or death situation wielding only a heavy knife is an experience of a lifetime, to be relished and told time and again.

Lindsay Denison was a journalist allowed to go on the Roosevelt bear hunt. He described the relationship between the men and animals in a 1903 issue of *Outing*: "There was the instructive picture of Holt Collier and some of the white men, too, on bended knee, dipping their horns in the water hole where the first bear had died, and drinking their fill of a puree of bear and dog and mud, all held in a solution of water that had been standing for eight months." Such is a good description of the bear hunters of yesterday and the hog hunters of today.

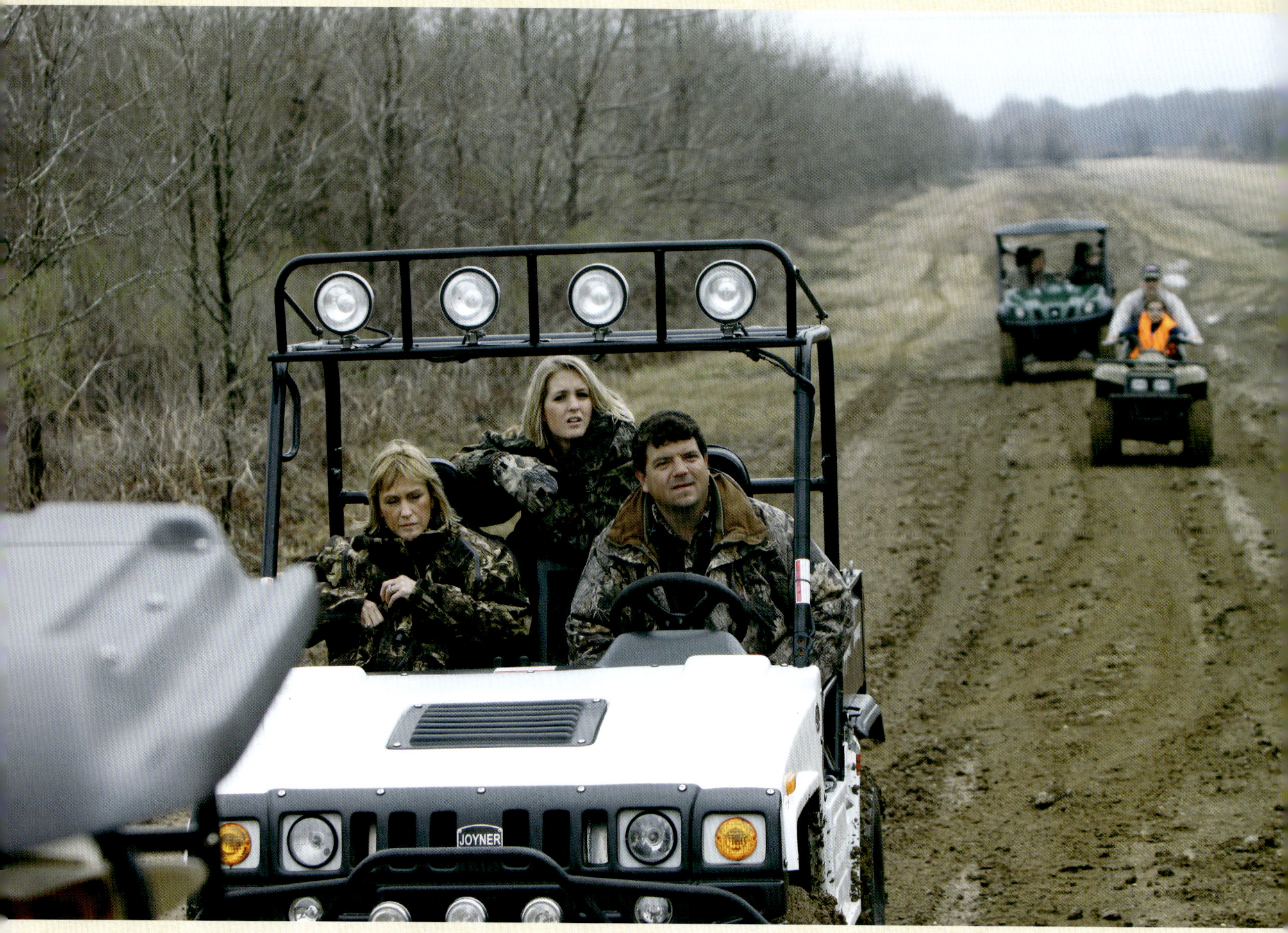

Mr. and Mrs. Mark Tate, daughter Ashley, Philip and Reeves Merideth.

Philip Merideth and his son, Reeves. >

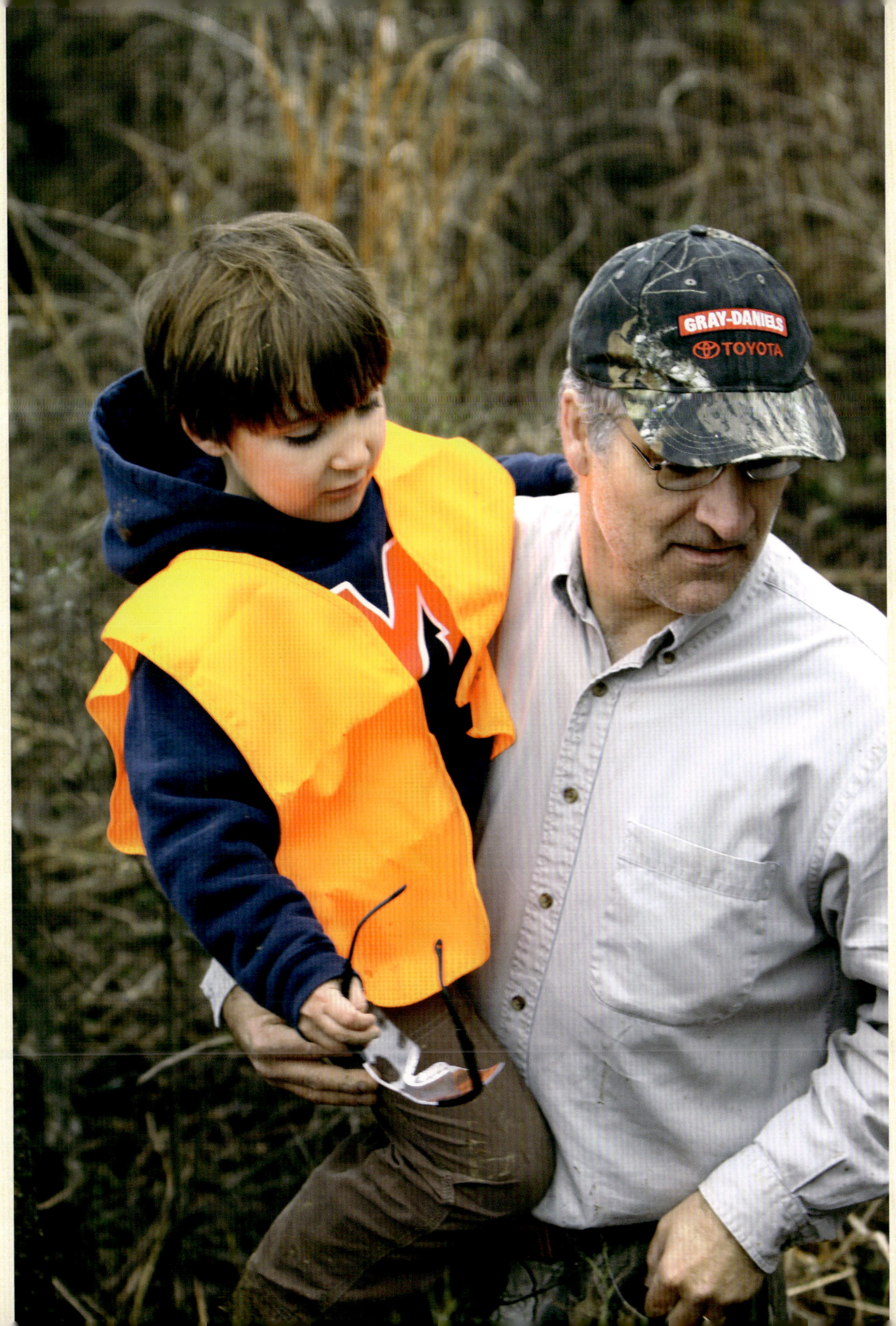
GRAY-DANIELS
TOYOTA

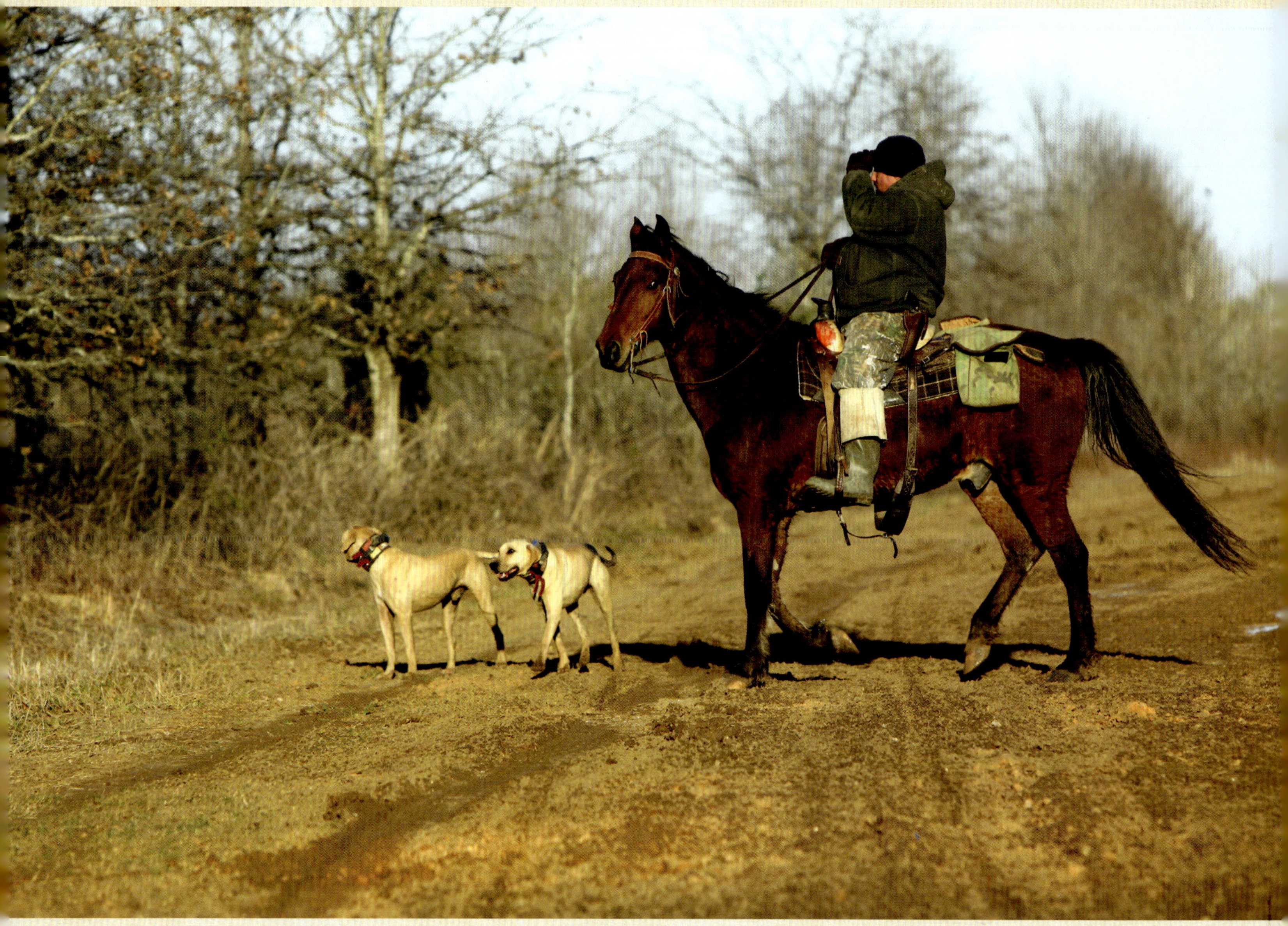

Hunter and dogs on the lookout.

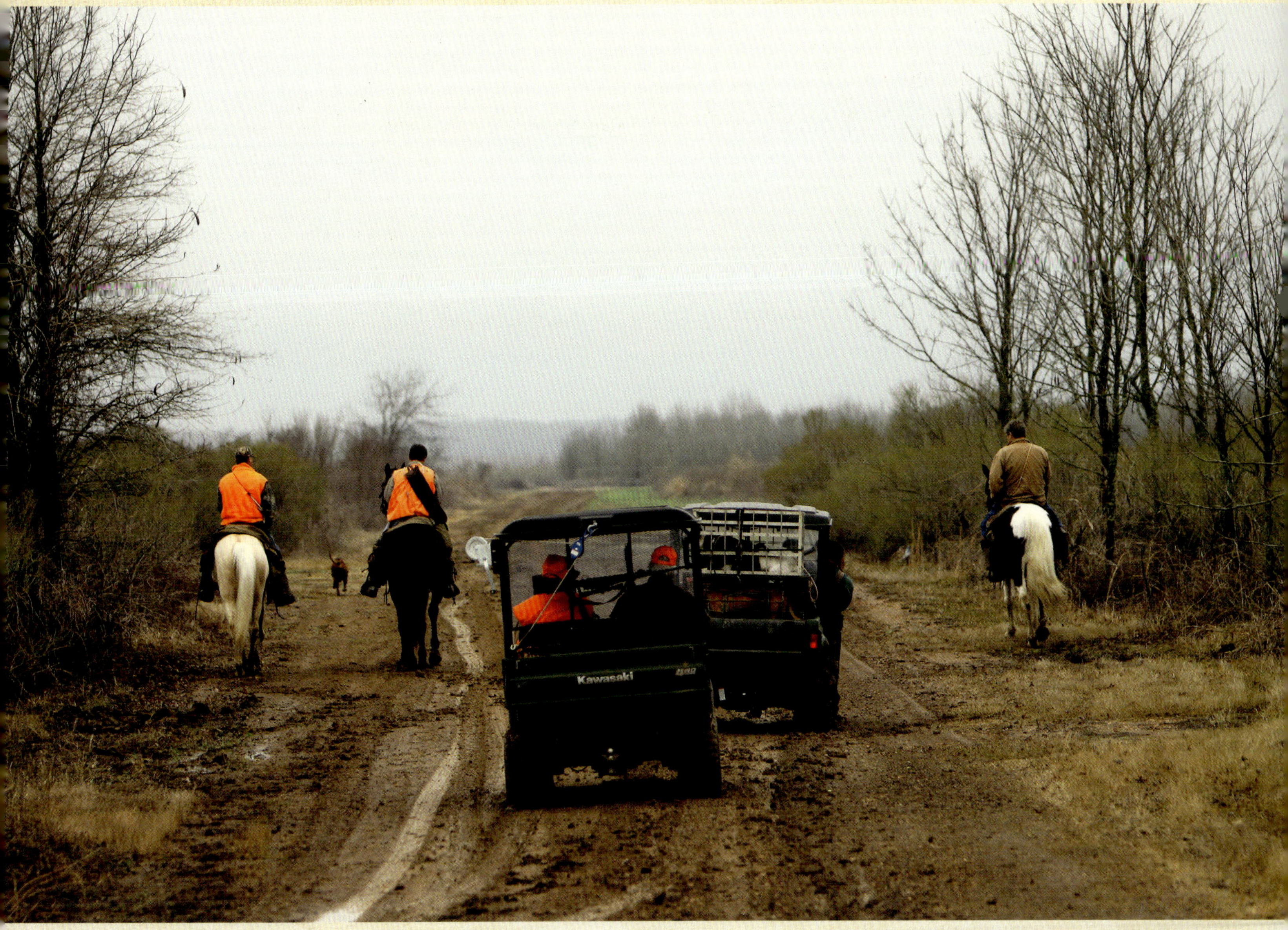

Headin' home.

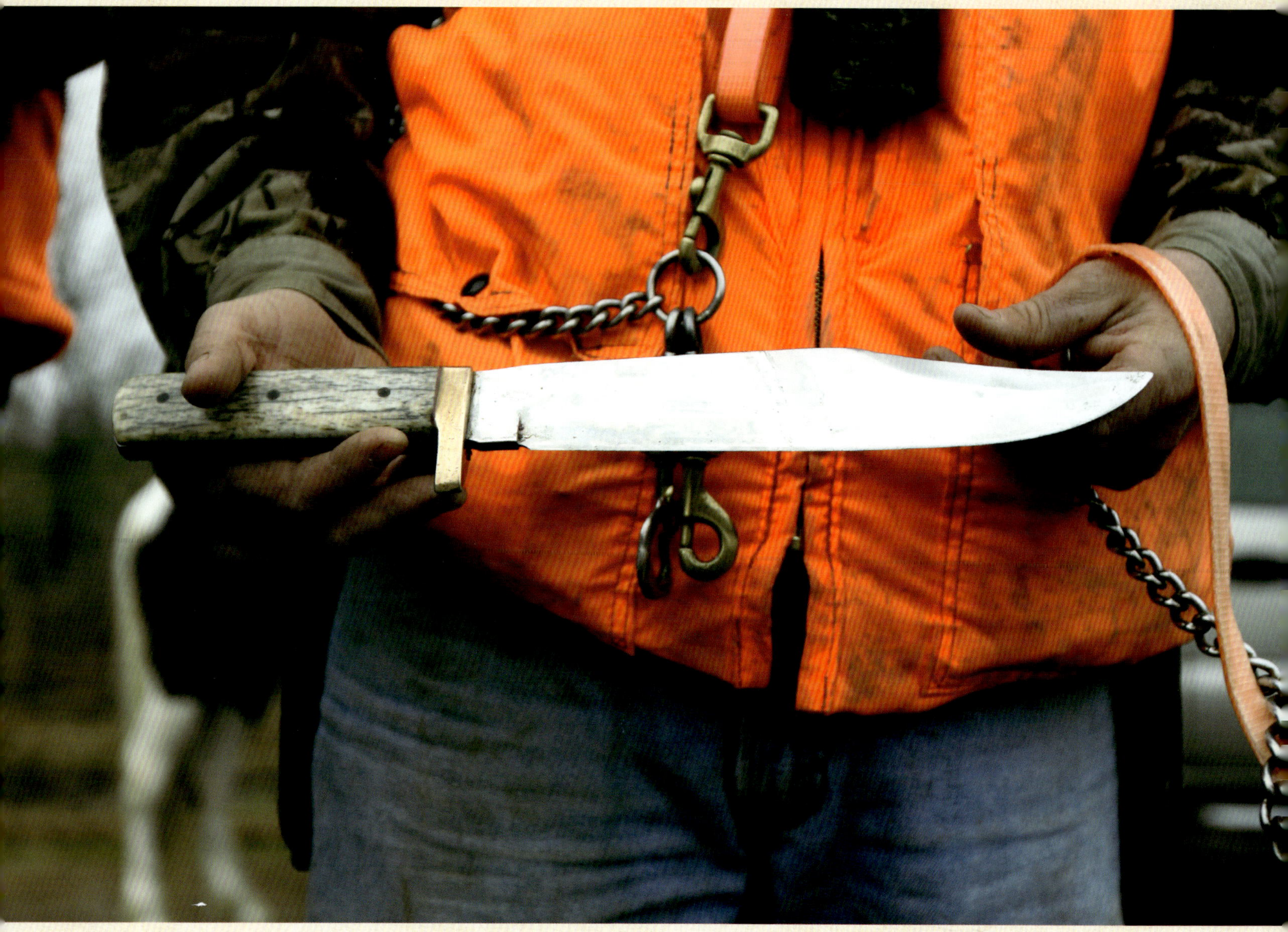

Bowie knife, 10¾-inch blade, made by Ed Jones.

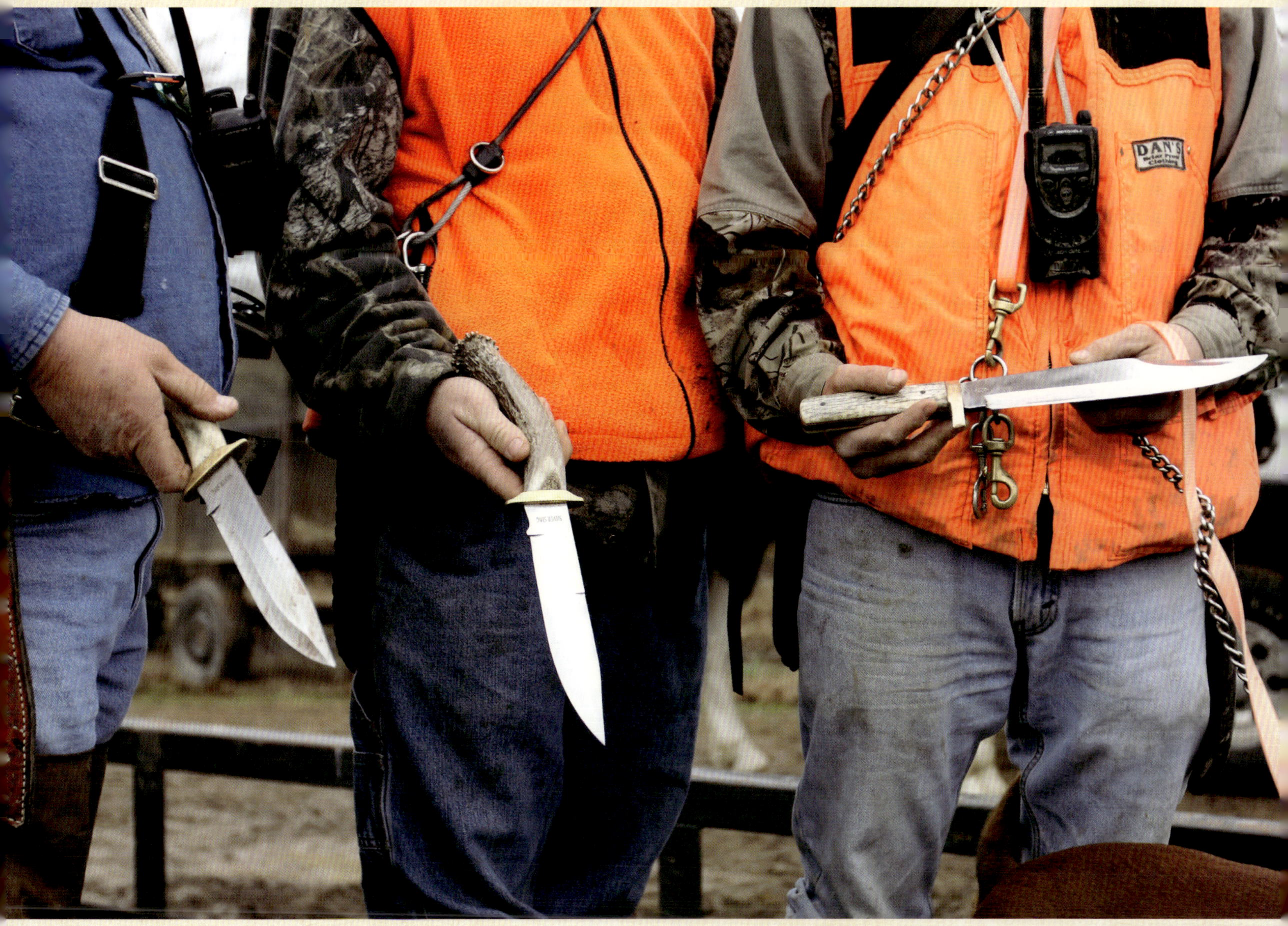

Braggin' rights.

Overleaf left: Sally Drinkhouse.
Overleaf right: Sonny Merideth.

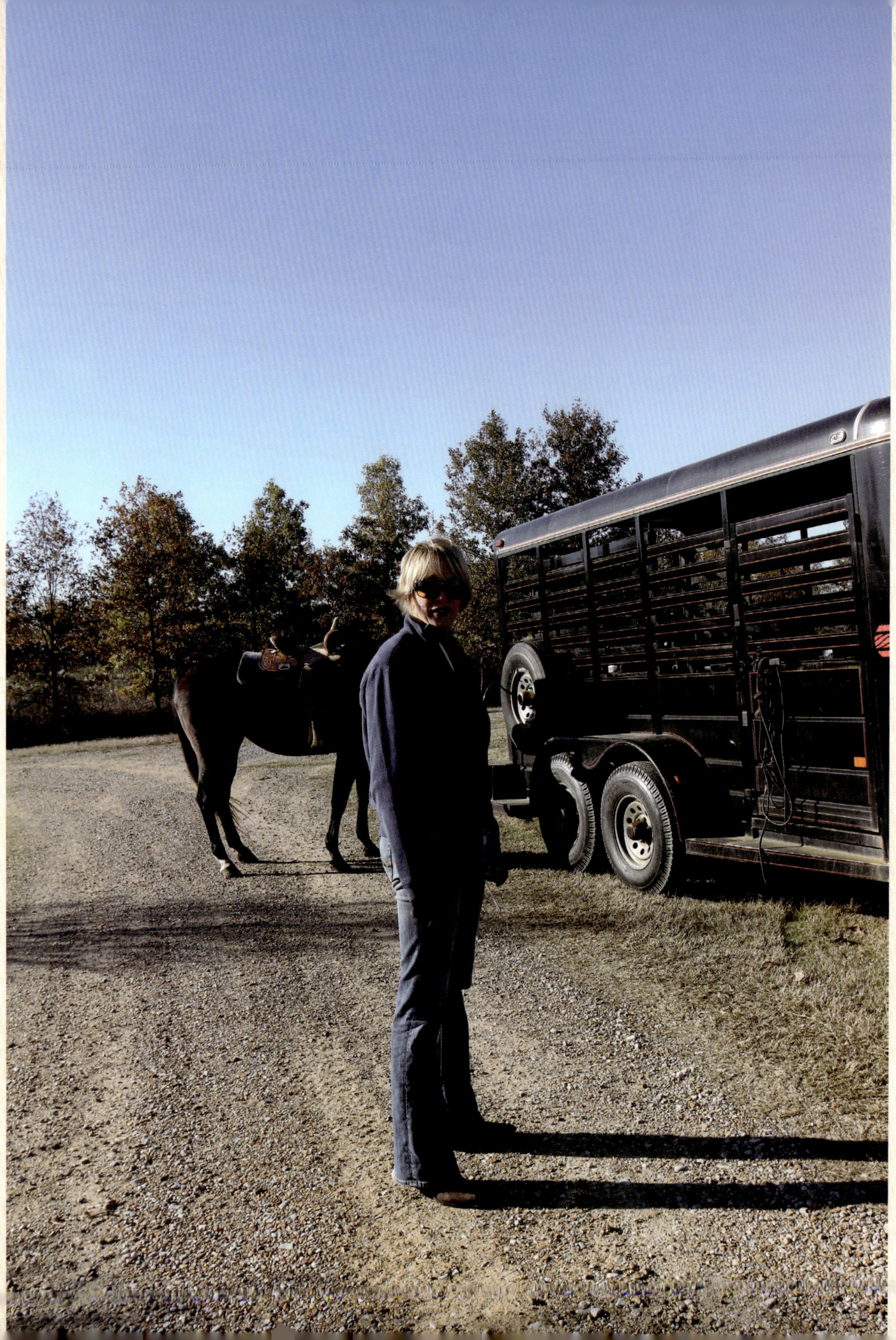

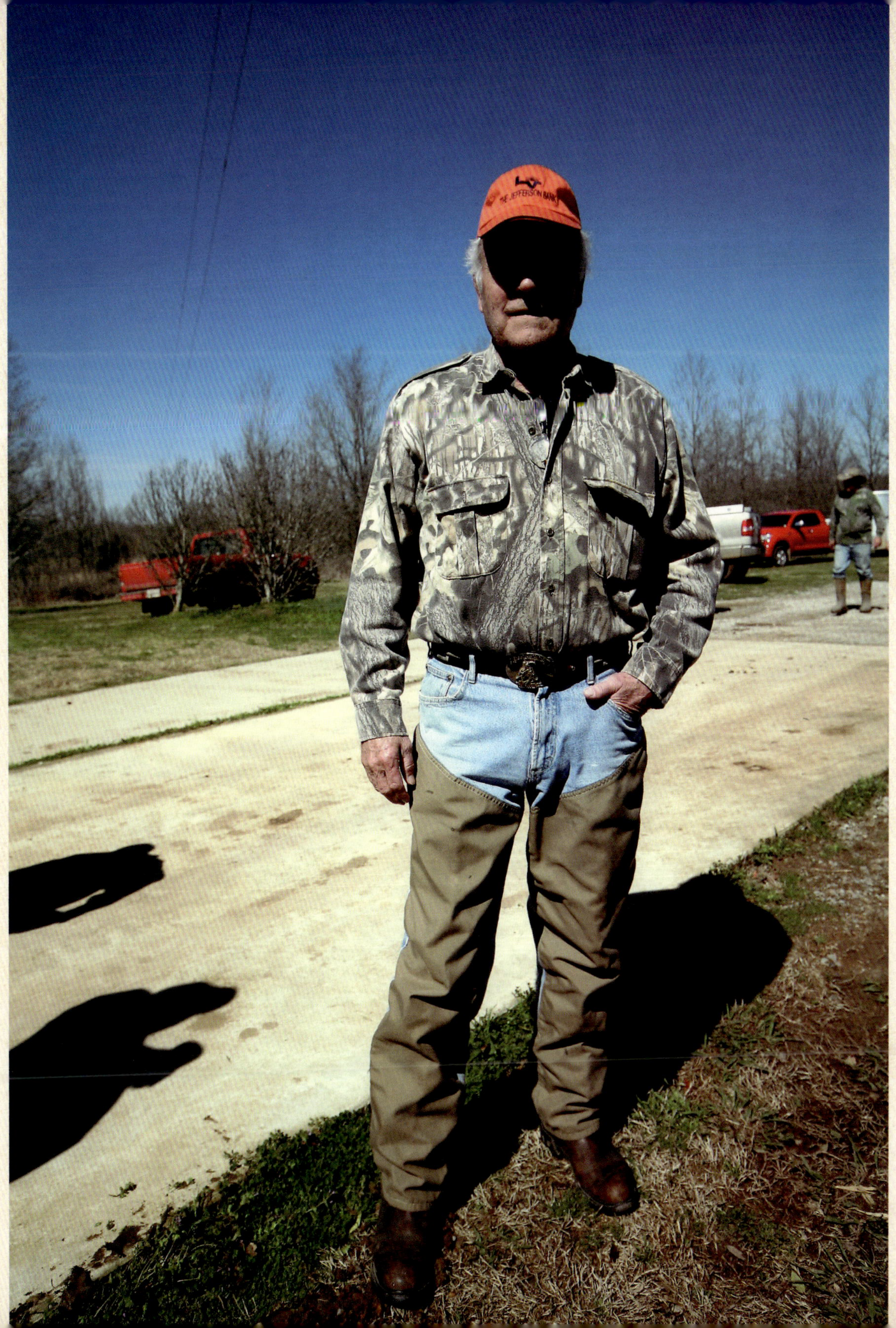
THE JEFFERSON BANK

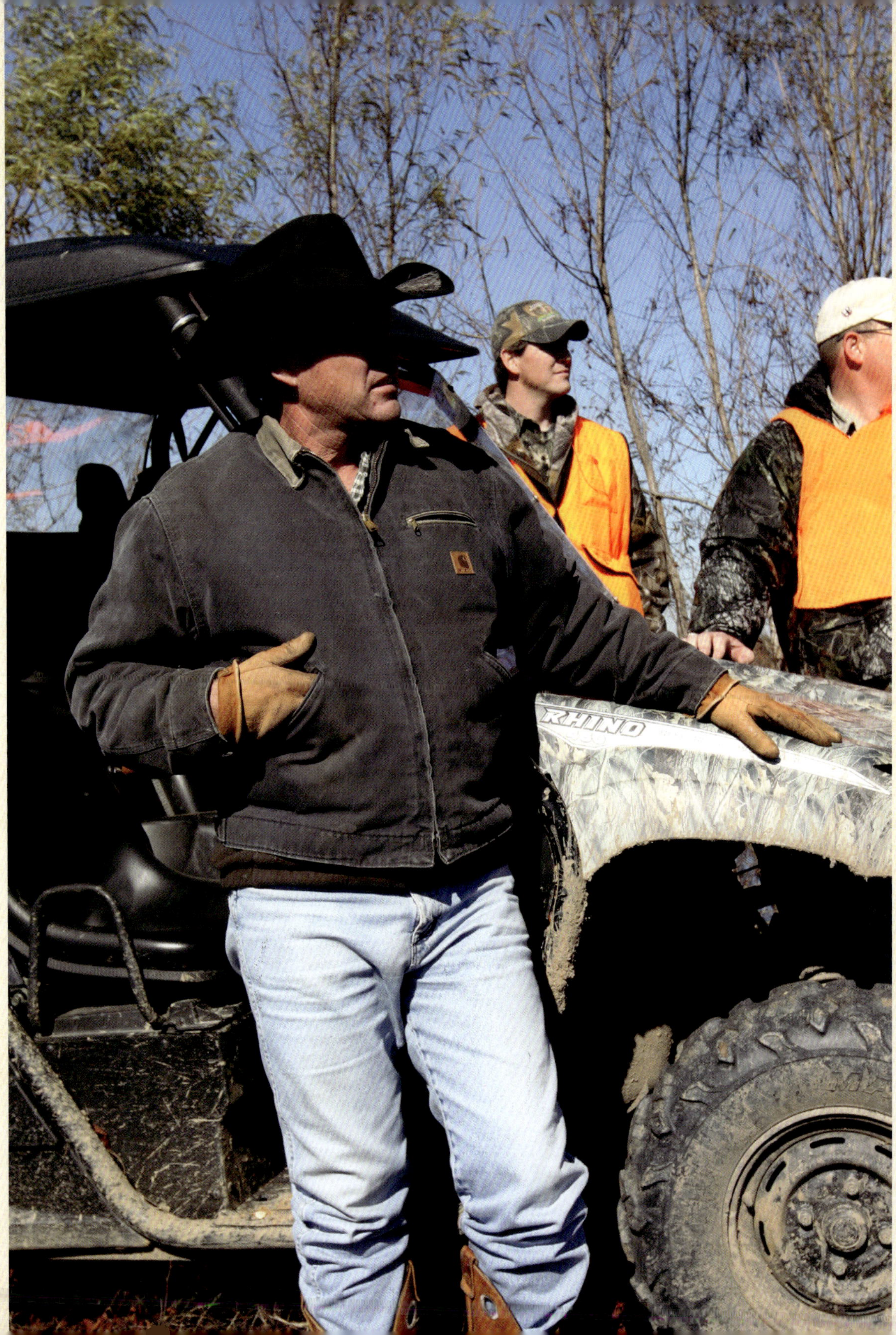
RHINO

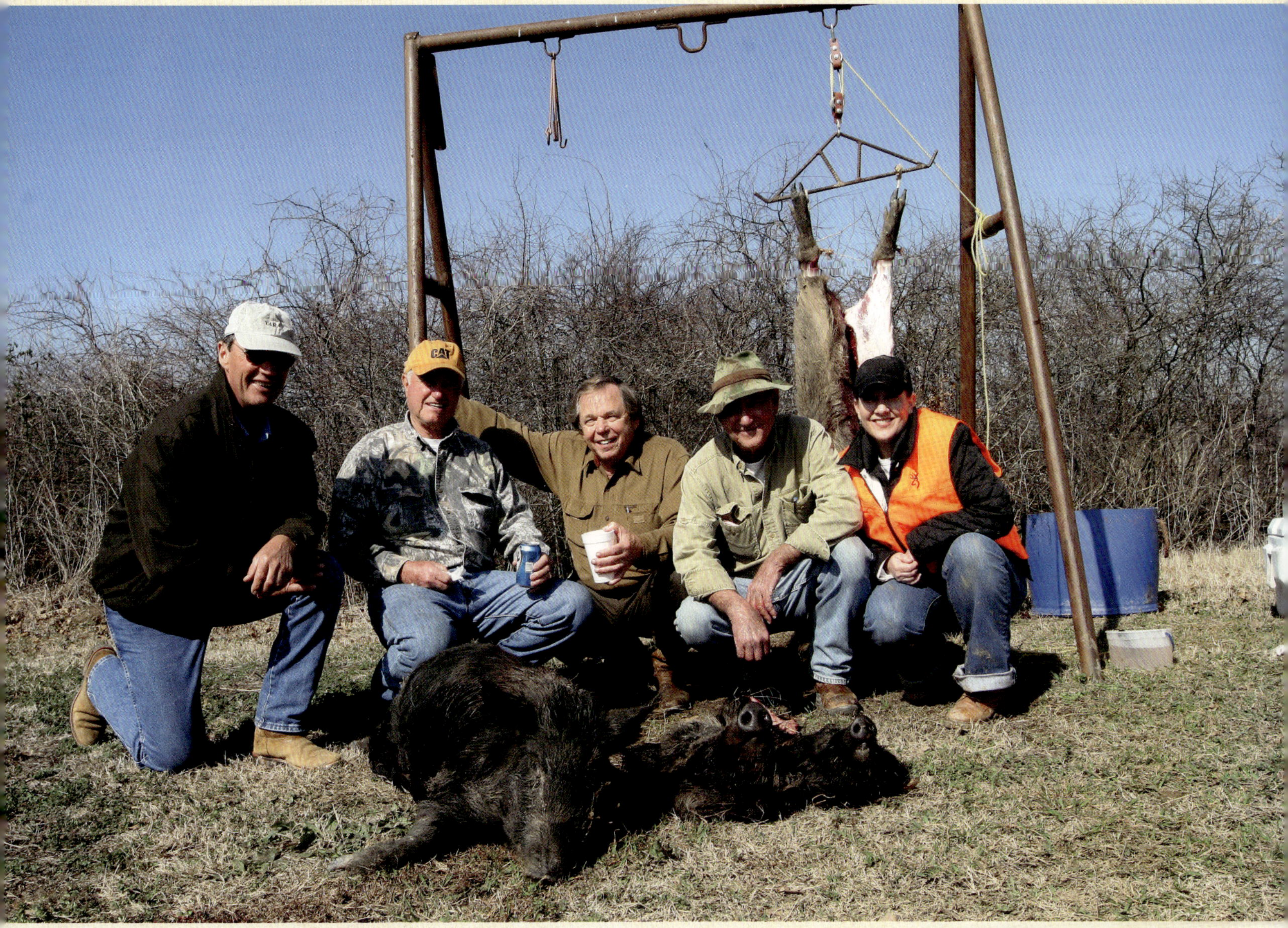

Gilbert Rose, Robert McConnell, Howard Brent, Charlie King, and Carrie Cooper.

< Bubba Bradley.

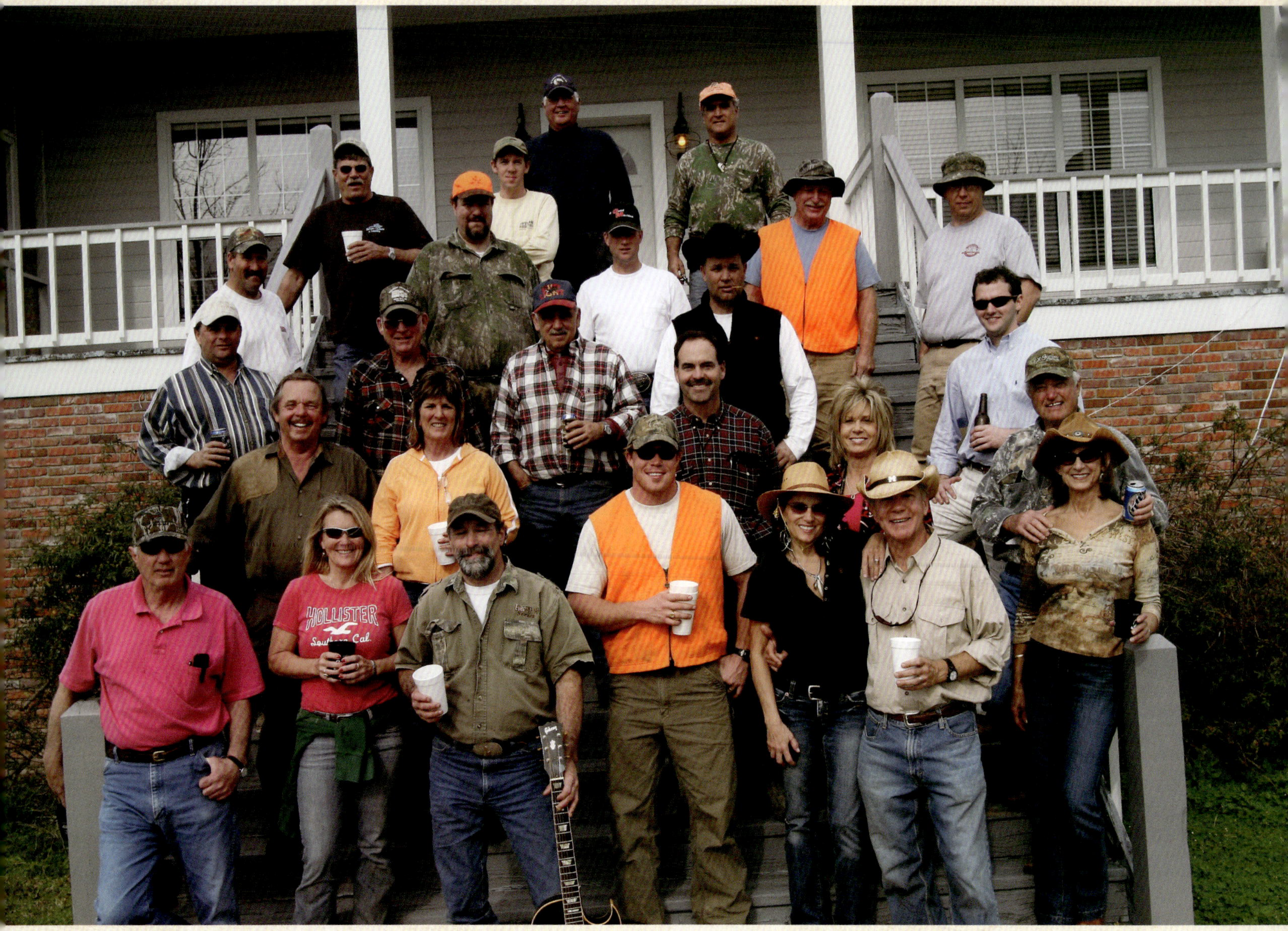

The pack at Panther Tract.

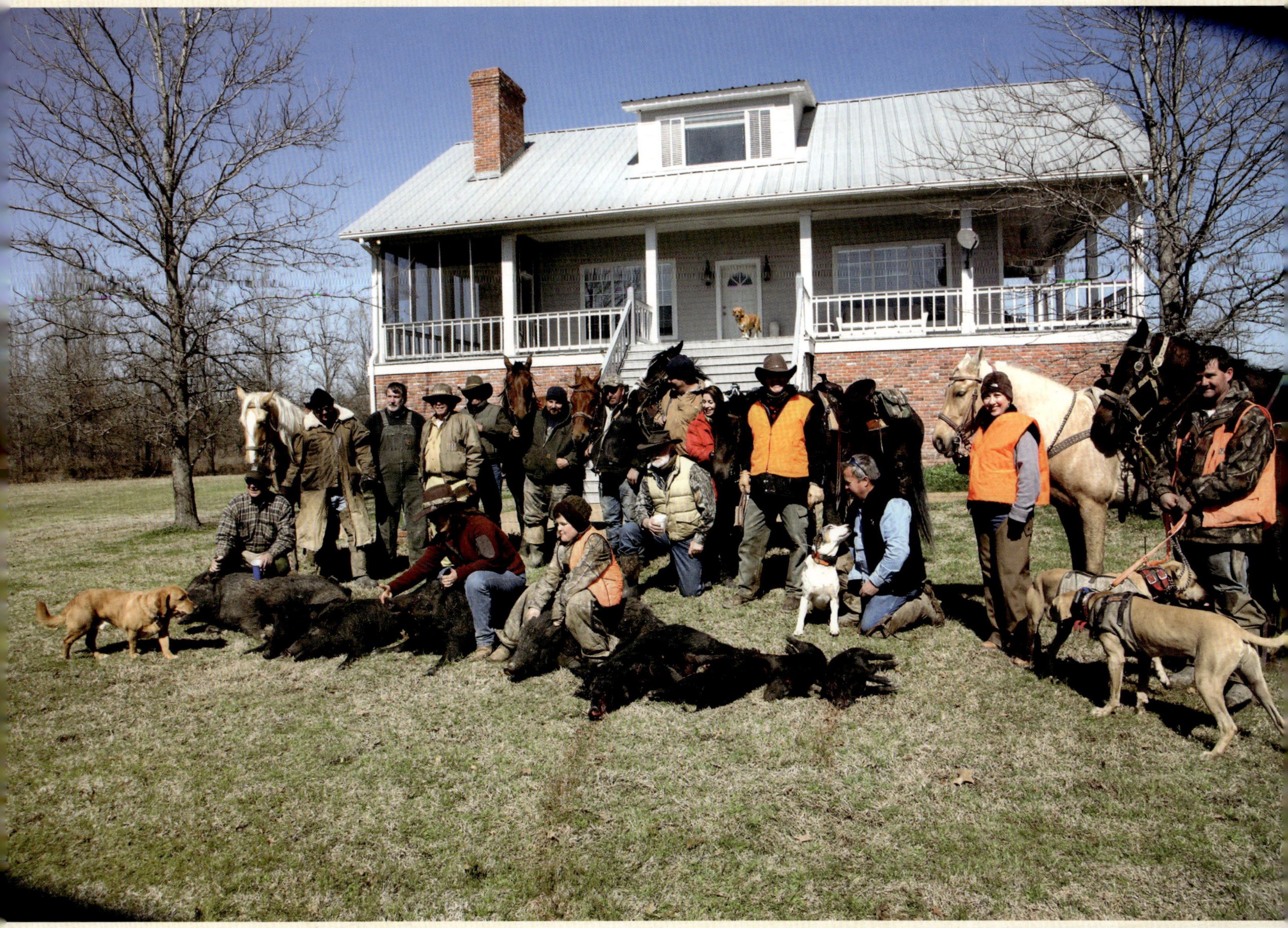

Brandon Yeager, Billy Vandever, Henry Coghlan, Howard Brent, Robert McConnell, Jeff Cremeen, Jesse Bates, Hank Burdine, Steve Ratcliff, Wendy Gentry, Steve Golding, Rickey Lowery, Carrie Cooper, and John Lowery.

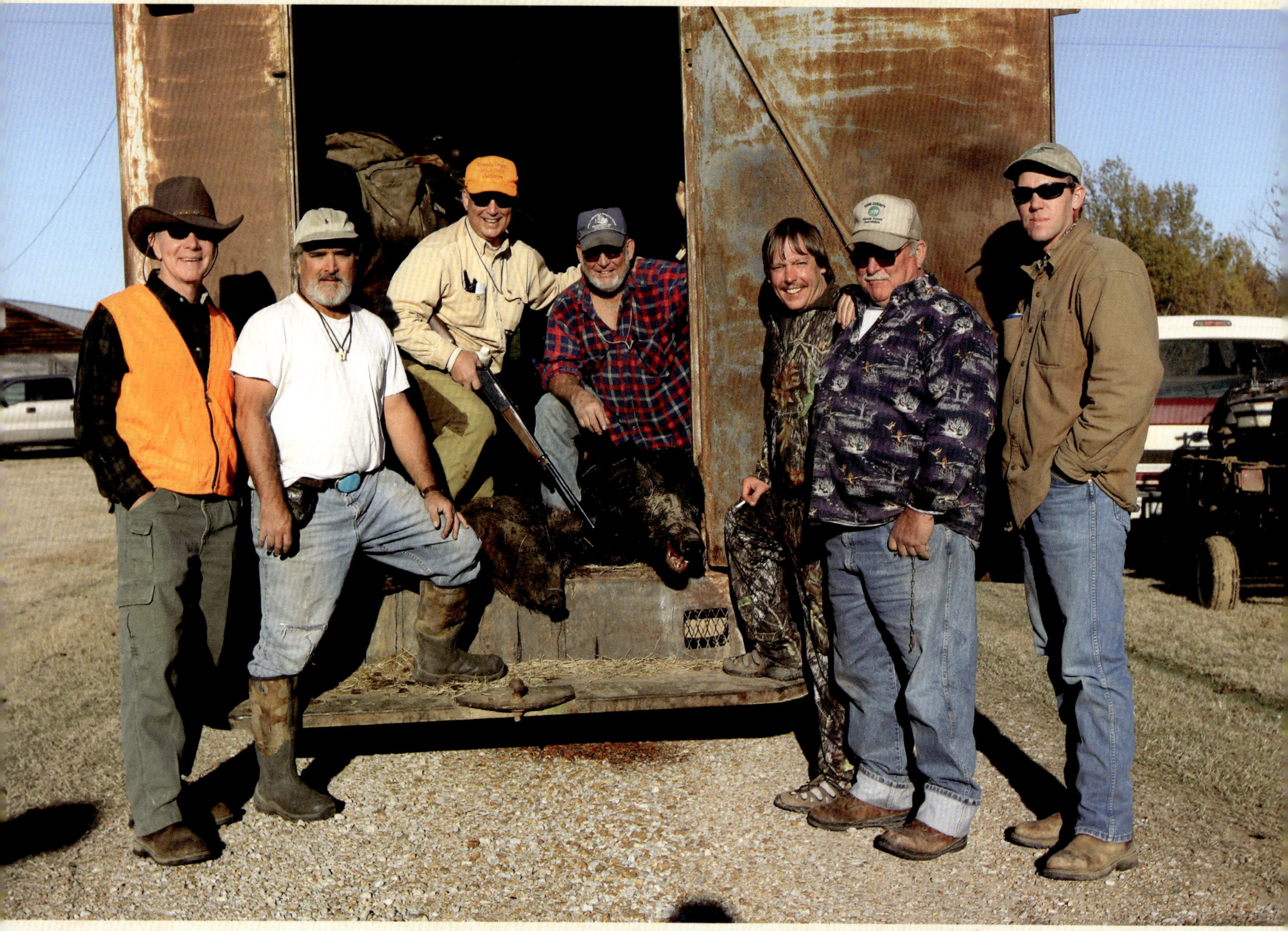

Steve Golding, Doc Merideth, Frank Dantone, Melvin (Bubba) Weeks, Lincoln Brent, Emmett Weeks, and Hines Outlaw.

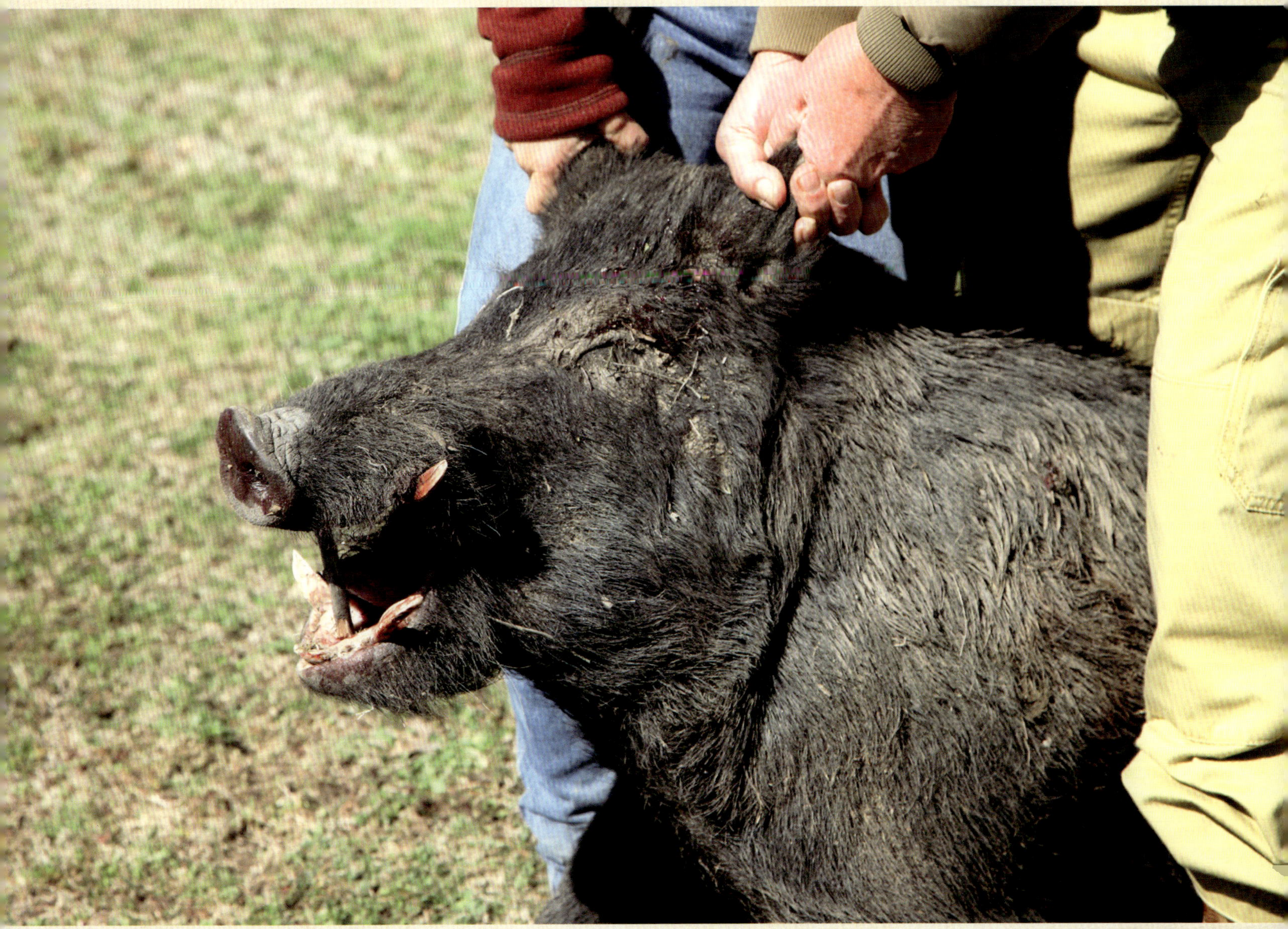

Harvested wild boar.

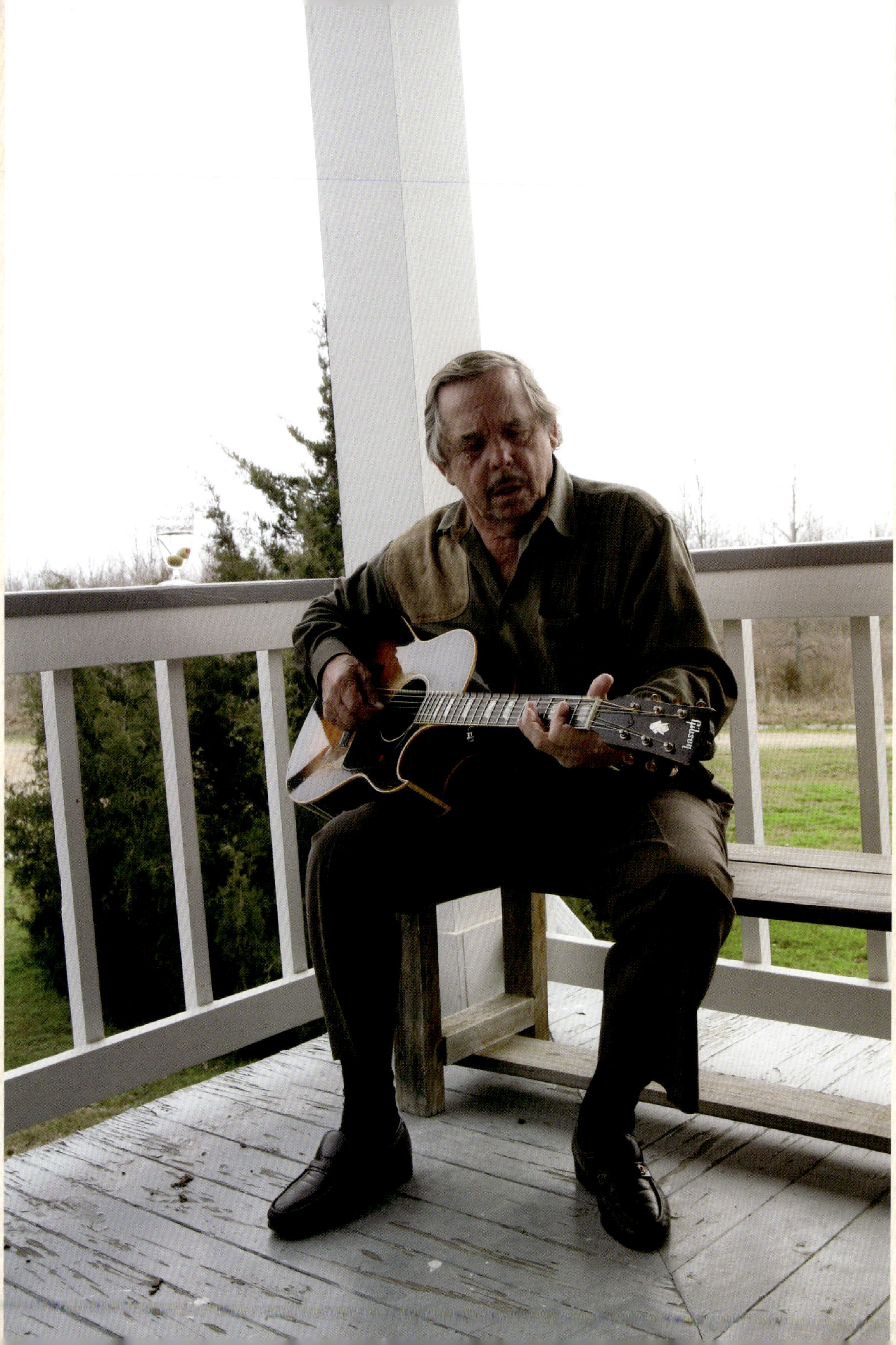

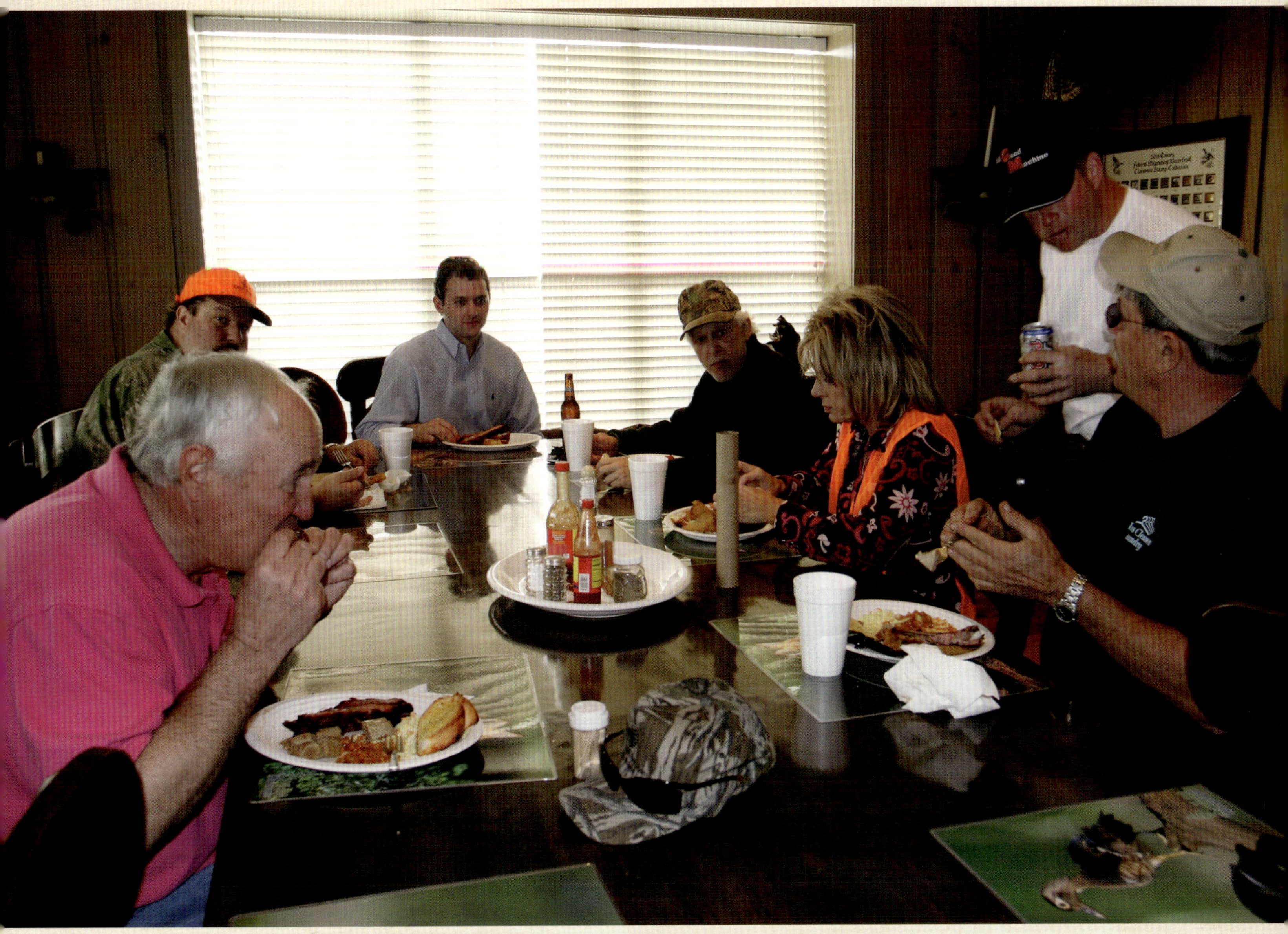

Wick Eatherly, Thomas Eatherly, Nancy Fulgham, and friends.

< Howard Brent playing his guitar.

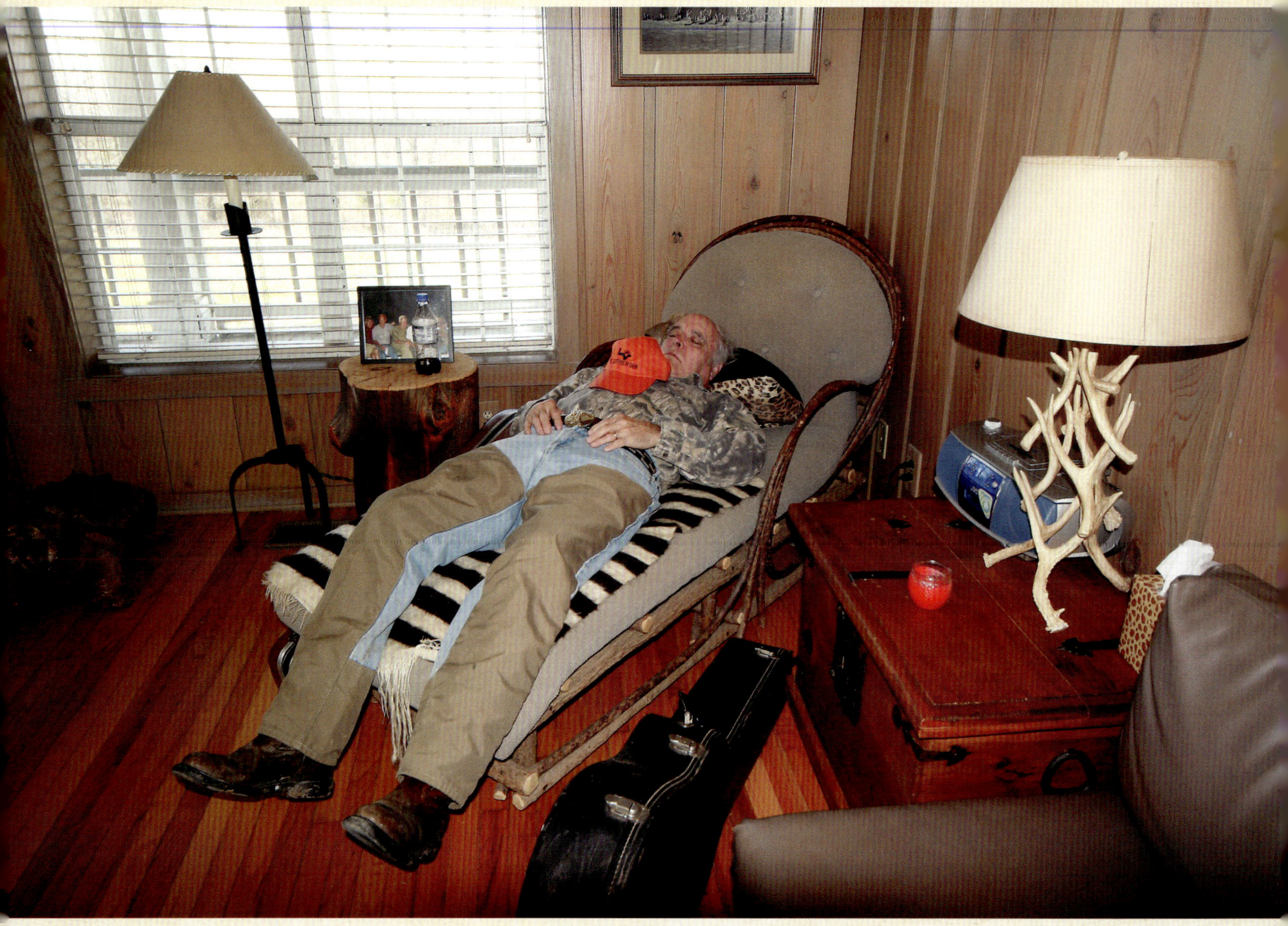

Sonny Merideth after the BBQ.

Old Great Mississippi Delta map. >

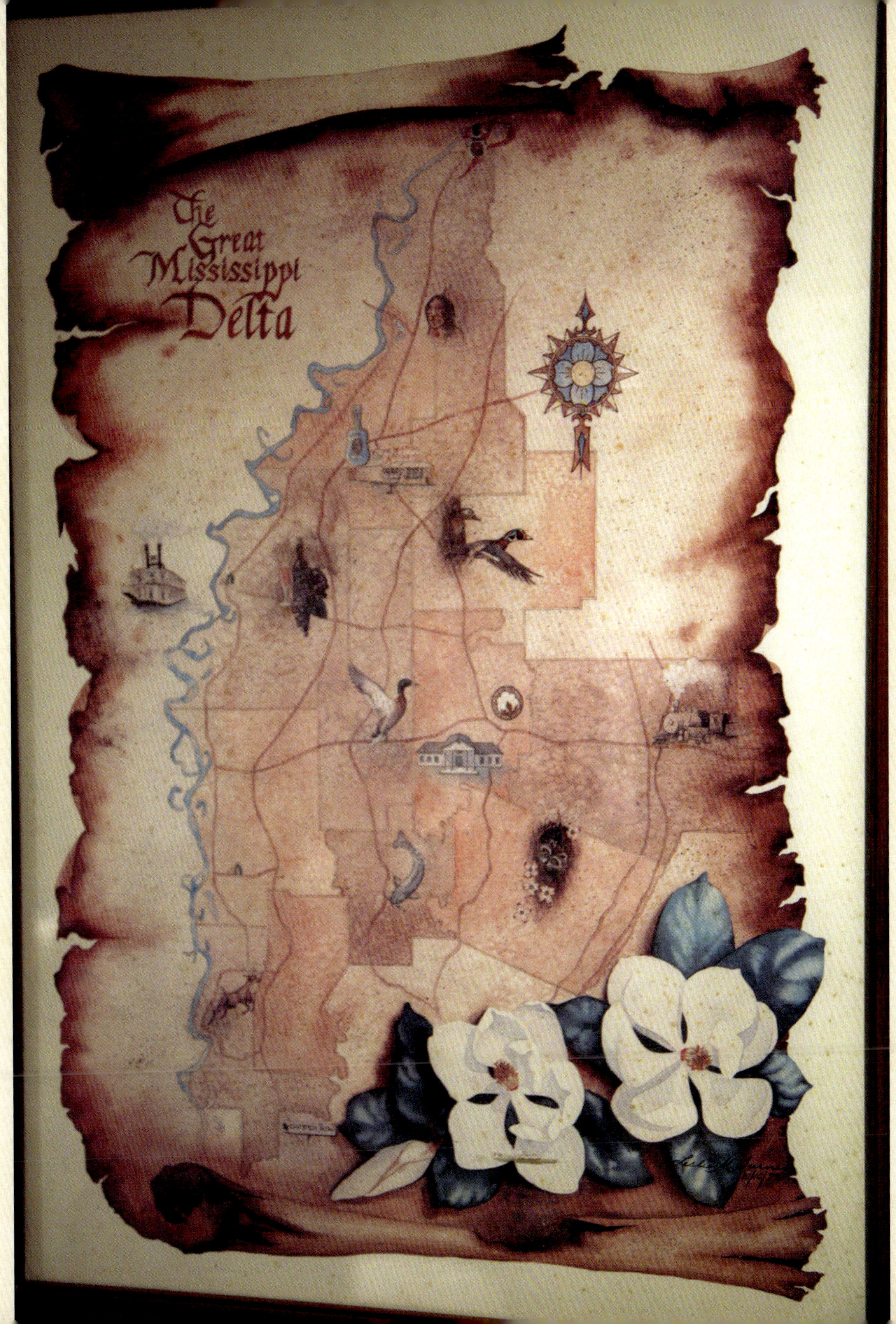
The Great Mississippi Delta

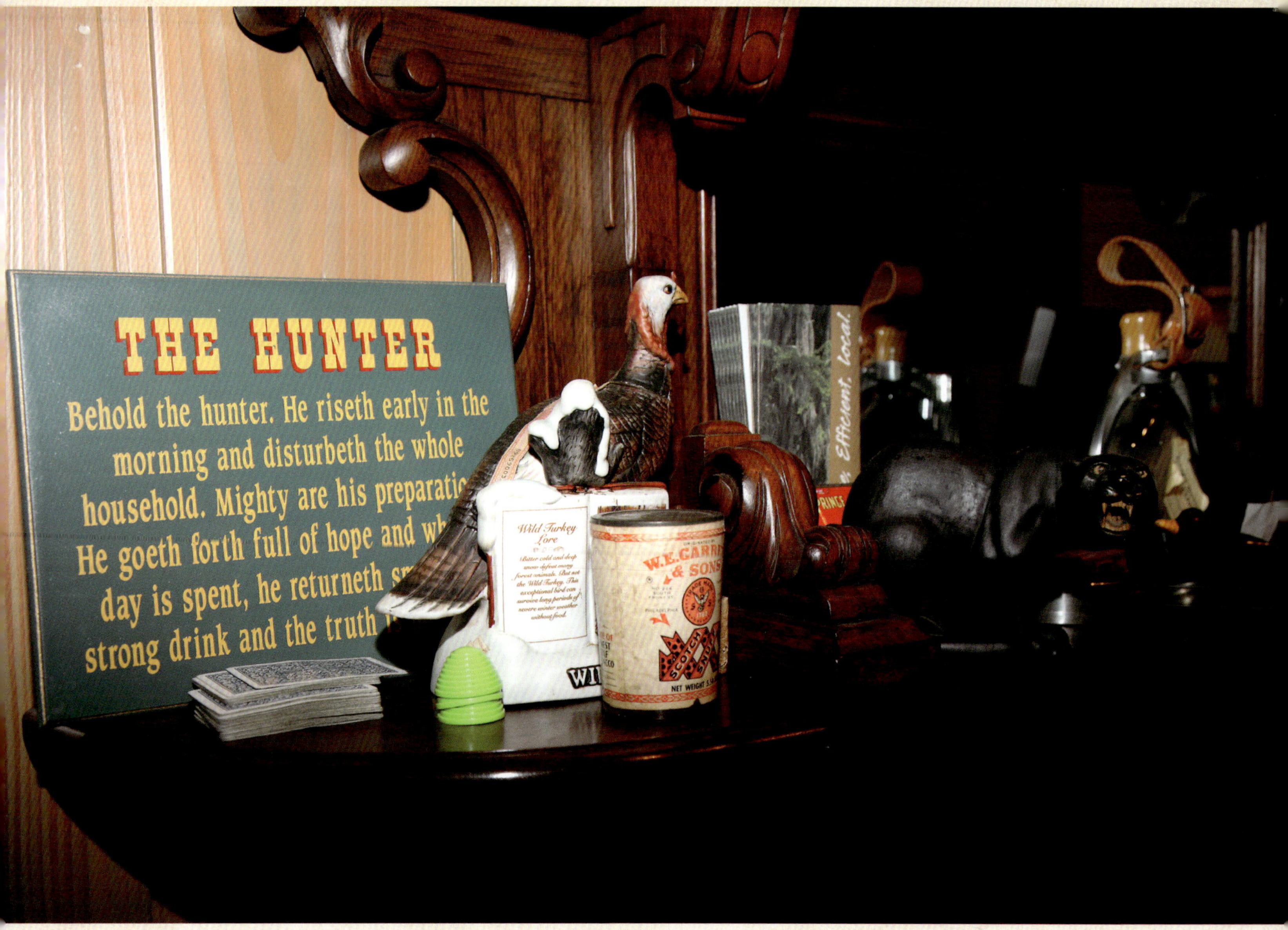

Shelf of hunting lore.

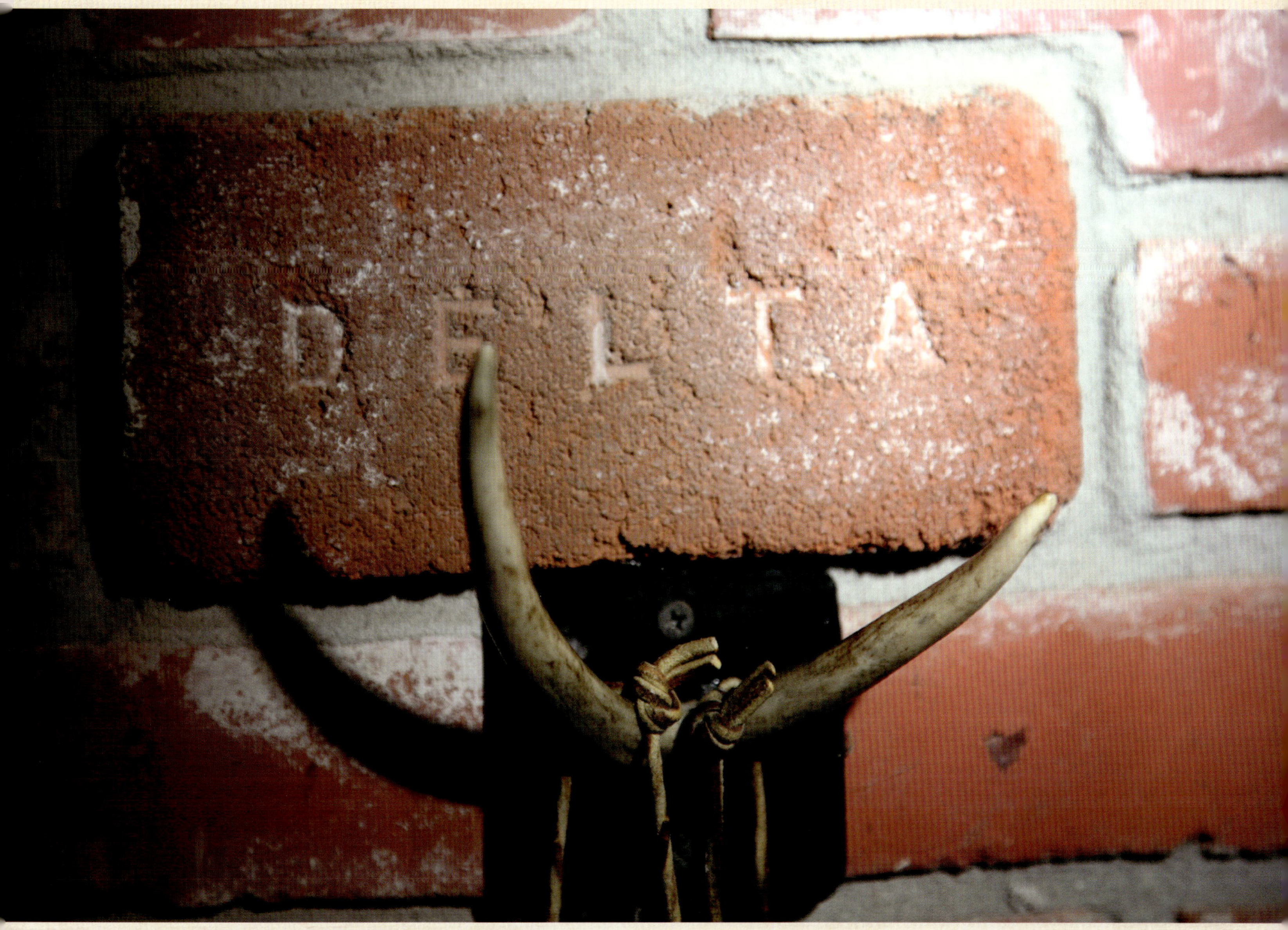

Brick from old Delta brick factory and deer antlers.

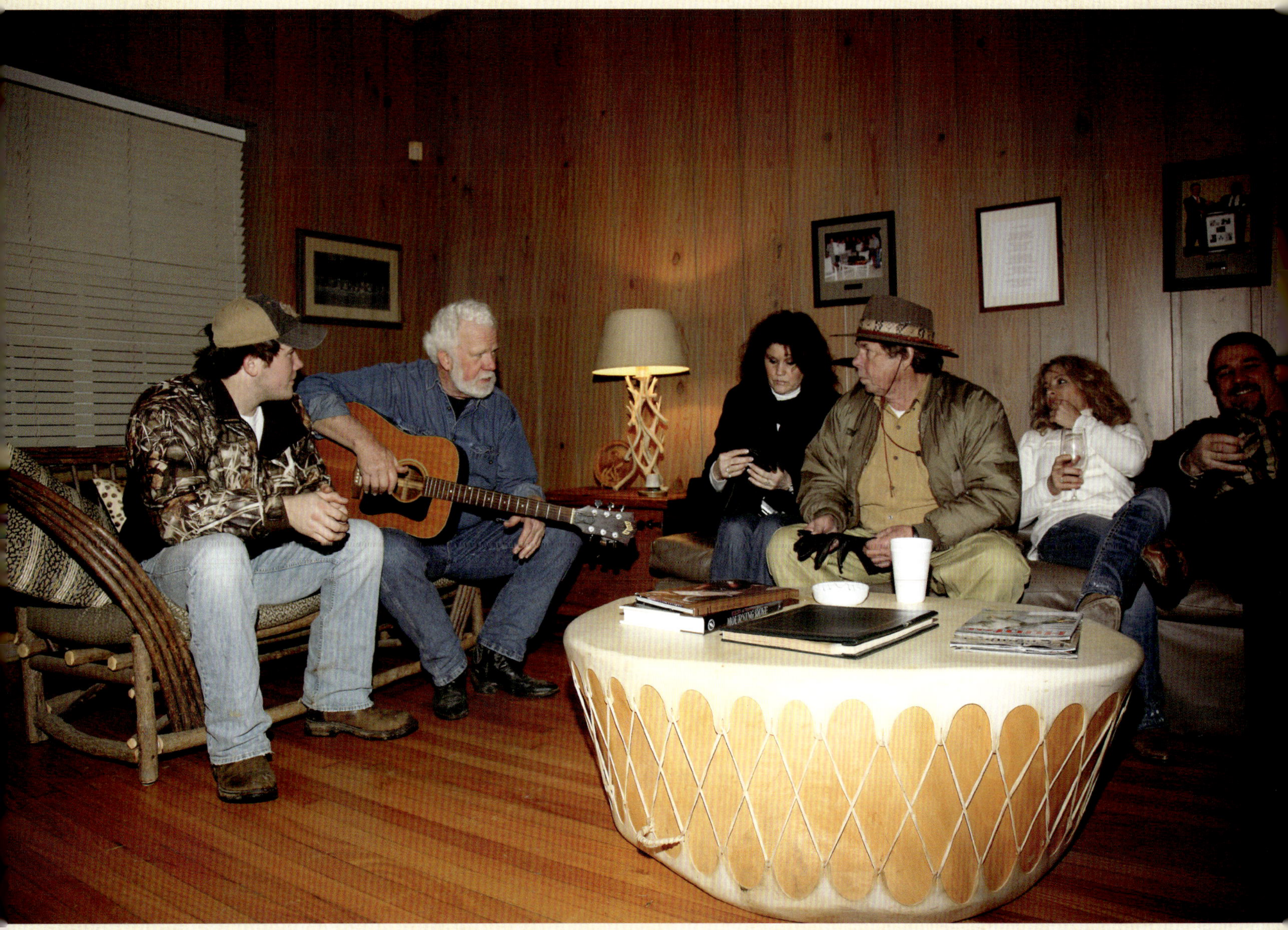

Jarrad Brown, Hank Burdine, Evelyn Brown, Howard Brent, Sam Zeponni, and Emmit Goodrich.

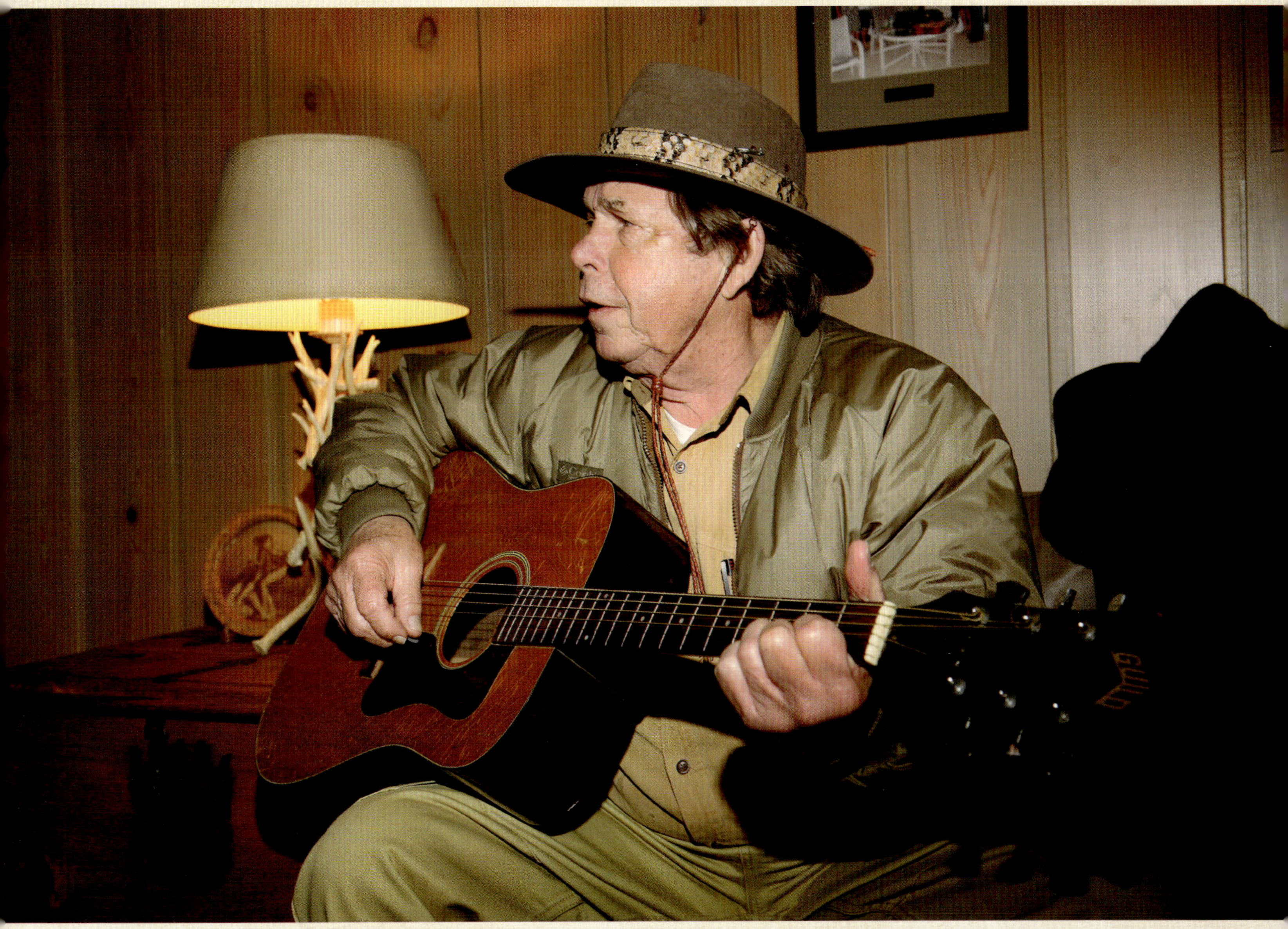

Howard Brent singing the blues.

Overleaf left: Hank Burdine and Carrie Cooper bustin' oysters brought up from the coast.
Overleaf right: Hank Burdine, Howard Brent, and Steve Golding.

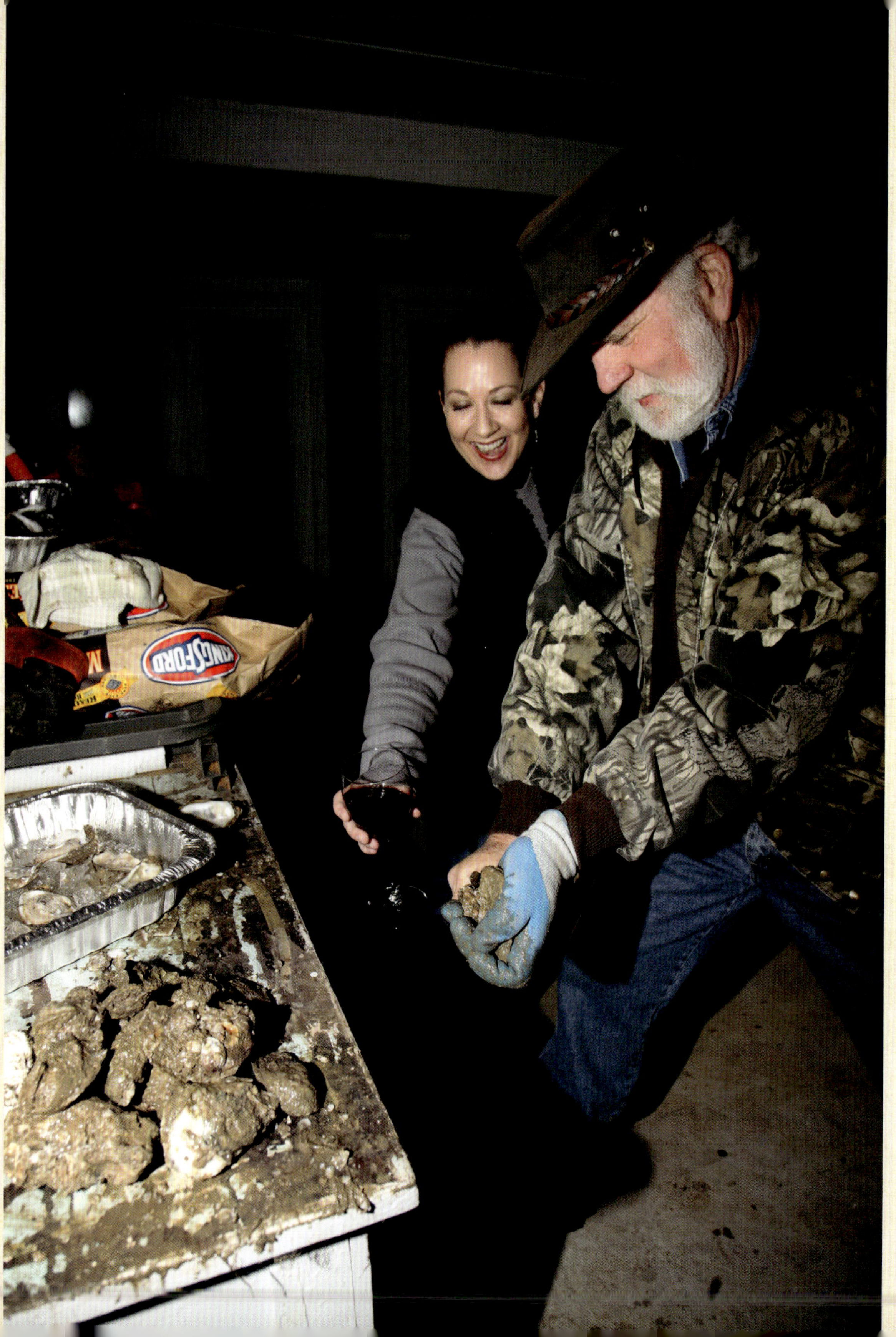
KINGSFORD

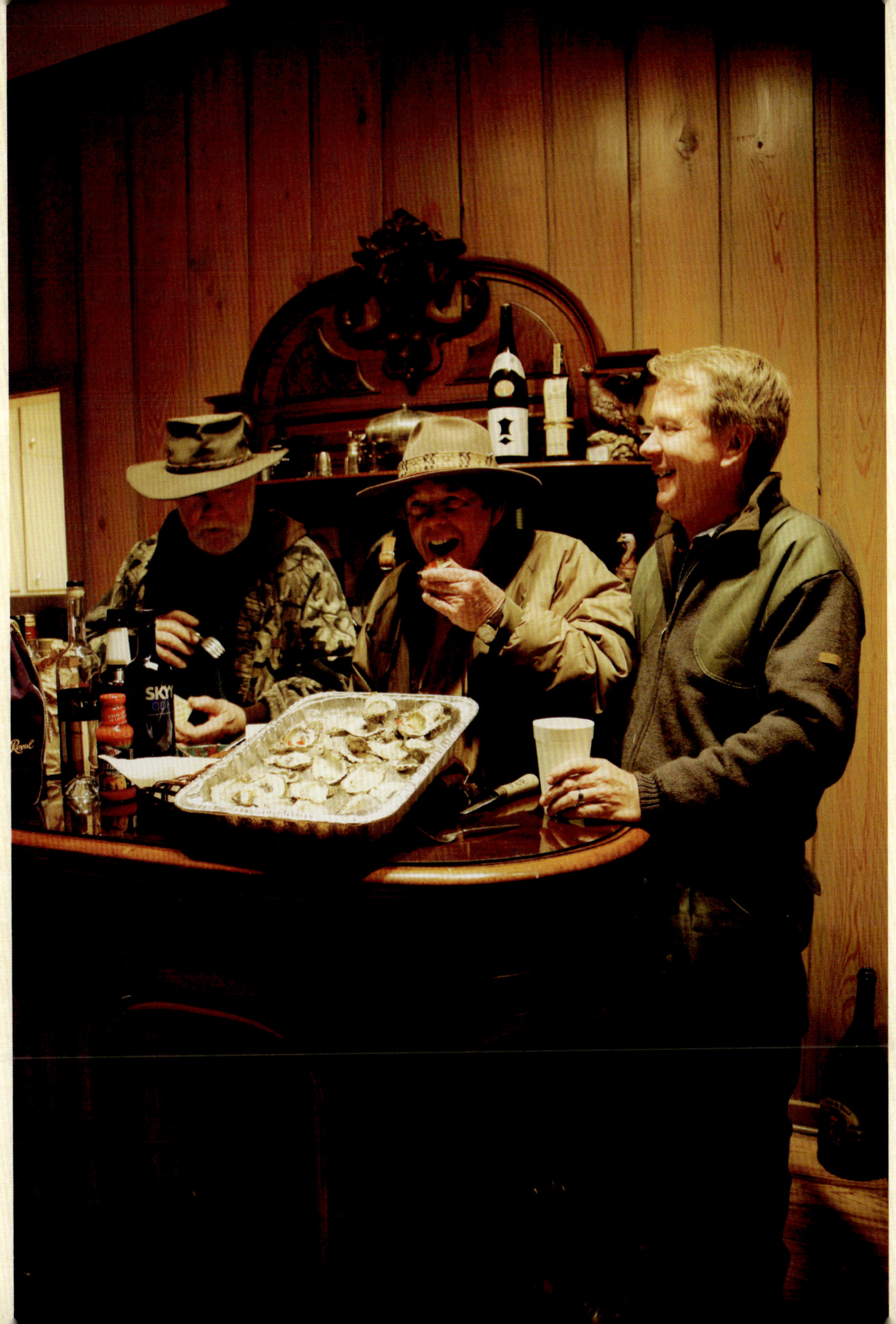
SKYY

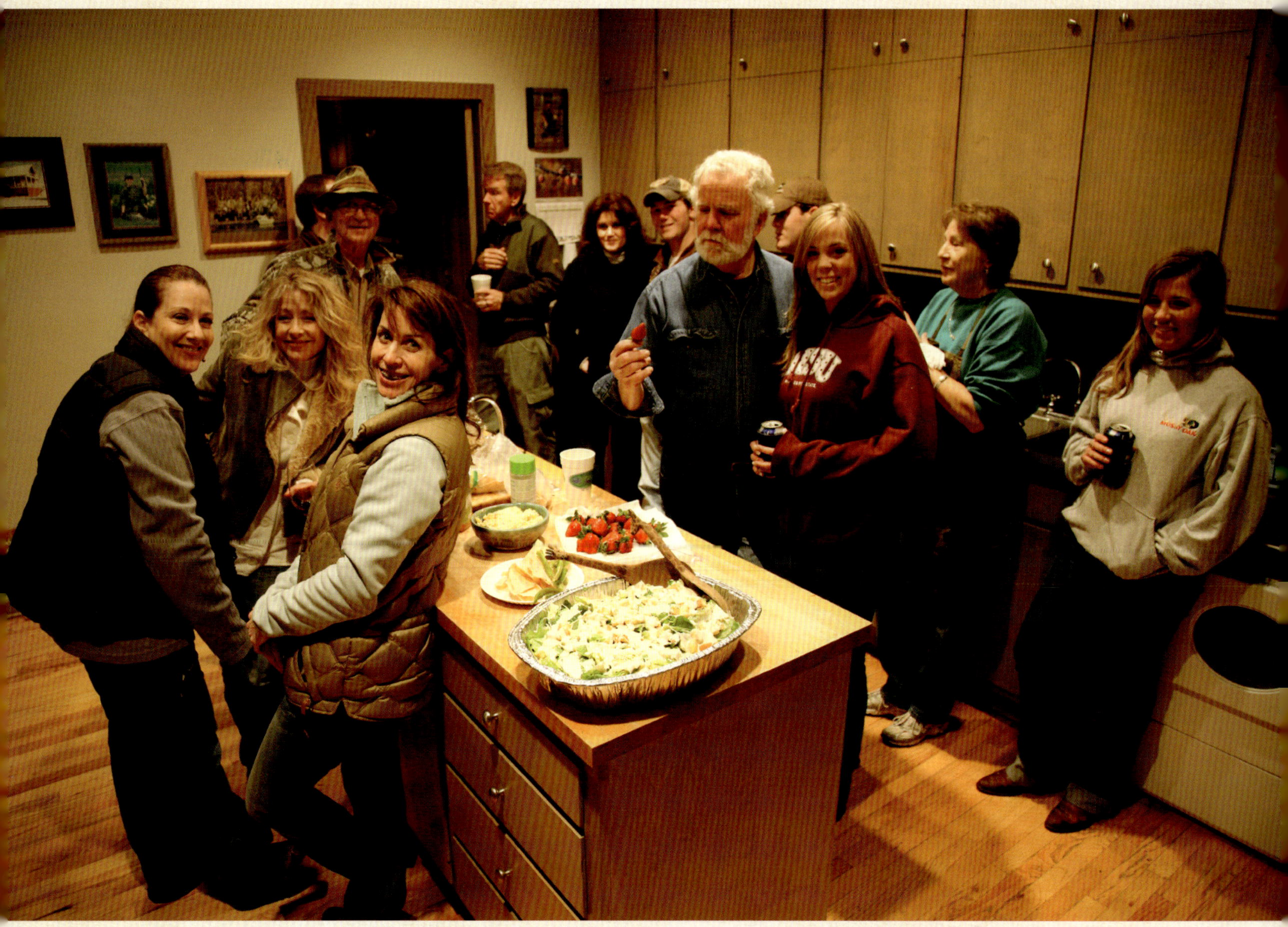

Carrie Cooper, Sam Zeponni, Wendy Gentry, Charlie King, Steve Golding, Evelyn Brown, Jarrad Brown, Hank Burdine, Patrick Brocato, Jane Alden Burdine, Betty King, and Mandy Mook.

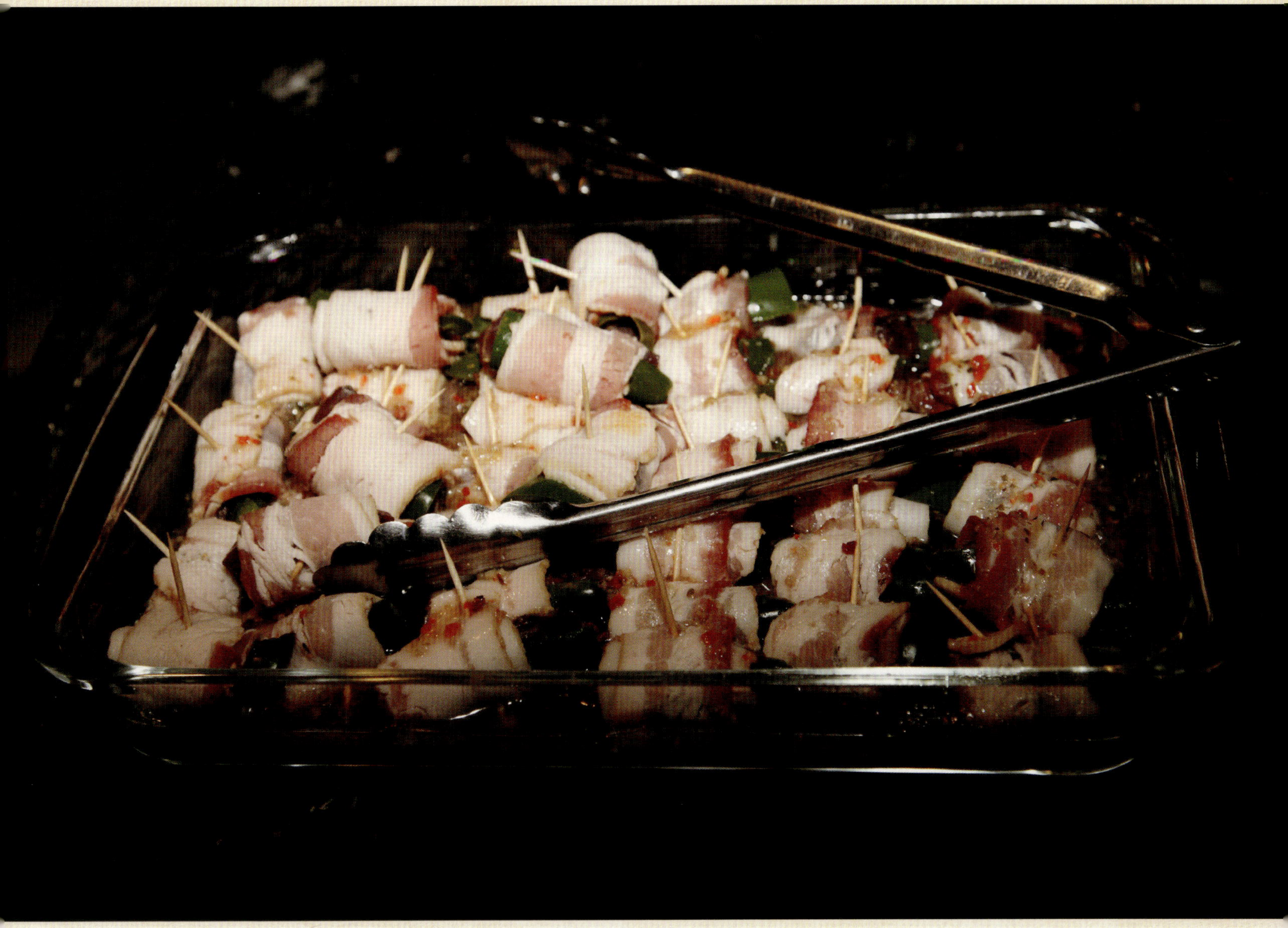

Mississippi Gulf Coast shrimp "poppers" with jalapeno wrapped in bacon.

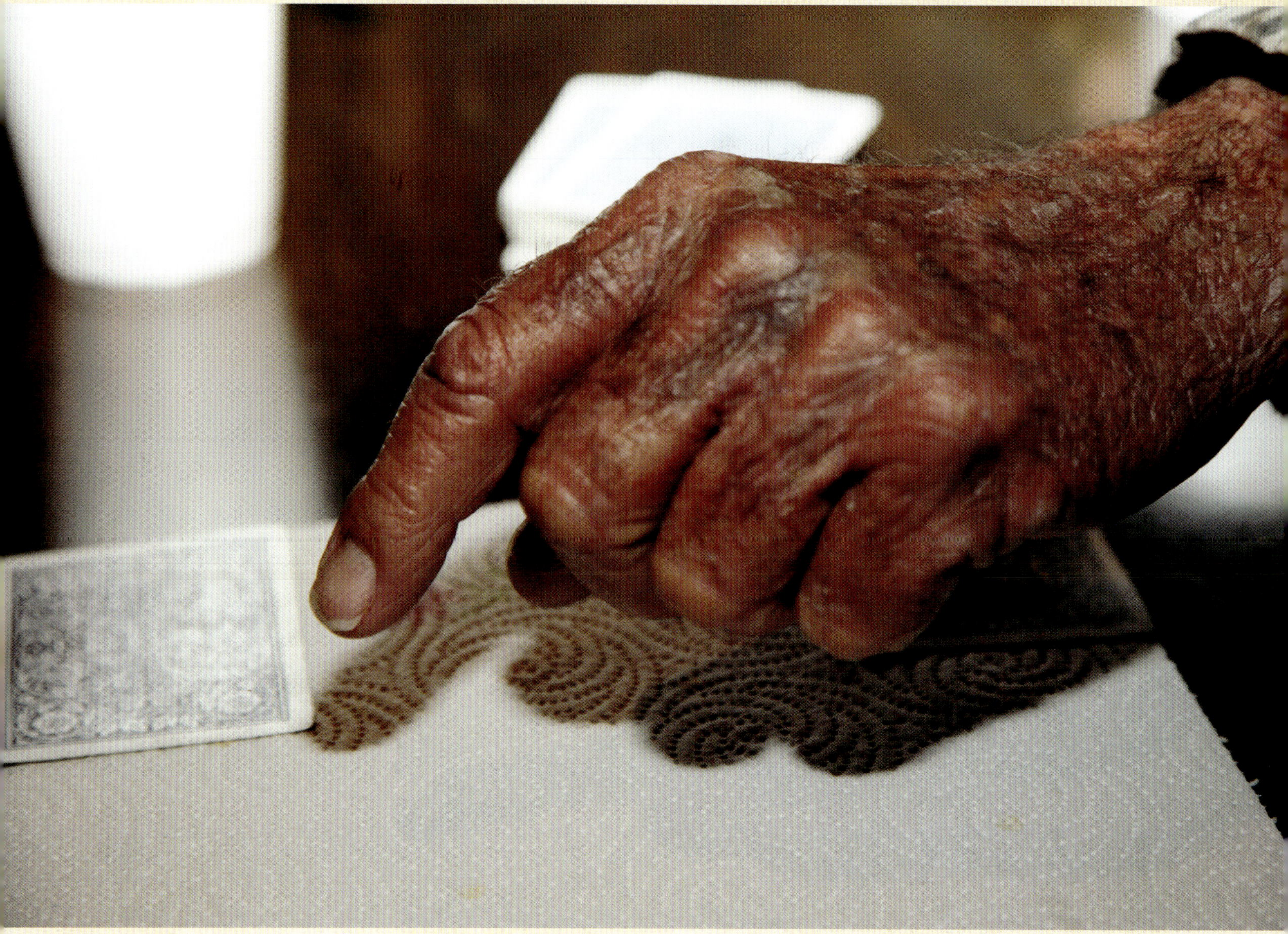

Charlie King playing cards.

Charlie King advising on Missy Webb's MHP ticket. >

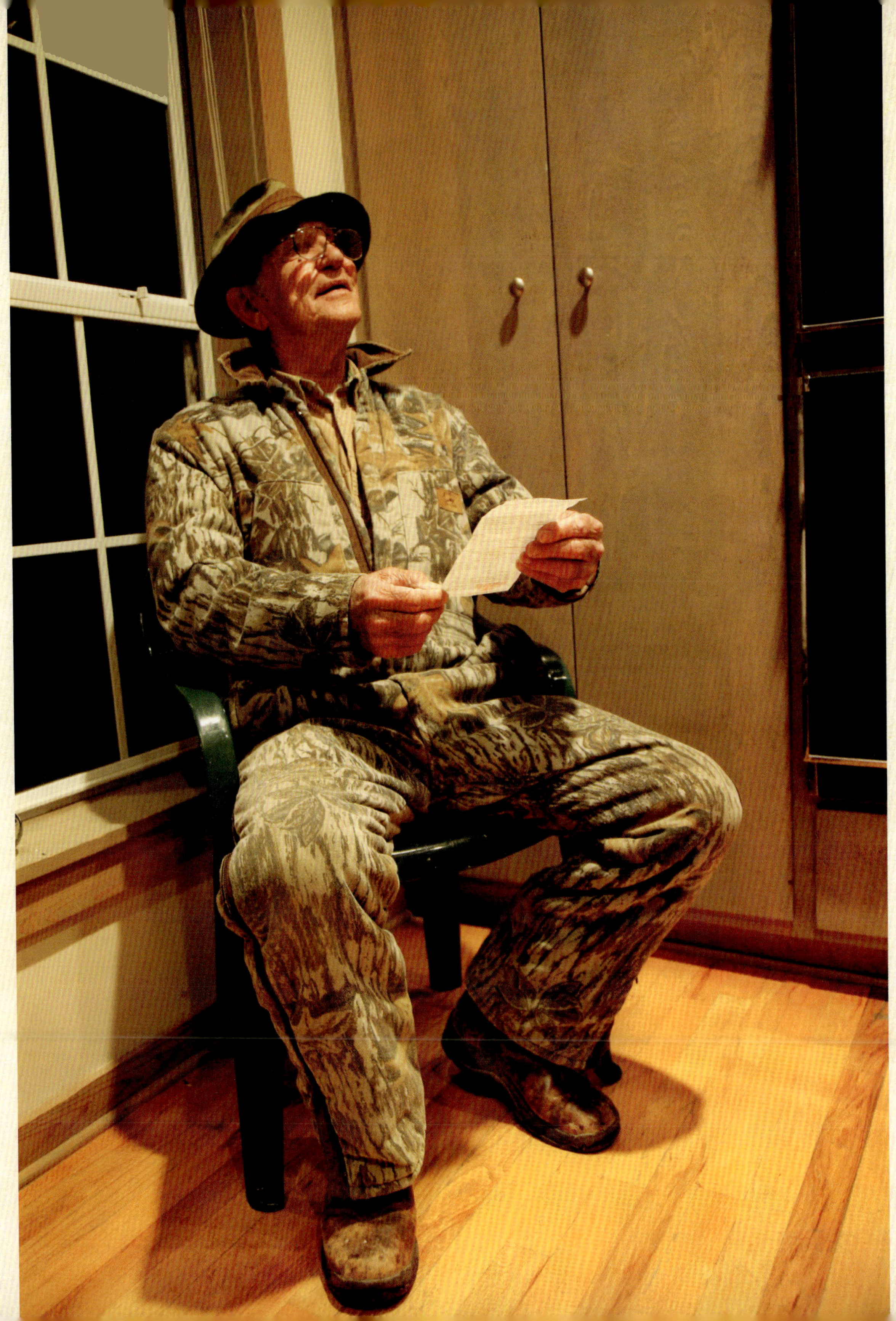

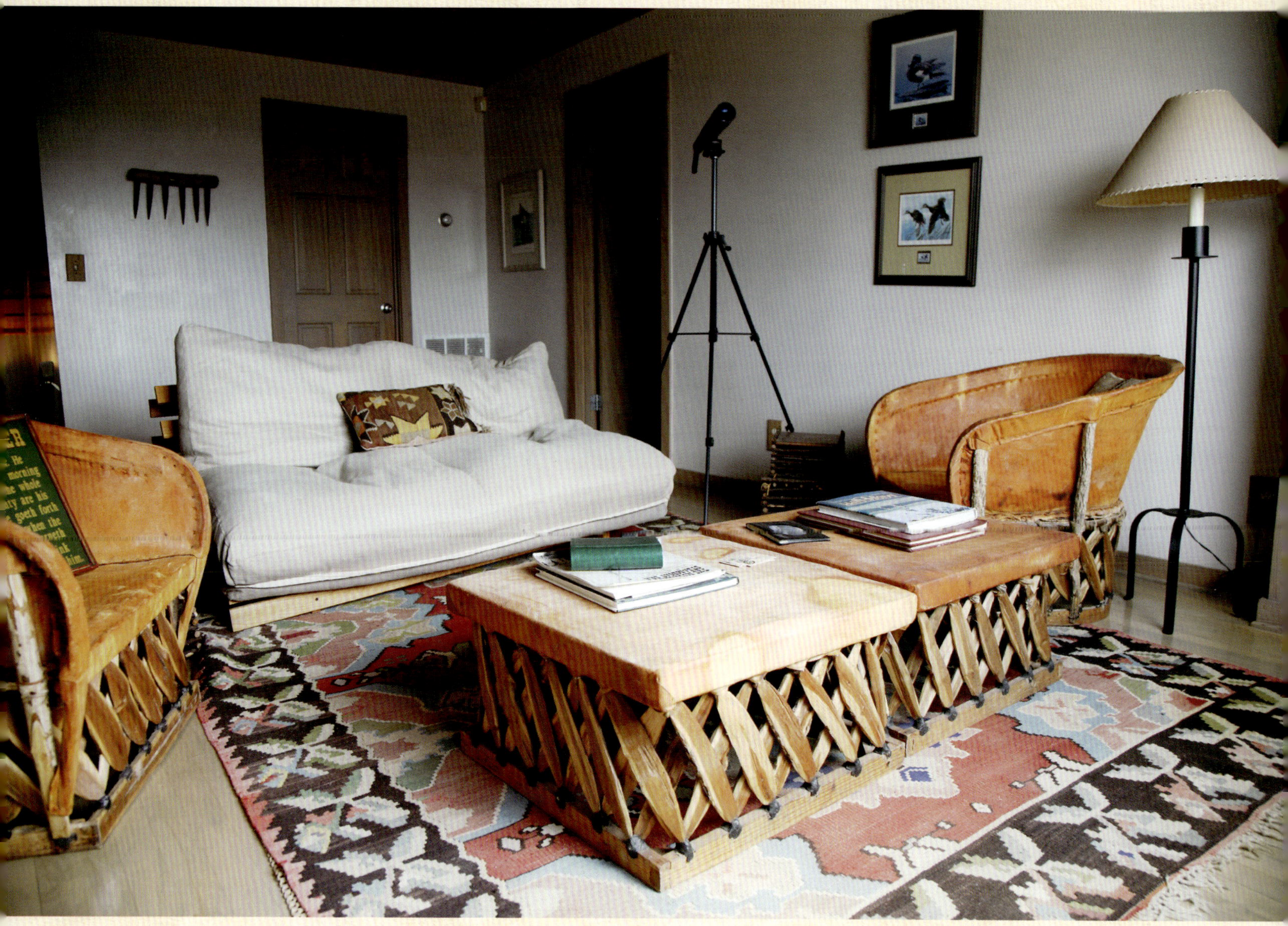

Upstairs living room at Panther Tract.

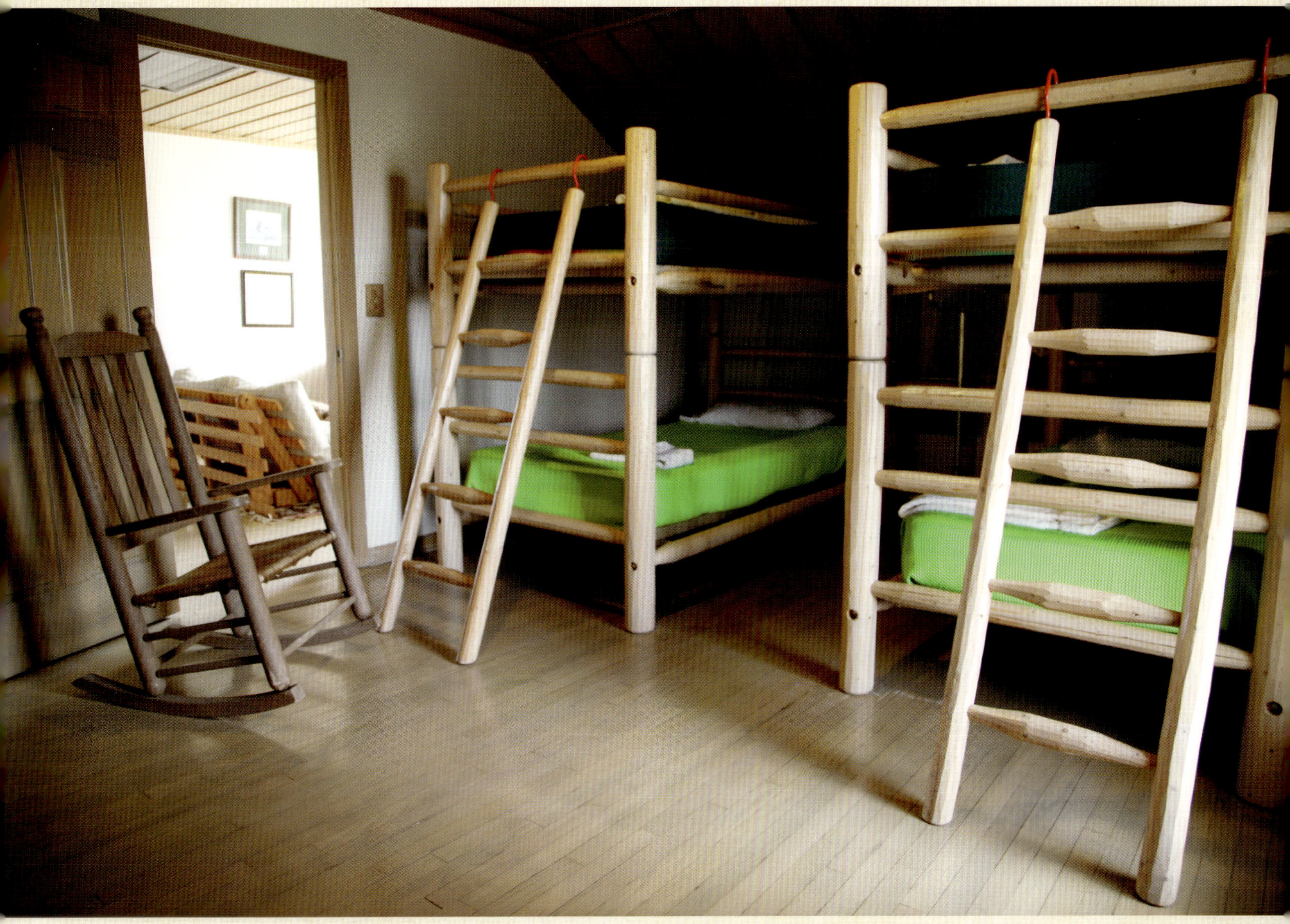

Upstairs bunkroom.

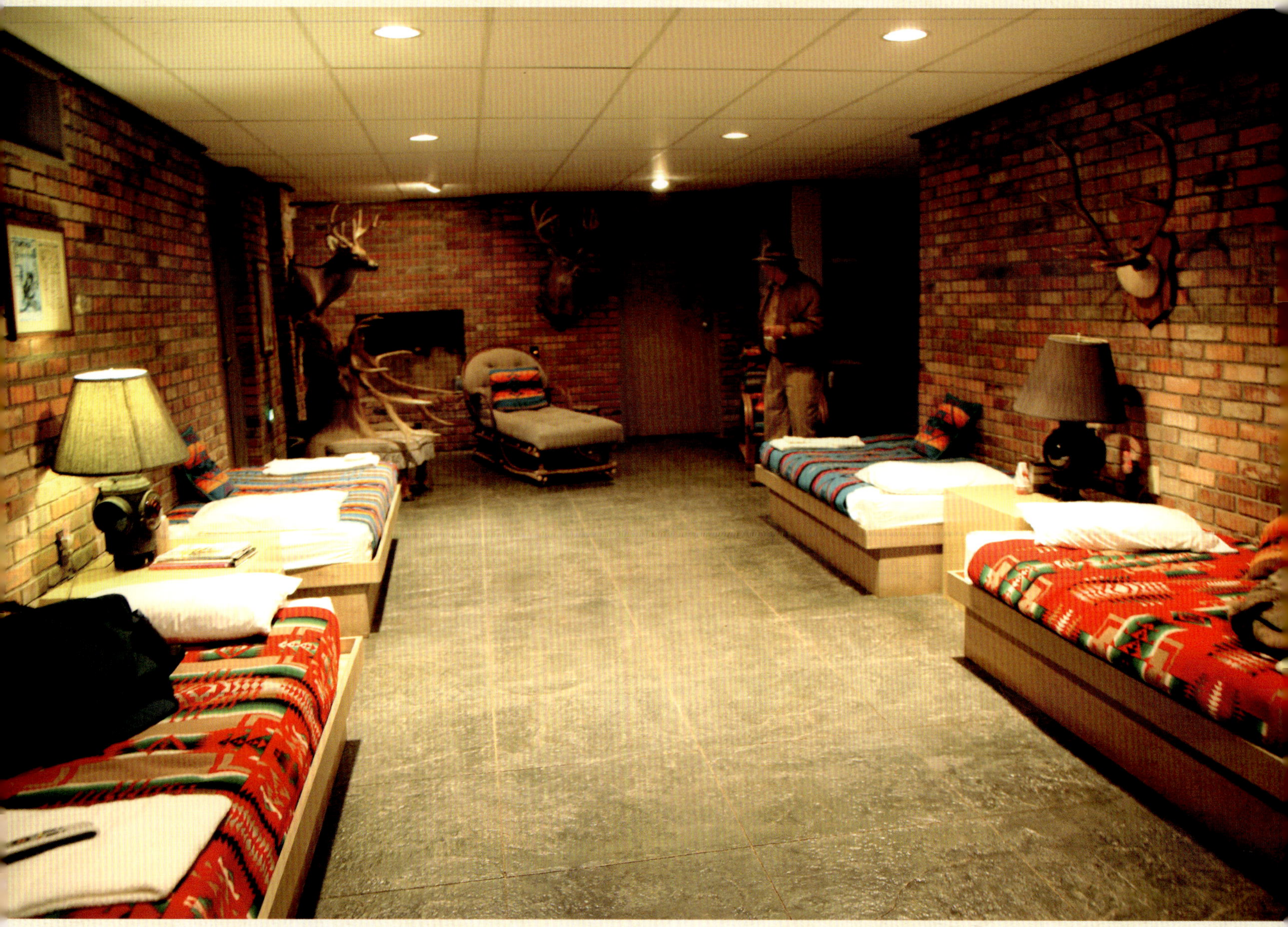

Downstairs bunkroom at Panther Tract.

Cordials. >

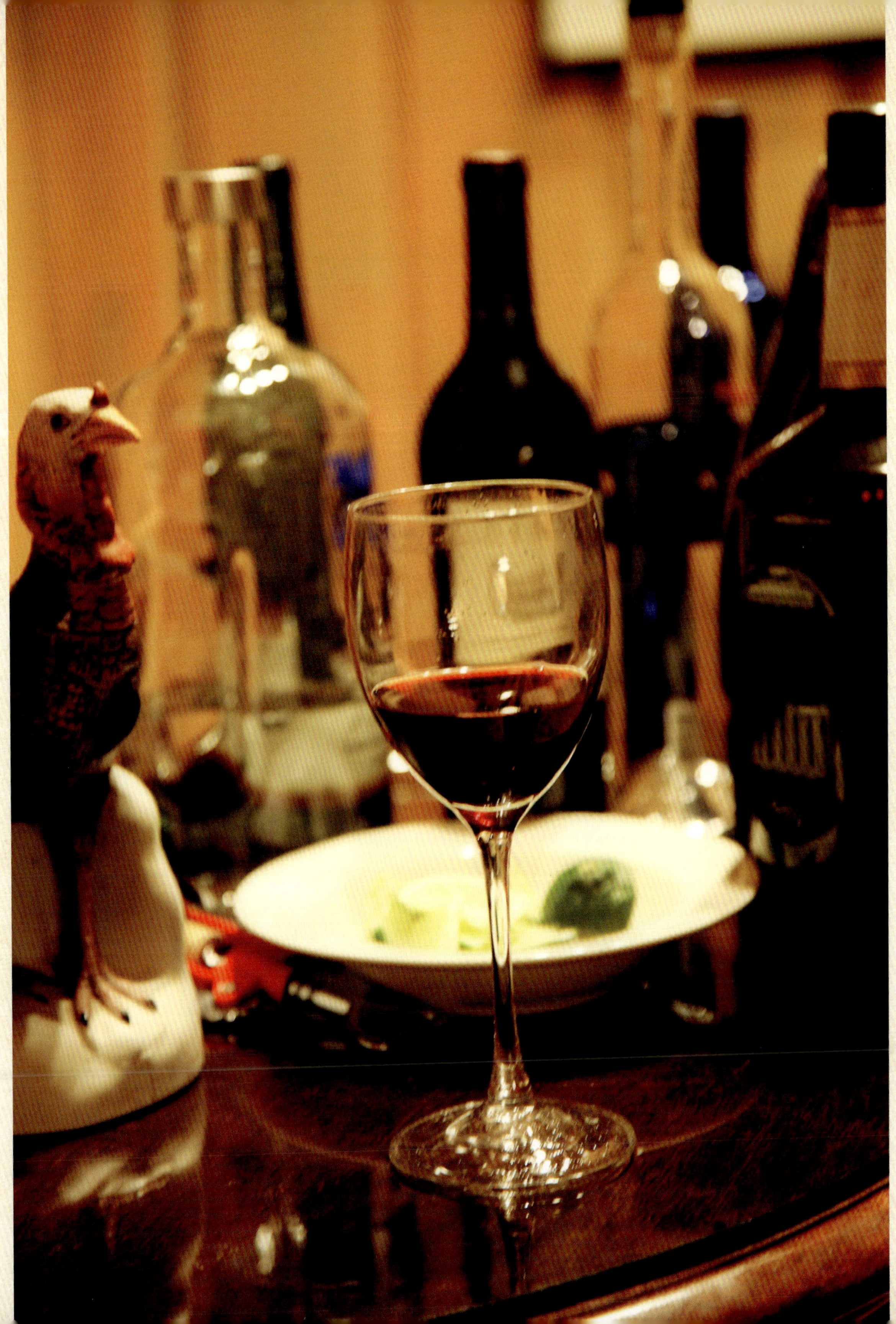

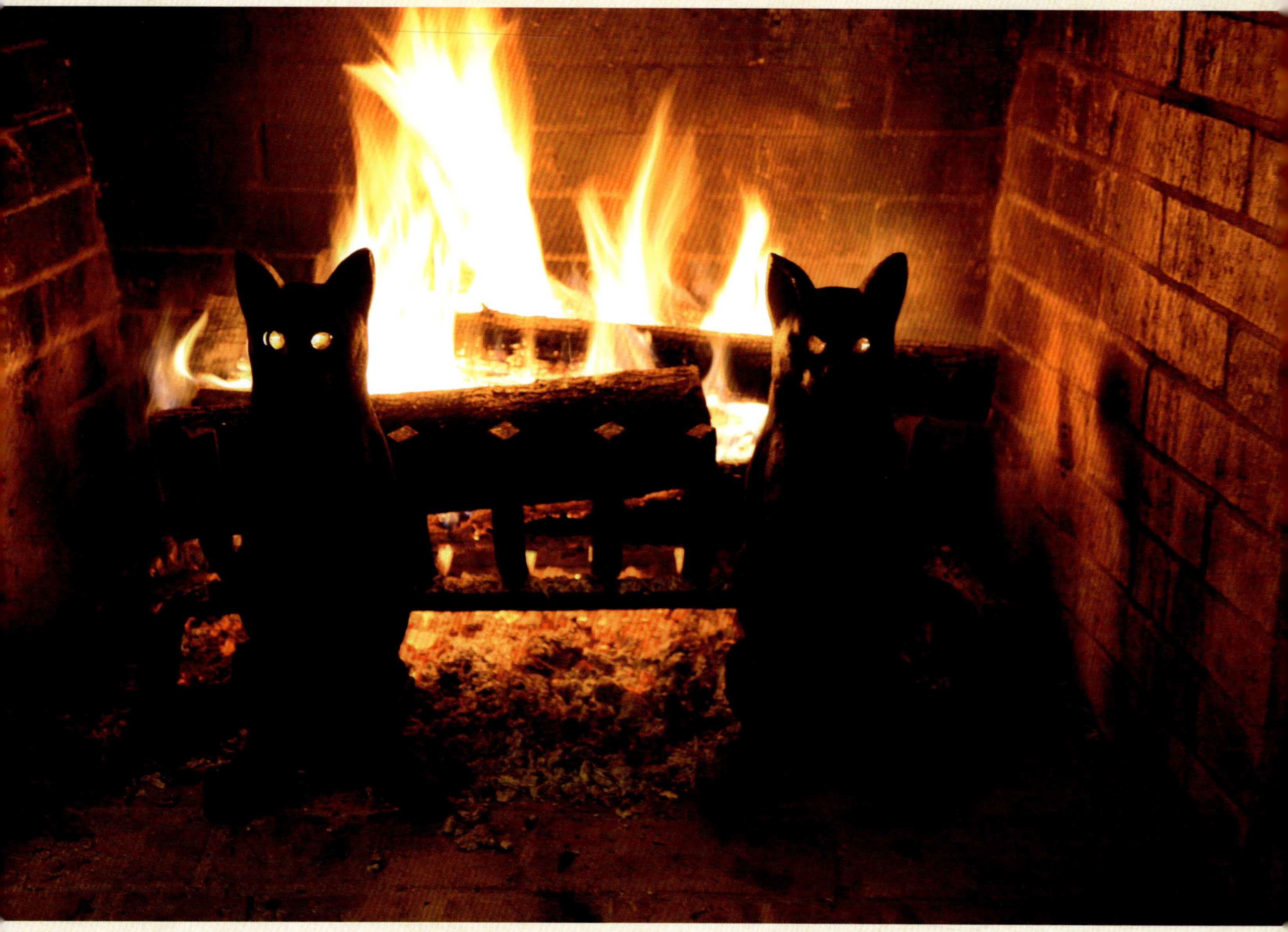
Panther andirons.

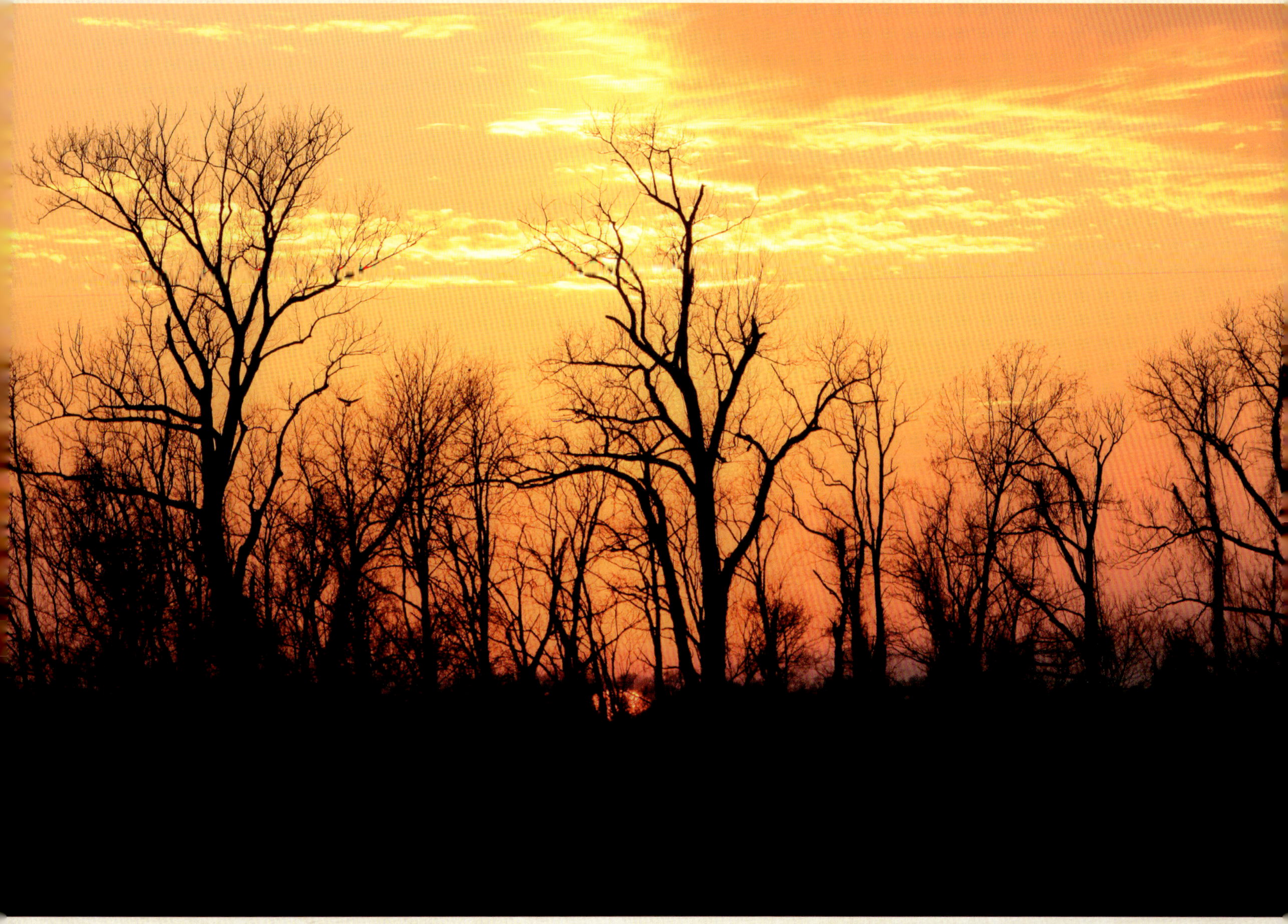

Delta sunset.

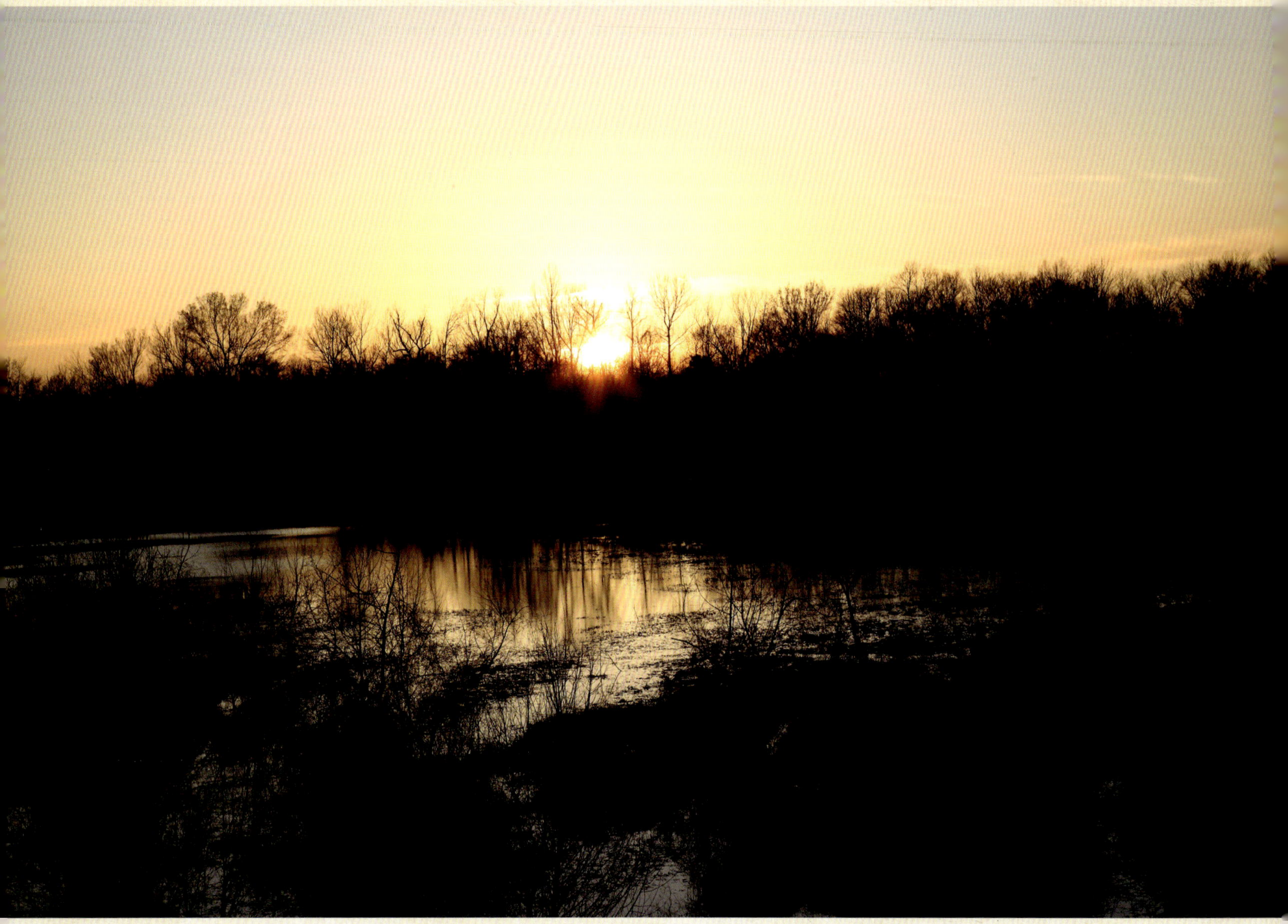

Lake Discovery at dusk.

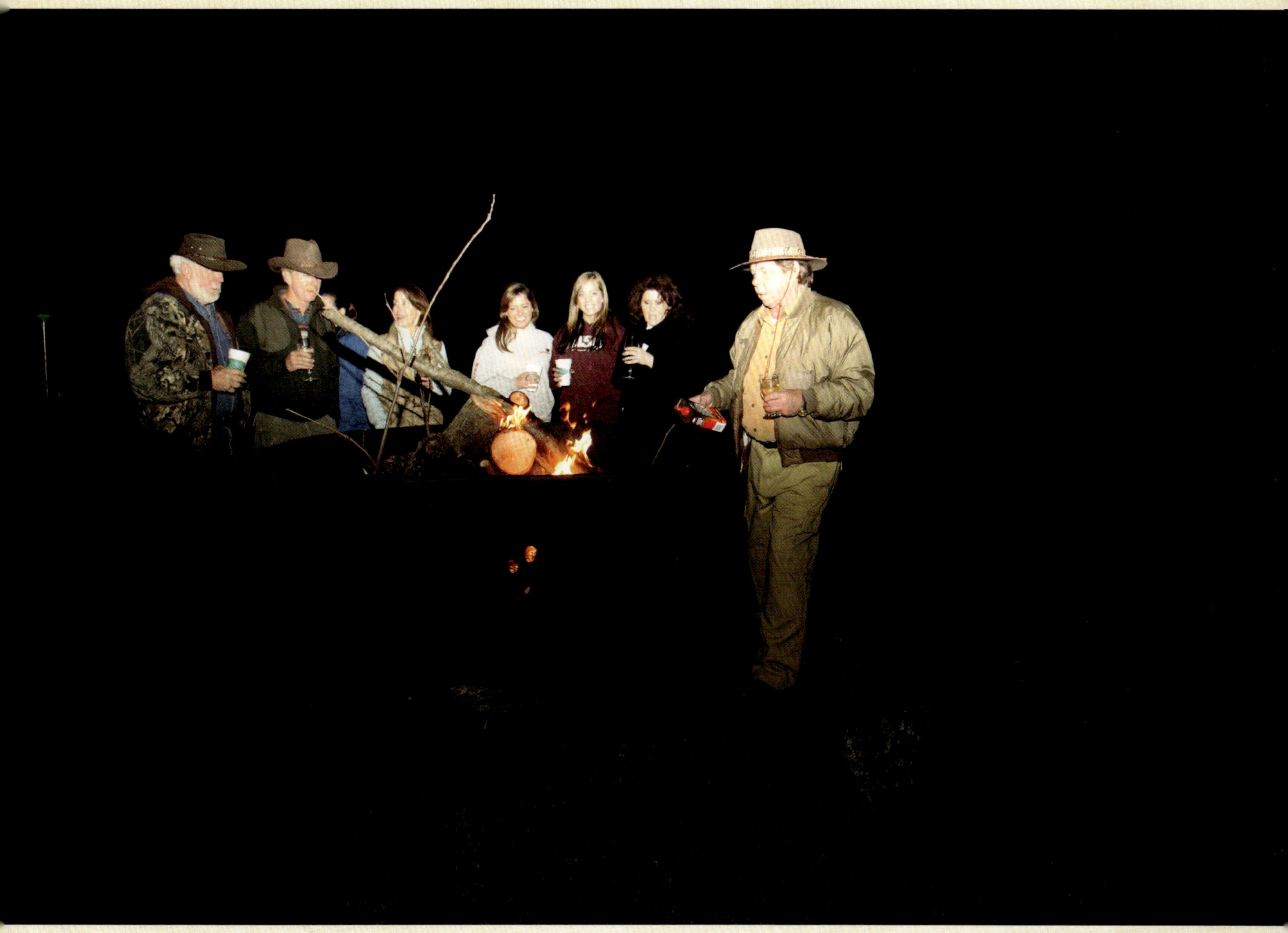

Hank Burdine, Steve Golding, Carrie Cooper, Wendy Gentry, Mandy Mook, Jane Alden Burdine, Evelyn Brown, and Howard Brent.

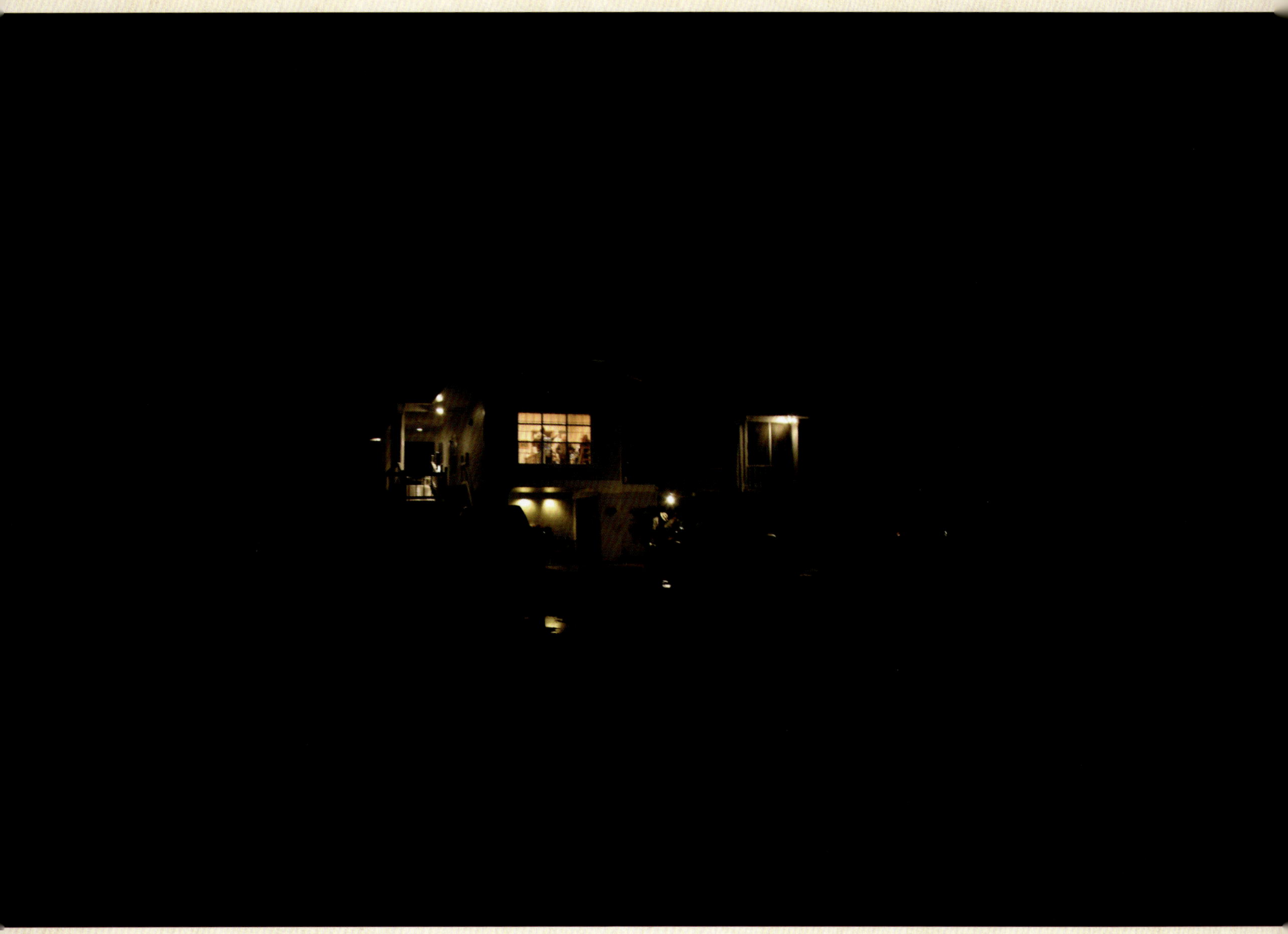

End of the day.

Tom's First Hunt

DOUGLAS MAULDIN

Somewhere, sometime back in the mid-1980s, I was contacted by Thomas McIntyre, editor of *Sports Afield Magazine* at that time. You see, Tom needed a favor. He had been given an assignment to write an article on wild hog hunting with dogs, and he thought I might know of someone in Mississippi who could arrange such a hunt. Tom and I had hunted together for years all across North America, and I deemed him to be a good friend.

Chris Marley, a prominent Delta farmer in those years, had become totally infatuated with hog hunting using dogs. His group was known for a lot of famous catch and release expeditions. Many attacks by raging boar hogs would end up with badly injured dogs and even hunters. A hunt with Chris Marley sounded too good to be true to ole Thomas.

Thomas arrived in Memphis on a direct flight from LAX. As soon as he got into the truck he wanted to know more details. So I began the story. "Chris," I said, "is a very colorful fellow—red-headed and full of BS, *but* he is a real hog hunter—a little crazy but real!" I told Thomas how the hunt would go down next. We'd all be on three- or four-wheelers and Thomas would ride with me. Chris would release the dogs on fresh tracks. As soon as they'd strike we'd all be "off to the races"—after dogs and hogs! "Don't know how long it would take," I told him, "but when the dogs bay a hog we must get there quickly to prevent injury to the dogs." At this point a capture would be attempted unless the hog was too big. If so, that called for a big knife to the throat of the boar. Only Chris would be allowed this honor! Thomas seemed satisfied with the plan and explained that if at all possible he'd like to be the knife man. I told him he'd have to be very skillful for that task or totally insane! Thomas thought for a moment and then replied, "I'm in."

After a good dinner and drinks we were off to bed. The hunt would be an early morning event to beat the Delta heat. The next morning Thomas and I met up with Chris and the dog handler. The location of the hunt was just south of the Helena, Arkansas, bridge along the Mississippi River, a hog-rich area full of big cane thickets and swamps.

Once on location Chris went over everything, including bad events that could occur as well as the good ones. He told us if the hog is small it will be caught and transported to another farm. If the hog is a medium sow it will be taken to eat, while the medium boars will be cut and released to grow into monster boars that only think about food. Big tusked hogs will be given the knife treatment if not too dangerous. Thomas had a few questions about injuries to animals and humans. Chris quickly displayed the first-aid kit—ready for stitching up of man or beast. "Enough questions, boys," Chris said. "It's time to hunt!"

Six very mixed-looking dogs called hog dogs were released on a muddy logging road. With Thomas on the back of my Honda 250 CC four-wheeler, we were ready for the hunt! Did I mention that Thomas was all of five feet ten inches and three hundred pounds? He was still very agile as a hunter but a large fellow nonetheless.

The dogs struck immediately and four-wheelers

raced off through the Delta swampland. I also forgot to mention that I was an experienced motocross racer and sometimes got lost in the moment. As the race extended, Thomas and I found ourselves well in front of the others, having ranged over ridges and treetops. At that point I thought I heard Thomas screaming, "You are gonna kill me!" It's a statement I had heard before from my passengers. As I flew over the next sand dune, Thomas was airborne and upside down on the trail. He bounced, rolled, and cruised but was OK and quickly back on the machine. Chris was now up with us and close to the hogs, the dogs, and the cane thicket. Soon we were all in the thicket and spotted the dogs working in a frenzy with hogs of all sizes moving in and out of the cane. The dogs focused on a very angry boar—two hundred pounds plus—and backed and bayed against a large fallen cottonwood tree. Chris signaled for all hunters to move in and help the dogs catch the hog. I charged forward with Chris and the handler as the boar faced his enemy. I was amazed at how fearless the dogs were as we moved forward onto the battlefield. The dogs went to work like one has never seen—darting in, biting at, grabbing hog legs, tails, and noses. It was wild and I was sure Thomas was getting his story along with plenty of action-packed pictures!

But where was Thomas McIntyre? I shouted, "Tom, move quickly—great photo op here!" Thomas replied, "I can't." I looked up and spotted Thomas about ten feet up a small willow tree trying to hang on. Don't know why the editor was up a tree—just go figure.

The boar was caught being held down by man and beast. Thomas was out of the tree and nervously taking a few pictures. Now came the boar's worst nightmare—Chris removed his surgical knife from a leather case and quickly took the hog's manhood. Then he stitched him up like any good doctor and said, "Let him go, boys!" Dogs tied up and men backing up quickly, the boar stood up with a puzzled look on his face. He turned and marched off into the canes. The hunt was complete and considered a grand success.

Now we were ready for the trip back to the trucks. Thomas said, "I believe I'll walk out." After a moment of explanation, Thomas understood that the ride back would not be nearly as exciting, dangerous, or whatever. He hopped on and we were soon home safe and sound. That night at dinner we all made toasts to the successful hunt and to the friends who were there to share in the excitement. We told the usual old war stories about giant wild hogs, went to bed slightly dazed, and arose in the morning fresh and ready for the next great adventure!

The Wild Boar Hunt

LOIS SWANEY SHIPP

In 1934, when the Natchez Trace was nearly finished, wild boars from Russia were imported to mix with our few and not-as-wild hogs. They were imported again in the 1950s and boar hunting was considered a new sport for Mississippians. Hogs from Louisiana and Arkansas were "imported" to improve our stock.

Pigs aren't native to North America. Hernando de Soto brought the first pigs to America on his trip in 1539. He needed Indian guides, as he was searching for the mighty river and there were no maps. He sponsored a feast and invited the Indians to be guests, as he needed them to be friends. Pigs were the fare at the banquet that night and so delicious that after midnight when the party was over the Indians came back and stole the pigs. From then on Soto and the Indians were enemies.

My son, Walker J. Swaney, and his friend Lee Brett killed a wild hog on a black moonless night in March on Caulk Island, in the middle of the Mississippi River. The hunt on Caulk Island was a very exciting one and impossible to ever duplicate. Pit bulls were used to find the hogs. The dogs matched the hogs in ferocity. Thirteen hogs were killed that night and distributed among the members of the hunting club to be roasted, barbequed, or boiled, and they said that they were delicious.

When God made the hog, I think he made them last, and as a joke tried to make them the ugliest animal on earth and the wildest. They have curly tusks protruding from their mouths that are used for ripping and digging. They can't see so well, but have an uncanny sense of smell and of hearing. They blend very well into the background because they are so well camouflaged. They have no fear of anything. If you are in the woods, they will turn and chase you. When you flee from them up the nearest tree, don't drop your gun. Can you climb a tree while holding a weapon? I can't either.

Country ham is one of the best things going for people who used to eat it all the time. We lived in the mountains of West Virginia for a while. Up there, my contemporary friends made their own country ham, so I asked for the recipe. It was a family secret, but since I was coming home to Mississippi, they figured I would not be sharing it if they gave it to me. Now, of course, I can't remember the recipe, but I came home to Mississippi and brought the "green" (fresh) hams. I remember rubbing the seasoning and salt into the hams. Then I had a needle and I injected the liquid seasoning in several places around the bone. Next, I wrapped them in cloth and hung them on the rafters in the cellar of Gray Gables. During Civil War days, people were craving salt and couldn't buy any, so they would sift the salty dirt that had dripped from their hanging hams and use the much-prized salt again.

My mother, Bertha Bonds, made her own country ham. After my daddy died, my mother made three hams and hung them in the basement to "ripen." That would take six months. She hung them in the cellar from a rafter. Her bedroom was directly above the hams. In the night, she heard a noise downstairs and she thought a varmint was after her hams! She couldn't find a flashlight, and so she lighted a candle and went downstairs to check on her hams. The cellar had an outside door and there was a big chimney that went up to her bedroom. Yes, her hams were all there in the flickering light. However, the next morning, they were all gone. A thief had been there and stolen her hams.

I haven't ever been on a wild boar hunt and I would be out of place on one. I am scared to death of those creatures!

Home from the Hill

JORJA LYNN

In the summer of 1959, I was twelve and I was having supper with my family (of seven) when the phone rang. My father answered it and when he returned to the table, he was very excited. He announced, "That was MGM Studios and they want me!" Now, he was a very handsome man, a dead ringer for Clark Gable, mustache and all, but this seemed unlikely. The movie *Home from the Hill* was being made in Oxford, Mississippi, that year. It starred Robert Mitchum, Eleanor Parker, George Peppard, and George Hamilton. In the movie, Col. Hunnicutt (Mitchum) sends his son Theron (Hamilton), with the help of his half-bastard brother (Peppard), to hunt down a rogue, filthy, red-coated, smelly, and covered-with-flies hog that is terrorizing the county. Now, that is where Papa came in. He had always raised bird dogs to hunt and show at the field trials and some Hollywood type decided he would be the go-to man to find some Catahoula hounds for the hog hunting scene. My father never admitted to having not ever even heard of a Catahoula hound before, but, being a very resourceful man, he told them that he would provide them, which he did, finding them down in Cajun country in Louisiana.

The hounds were a very important part of the scene in which they lead Theron into the forbidden swamp while chasing the hog, which, after throwing off the dogs, charges the boy, who at the *last* minute kills it! So, my father provided the dogs and my mother, father, and I got to watch the "fight scene" between the two Georges there in the yard of Annesdale in Oxford and I got to meet and greet all four of the stars—all because of the hog!

Incidentally, a few nights later, the ca. 1836 Holly Springs jail was the overnight host of Mr. Mitchum, who had been booked there for possession of marijuana!

Twenty Years of Hogs

JUSTIN BRASWELL

I've been hog hunting for probably twenty years. One time we went hunting across the river and my brother-in-law, who isn't much into hunting or the outdoors, joined us and brought two of his friends. We bayed some hogs up and Stephen and I got down off our horses and caught the hog and shot it with a pistol. The dogs had bayed up another hog and it was a big sow. A sow will bite you. By the time we got to the sow the dogs let go and she came right at us and I grabbed her by her ear and Stephen grabbed her by her back legs and flipped her over and killed her. My brother-in-law's friends said, "Your in-laws are some crazy people!"

Ninety percent of the time we hunt off our horses. Sometimes we take four-wheelers for people who aren't "cowboys" and don't know how to ride, but we like to ride the horses. We keep .22 pistols in the pockets of our chaps when we are riding horses and we keep our guns in the gun scabbard down by the saddle.

We were hunting in January of 2008, and the dogs bayed up a bunch of hogs up in a break we call dishpan; Stephen was on one side and I was on the other. Well, I walked out on a log and the water was probably about four feet deep and we could see the hogs and the dogs out there. Mr. Wick got on the radio and said, "Stephen, you need to go help Justin get those hogs! He's already waded out there!" Stephen took off through the cold water but what he didn't know was that I was standing on a log. Stephen got all wet and wouldn't talk to us for the rest of the day, he was so mad! But we got three hogs that morning.

The biggest hog we ever got weighed four hundred pounds. He was a big boar hog. That's a hog that's been castrated.

We like to keep our hunting group to six experienced hunters. It can be really rough sometimes going through that brush. I've had a perforated eardrum where a limb went into my ear, and bumps and bruises from going after the hog. I always wear safety glasses.

My favorite place to hunt is Delta Wildlife. It is twenty-five thousand acres. I grew up hunting there and know it well. I had good teachers and probably learned it well because of covering it on the horse. You can cover a lot of ground on a horse.

My Most Memorable Hog Hunt

FRANK DANTONE

The hog hunt I remember most occurred several years ago at Black Bayou Hunting Club located just north of Greenville, Mississippi, inside the Mississippi River levee.

Jimmy Donahoo, a close friend of mine, invited me and about fifteen other men on a hog hunt at Black Bayou. Deer season had just ended and Black Bayou, at that time, had a large population of hogs.

The hog hunt began, as most of them do, the night before. We had a wonderful meal followed by numerous hunting stories that were certainly exaggerated by the amount of alcohol that was consumed. Every time one hunter would tell a story, another hunter would come up with one a little bit larger and a little bit taller than the previous story. It reminded me of a quote from an old friend who has since passed: "The first liar don't stand a chance."

Jimmy had invited Mike Braswell, who primarily was responsible for the dogs, and he brought his longtime hog hunting partner, Wick Eatherly, with him. At the time, though I did not own a horse, I owned everything but a horse. I brought my saddle and saddle bags and spurs, and I was told that a horse would be waiting when I got to the camp. Austin Jones, another invited friend of Jimmy's, brought a big, black saddle horse, Dante, supposedly named after Dante Jones, the Mississippi State basketball star who passed thirty-six hours of summer school prior to his junior year at Mississippi State. The next morning, we enjoyed one of the best breakfasts that these lips have ever tasted. Jerry Ferguson "Kikie," another good friend of Jimmy's, cooked a breakfast which consisted of thick sliced hickory smoked bacon, venison sausage, homemade biscuits, hashbrowns, fried eggs, freshly squeezed tomato juice, and red-eye gravy, white sausage milk gravy, and tomato gravy.

After breakfast, Austin had one of his workers saddle up my horse and I was told by Jimmy to stay close to Braswell and Eatherley because that's where the action would be.

We took off down the road and it wasn't too long before the dogs stuck and all hell broke loose. Braswell and Eatherly started traveling at breakneck speed through the middle of the woods, and it was during one of the turns that my saddle slipped over; by the time I got the horse stopped, I was hanging out perpendicular sideways from my horse in between two trees. My spur was caught in the saddle and it took four guys to extricate me from that situation. By the time I got back on the horse, the dogs had bayed the hog in a thicket. Prior to the hog men getting to the hog, it had ripped the side of one dog from front shoulder to rear leg and was working on another dog when it was finally dispatched by one of the hog men walking up, quickly placing a revolver behind the hog's ear, and pulling the trigger, killing the hog.

It was at that time that I first realized how tough hog men and hog dogs are. Mike Braswell picked up his wounded dog and went back to his horse and got his medicine kit. He gave the dog a shot of penicillin, cleaned the wound, which was approximately sixteen inches long, and applied salve to it. He then used a staple gun and stapled the dog's skin back together.

Later that day, we came up to a big sow hog with a litter of approximately twelve pigs. The little pigs weighed about 10.5 pounds each and one of the guys caught a pig and put it in his saddle bag. He brought the pig home, raised it, and named it Squeal. When Squeal got up to about 250 pounds, he let it loose in the national forest.

Needless to say, it was a memorable hunt and by the time we stopped hunting that afternoon, Black Bayou was short twelve hogs.

It should also be noted that during the course of the day, there were many occasions to celebrate and we celebrated each kill by consuming adult beverages consisting mainly of moonshine, beer, and an occasional drink of wine.

All in all, it was a day to remember, with memories that I will always cherish.

Hog Trap

KEVIN KEEN

Hunting in Mississippi has always been a passion or even a way of life for me. As a young child, I experienced the excitement and anticipation prior to the fall hunting season. Over the years, the majority of my time in the woods was spent in pursuit of the whitetail deer in the Mississippi Delta. Several years ago, we started to note signs of the presence of wild hogs in the area. It did not take long to see my first hog. I was excited about the possibility of harvesting one of these strange-looking creatures. They should be an easy kill, I thought. After all, I had hunted for deer for many years and there is no way these animals could be as wise as the elusive whitetail.

Approximately a year later, I shot my first hog with a .45-70 rifle. In the years that followed, our hog population seemed to exponentially increase. Though we were able to shoot several of the critters, we were quickly realizing that these hogs were not as stupid as we first believed. Being the wise hunter, I quickly decided to enact "population control" measures. It seemed like the more I hunted the hog, the smarter they became. I realized that, as many as were there, I had never actually seen a hog while walking in the woods. I only saw the hogs while on a stand sitting quietly. After a hog was shot in a particular location, it seemed the others made a mental note and told their buddies to stay away from that area.

Were they that intelligent? Were they that elusive? It couldn't be that difficult to outsmart a hog. I just needed a different strategy. A quick Internet search and I came up with several answers. One answer included hunting the wild hogs with dogs and knives. What? Much too dangerous for me. Then I found it. I could build a trap!! How simple. Build a cage, add food, control the population. We quickly constructed a trap strong enough to contain an elephant and assembled it on the property. It did not take very long to catch our first hog. We shot him through the trap and quickly removed him so as not to spook the others. The trap was reset and baited. Weeks went by and the hogs would not even attempt to get access to the food. Did they know that one of their brethren had met his demise in this trap? Was the smell of death still present in the area? I figured it must have been because of the blood left in the trap from the first hog. The trap was relocated to an area far from the kill site and new bait was added with a more "advanced" trigger mechanism. Three days later, we had trapped a large boar in the new location. This is where the adventure began.

From the previous experience, we knew that we could not shoot the hog in the trap and expect others to return. But what could we do? My cousin and I started to devise a plan to extract the hog live to prevent spoiling our new location. While we were planning our capture, we noted that not only was the hog not scared of us, but was very aggressively trying to get ahold of us. We finally decided on a plan. I would lasso the hog while he was still in the trap, gently lead him out (like you would Wilbur the pig on the farm) to an area far from the site, and dispose of him. It seemed foolproof.

Several attempts to secure the rope around the hog's neck were unsuccessful. Not only did it make him more pissed off, but he was grabbing the rope with his teeth and shaking and pulling on it like a dog would a chew toy. At that point, I was getting as mad as the hog and I was not going to give up. Finally, while my cousin had him distracted, I was able to lasso the hog. He went as crazy as a Tasmanian devil but I was not going to let go of my hold. My cousin opened the gate of the trap and I pulled him out. At that time, we both realized we had a maniac on the end of a rope and that the only direction he could run was towards us. I ran to the edge of the tree line, keeping hold of the rope, with the pig a close distance behind. I got behind a tree and the pig made about three circles around an adjacent tree and hog-tied himself.

Our goal was to slay the boar a safe distance from our existing trap as to not scare away future catches. We had gained less than twenty feet in that fiasco. In a last-ditch effort, we tried to untangle the hog yet were unsuccessful due to the pig's desire to kill us using only his teeth. The disaster was finally ended with a close-range gunshot to the head. What started as an efficient, clean, and tactical way to remove a hog from a trap ended in what looked like a crime scene right next to it.

We untied the dead hog and loaded it onto the ATV. We tucked our tails and headed back into camp in disbelief that a pig had made such a fool of both of us. We both gained a new appreciation and newfound respect for the hogs. I don't know about my cousin, but from now on I wanted to hunt only the more timid creatures of the woods, like deer and rabbit.

Deer Hunting for Hogs

JOEL HENDERSON

I have seen grown men crawl on their hands and knees, run to try to keep up with a pack of dogs, ride horses through undergrowth taller than their horses, strip down and swim ditches and bar pits (a borrow pit is a small lake created when the U.S. Army Corps of Engineers built the levee system along the mighty Mississippi River), and forego meals and spousal responsibilities to participate in a hog hunt. I am not one of those individuals who is overcome by a passionate desire to hunt hogs. My hunting passion is watching my black Labrador retriever, Lagniappe, retrieve mallards, gadwall, teal, widgeons, wood ducks, and an occasional strange duck, such as a black-bellied tree duck.

However, I have always wanted to "harvest" a hog, which is excellent table fare, and entitles a hunter to a greater standing among his/her peers. Good dogs make hogs fairly easy to find, but taking a hog without a dog is difficult as the hog has to be located and has to be in a position for a clean shot with rifle or bow.

Generally, I hunt on land adjacent to the Mississippi River below Greenville. Hogs in the areas where I hunt normally stay in the bottoms close to water. They usually travel in cover so thick that, on several occasions, I have had hogs pass within twenty yards of me while I was eighteen to twenty feet off the ground in a climbing deer stand, and yet I could only hear the hogs and see the grass move as the hogs grunted past me.

My nephew, Brooke Henderson, is an avid deer hunter and could quickly become a passionate hog hunter. For several years, while deer hunting with me, Brooke seemed to see hogs every time he climbed a tree to deer hunt. He took several with his gun and attempted to take some with his bow, to no avail. As an experienced deer hunter, I began to feel somewhat intimidated by my young nephew as I could never take a trophy hog, despite my self-professed greater experience as a deer hunter. I declined Brooke's offer to take me with him to let him serve as my hog guide much as I had done when he was learning to deer hunt.

Finally, I made up my mind to "deer hunt/hog hunt" with the intent to take either a trophy buck or trophy hog. Typically, I use a tree climber and climb in areas so thick that rarely can I see more than thirty yards. On this particular trip, I had scouted an area and observed where hogs had been rooting, as well as several buck rubs and scrapes in the area. I climbed a tall hackberry and because of the thick undergrowth consisting of briars, grass, and pawpaw trees, I could see no more than twenty-five yards in any direction, except to the northwest where a wheat lane opened up which allowed me to see a distance of approximately seventy-five yards with a "window" of approximately ten yards. I anticipated that if I saw a nice buck or hog, my only real opportunity would be for him to show up within the twenty-five-yard radius of my tree and not the wheat lane "window" because a buck or hog moving through the wheat lane would only have been visible for a few seconds.

Nevertheless, I climbed the tree about 1:00 p.m. on a clear, calm, and cool December afternoon. Fortunately, I could move around the tree in my tree stand to take advantage of the direction in which the wind was blowing. I knew that the bucks were in rut and I had little concern about a buck seeing me. If he smelled me, I would not have a chance to take him. Similarly, a hog has a keen sense of smell and of direction, and seems to sense any danger. I actually had expected that if a hog came through, I would only have a few seconds to locate it and, I hoped, let him get in a position where I could take a clear shot just behind the front shoulder into the lungs, as hogs are notorious for being able to "take lead" and keep on going. In addition to recognizing their strength, I am always amazed at how they can survive despite droughts, storms, and floods and continue to populate an area. In fact, I absolutely believe that if someone was able to build a suitable fence around a Wal-Mart parking lot, a hog could not only survive, but find a way to produce more hogs. Hogs can, literally, live off the land, or the parking lot.

During the first hour of my hunt, I saw nothing but a few squirrels. The second hour produced an obnoxious armadillo which took forty-five minutes to stroll by me, creating as much noise as he could, knowing that on this particular day I would not remove him from the environment. In the third hour, I had a nontrophy (less than ten points and twenty inches inside spread) buck around me for approximately fifteen minutes. He was downwind and spotted me quite easily, but he had no fear as he could not smell me. In the last hour, the sun began to drop fairly quickly, and although the woods and the sunset carried the beauty of a heavenly design, no hogs or bucks showed. I stuck my hand in my left front pocket to make sure that it was empty. I always carry four shells and load my bolt action .30-06 with four shells. It will hold additional shells, but I have always felt that four is enough. When I put my hand in my left front pocket and it is empty, I know that my gun is loaded. If I reach in my pocket and count four shells, I know that it is completely unloaded. After giving a little pep talk to my trusted .30-06 and myself, I started to bolt my gun to unload it and prepare to climb down. As I reached for the bolt, for some reason, my eyes were attracted to the "window" of the wheat lane in the distance. To my absolute utter shock, I saw what appeared to me to be a small bison or buffalo standing on the edge of the wheat field about to step into a briar thicket. It happened so quickly that I did not have time to hyperventilate and, out of habit, I had already raised my gun to my shoulder. The crosshairs immediately centered just to the rear of the left front shoulder, and almost instinctively, I pushed the safety off and squeezed the trigger. I continued to watch through my scope and, to my amazement, the brute did not move. I bolted another round and raised the gun to aim in disbelief as I could not believe that I had missed. Before I could squeeze another round, he dropped dead in his tracks.

After climbing down, I quietly made my way to him, being fearful that he would get up and run, but he did not move. He weighed approximately 225 pounds and had two-inch tusks. He had long black hair and was generously endowed. After taking several photographs with my .30-06 appropriately laid across him, I had to use the winch on my truck to drape across a tree in order to load him into the bed of the truck. When I arrived at the skinning rack, I delayed that process as long as I could so that all could see my prize, knowing that the news would spread quickly. We had several tasty meals. I have since become a "certified" hog hunter as I have taken several others while deer hunting for hogs. None of the hunts were as special or memorable as this one.

Boss Boar

JOHN LOWERY

I was invited to go on a hog hunt at Panther Tract. It was a perfect morning. We had a light dusting of snow after a half inch of rain. On the hunt was my dad, Ricky Lowery, my best friend, Steve Radcliff, and many others. Jeff Cremeen and Brandon Yeager came with Steve. Jesse Bates and his son, Martin, came with me. Steve and I hunt together a lot. We hunt three dogs and a pup most of the time. On this morning Steve brought his dog, Molly, and a pup and I took my two dogs, Bo and Little Joe. After about thirty minutes of hunting Molly struck a hog in the thick brush. All the dogs went to her and tried to bay the hog but could not hold it. There were a lot of people around on horses and the brush was around nine feet and thick. The hog kept breaking and would only go about a hundred yards and bay again. We thought it was a pig around a hundred pounds. I just could not figure out why the dogs could not hold it. It was really odd: they usually catch everything under a hundred and fifty pounds, but the hog would not leave; it just stayed in the brush. After a short time I got tired of this and told my dad, who was on the Ranger, to turn my bulldog, Homicide, loose and let's get this one caught! Right when he turned him loose, the hog broke again and went about a hundred yards, crossed a road, and bayed on a big slough where it was not quite as thick and the bulldog caught for thirty to forty-five seconds. He reminded me of a paddle ball. This hog was much bigger than we thought. Steve got there just in time to see the bulldog catch and hollered to me to get there, that this was a bad hog! I ran in, dove off my horse, and caught the hog by his back legs. He was still throwing the dogs around so I rolled him and lay on his stomach holding his back legs over me! Steve jumped off his horse and came in to help me. I was still having a hard time holding this big hog. He came in and put his knee on the hog's chest so we could hold him. Steve pulled his .357, put it to the hog's chest, and pulled the trigger to kill it. What we thought was just a pig turned out to be a two-hundred-forty-pound boss boar with teeth a little over three inches long!

This was the beginning of a great hunt! We caught several more hogs that day and made a lot of new memories. To me that's what hunting is all about—friends, fellowship, and especially the memories.

The Year's Last Hog Hunt

HANK BURDINE

Howard postponed the hunt until Sunday because of the forecasted rain. That was okay with me because the sack of oysters and the Gulf shrimp I had just bought and iced down would easily last one more day. We were all looking forward to a great feast the night before the last hog hunt of the year. There would be much camaraderie and fun and good hunting. We weren't going to let a rain front spoil our fun.

We all gathered Saturday night and the party began. Oysters were being shucked while the shrimp étouffée bubbled on the stove. The rain fell and then slacked off; it was snowing in Memphis. The temperature abrubtly dropped in the south Delta and someone hollered, "Look yonder, it's snowing like hell out there!" Howard threw another log on the fire and as drinks were refilled, the guitars were tuned up. It was going to be another wonderful evening at Panther Tract.

An hour before the crack of Sunday's dawn, I smelled the coffee pot and walked upstairs to find Melody and Steve already in the kitchen. She was getting ready to go out and photograph the beautiful snow-covered vista for the documentary. It was one helluva day for a hog hunt and we were determined not to miss one bit of it! One by one, the revelers from the night before came into the kitchen for a little hangover helper, shaking their heads as they passed the bar, full of empty soldiers from the night before. Coffee mugs were filled and refilled and the excitement of the forthcoming day began to emerge. Ten pounds of bacon and four dozen eggs were prepared, along with two trays of cathead biscuits, and hungry hunters dove in as jars of homemade preserves were passed around the banquet-sized dining room table. It was cold outside and a full belly would help keep us comfortable until the noon feast.

At precisely 8:30 it seemed as if a land-based Spanish armada was coming up the driveway. Trucks and horse trailers were everywhere. Dog pens held yapping, whining, and shivering mutts of all breeds and mixtures. Off-road vehicles, four-wheelers, and Yamaha Rhinos were being unloaded and revved up in the freezing air. Thick Kevlar flak-jacket-type vests were being attached to huge bulldogs that are used for catch dogs and radio antennae collars were being fastened to the chase dogs so they could be tracked with a GPS transmitter. Handheld radios were looped around the dog handlers' necks as stout dog leashes were snapped on bandolier style. Velcroed backpacks and bags that were slung over shoulders held antiseptic syringes and medicines and medicinal staple guns to sew up any dog or horse that happened to get slashed open. The anticipation was rampant and adrenalin built up in those that knew what was soon to come.

Hog hunters and dog handlers are a unique breed of sportsman. Thinking that their pack of dogs is superior to any other, they refuse to hunt with other packs. And it is probably a good thing, as dogs are raised in the same pens and get used to each other. They know what to expect from their handlers and their penmates. Chase dogs are more lean and fleet of foot than catch dogs. Today, sitting atop a crate of chase dogs was a solitary white bulldog named Frog. He was huge and had a large Kevlar vest strapped securely around his body. He was quite docile in the cage and would let you pet his nose protruding out of the welded aluminum tubing. However, once on the ground and onto the scent of a wild hog, he turned into a hunting and attacking machine.

The dog handlers gathered and made a plan of where to go and how to pursue the hunt based on sightings of hogs and their fresh rooting areas. Leading the way, twenty horses and riders in all sorts of attire and riding habits proceeded down the road in front of the big house. Following at a respectable distance were five or six Rhinos with dog cages attached or pulling dog trailers. All were loaded with shivering and excited hounds, with the catch dogs separated from the chase dogs. Following the dogs was a small army of four-wheelers. Howard, of course, followed close behind in his Honda four-wheel-drive pickup truck with an ample supply of human antifreeze labeled peach brandy, and off into the swamp we headed. Twirling his thick, black handlebar moustache, Sid "Bo Weevil" Law III of nearby Berkeley Plantation commented, "Just look at that! Women and men, boys and girls, all hunters to the core, horses, dogs, guns, big knives, snow and mud, four-wheelers, and good whiskey! It just don't get no better. This is the Delta at its finest!"

A mile or so down the road, the two packs of hunters and dogs soon separated on either side of Lake Discovery so as not to get intermingled. The dogs were turned loose and the hunt was on! The oak trees that were planted eighteen years ago have grown substantially. The trees were planted in rows and make for pretty easy traveling on four-wheelers if you know how to drive with your head ducked down. The Rhinos could travel at high speed, knocking down everything in between the trees big enough to run over. The horsemen and dog handlers were hell bent for leather as they were the ones that kept right on top of the dogs and were the first on the kill, most of the time. A hog will go wherever it wants and will sometimes break and run after it has been bayed. At all times, he is ready to fight, and fight he will.

The first and biggest of the eight hogs to be killed that day was soon routed. Ripping and tearing through the swamp, the boar was soon at bay only to break and run again. Frog had been turned out, but missed the catch. The chase dogs caught up with the big boar hog again. Three dogs circled the hog, snapping and barking while one small rednosed pit bull charged in. He was expertly trained and knew exactly what to do. A good hog dog will not turn loose once he has hold of a hog and this was no exception. This particular dog had caught no less than one hundred hogs. The huge hog flipped that forty-pound dog over and over in the air like a ragdoll, but the dog would not turn loose. When the handlers got to the riotous fight, there was no time for a knife to be drawn to deliver the coup de grace. A .357 magnum pistol was pulled and the hog was immediately dispatched.

What happened next is burned into my memory and I will never forget it. Three grown men leapt off their horses in midstride and rushed to the dog that still had hold of the boar hog's ear. He was gashed badly in two places. While one man held the dog down, another cradled his head and rubbed his ears trying to calm him. John Lowery, a self-made vet, went to work on his prized dog, immediately assessing the wounds and deciding which to work on first. Gauze pads were pulled out of one of the backpacks and syringes of antiseptic were immediately injected. The wounds were cleaned professionally and a medicinal staple gun used to sew up the openings after vital organs were pushed back inside. I counted twenty-four staples in one gash and twelve in the other. As all gathered around to look at the massive, steaming boar hog with three-inch tusks and inquire about the dog, I realized the scene before me was one very few sportsmen ever get to see. There, on the ground in the mud and the blood and the guts and snow, lay three grown men fervently working to save a favorite dog's life. Tears were in their eyes, as they were in mine and many others. There was a reverence and a quietness that was deafening. One late horse rider dismounted and walked into the crowd exclaiming, "My, what a hog!"

Slowly, the biggest of the three men on the ground turned and looked up through tears and said, "Yeah, but this is one helluva dog. He got more heart than a country cur got fleas. His name is Homicide."

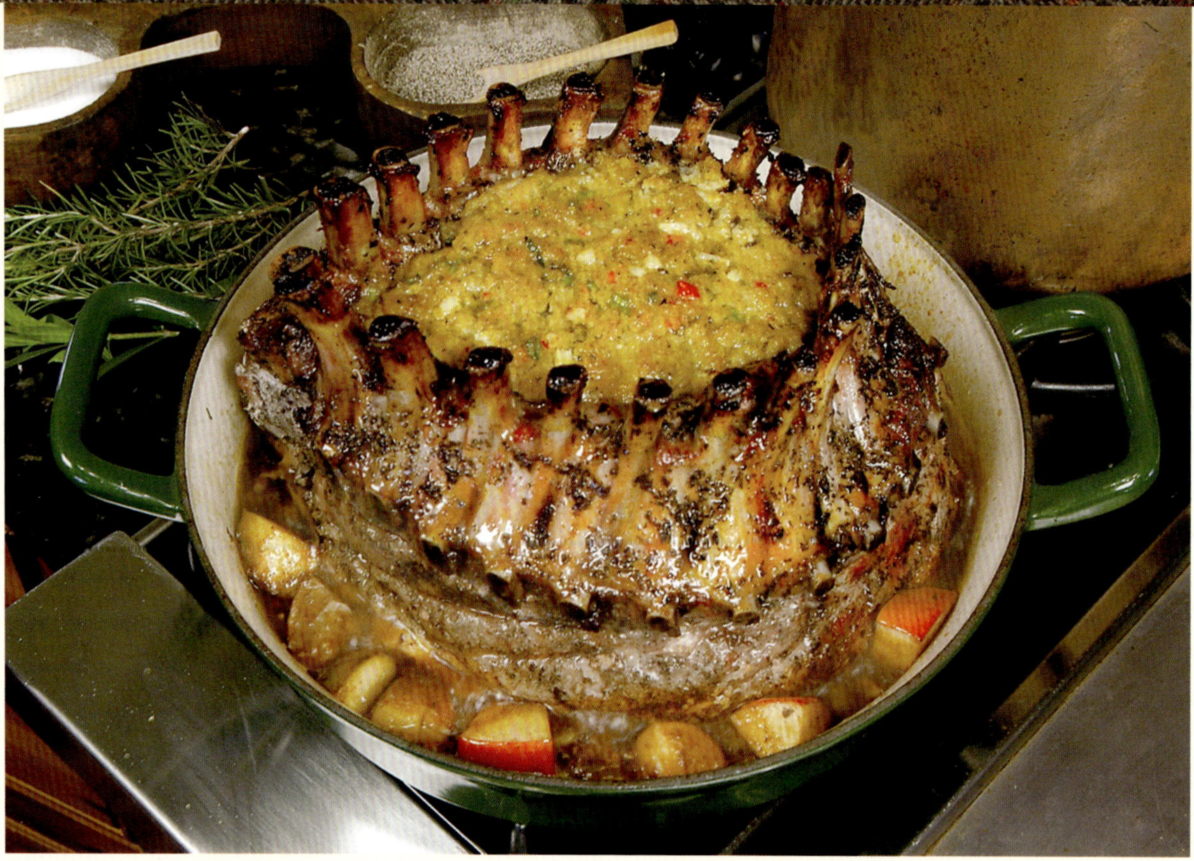

Stuffed Crown Roast of Wild Boar.

Wild Boar Recipes

CHEF JOHN FOLSE

The international success of Chef John Folse's Lafitte's Landing Restaurant in Donaldsonville, Louisiana, spawned the incorporation of several other properties. Chef John Folse & Company Manufacturing, in operation since 1991, is one of the few chef-owned food manufacturing companies in America producing custom-manufactured foods for the retail and food service industry. Other divisions include Chef John Folse & Company Bakery, White Oak Plantation, Bittersweet Plantation Dairy, and Chef John Folse & Company Publishing. He hosts an international television series through PBS and a radio cooking talk show, and is the cofounder and namesake of the culinary institute at Nicholls State University.

Folse has received numerous accolades, including being named National Chef of the Year by the American Culinary Federation (ACF) and Louisiana Restaurateur of the Year by the Louisiana Restaurant Association. Research Chefs Association (RCA) recognized Chef John Folse & Company as "Pioneers in Culinology." Folse is a recipient of the Silver Spoon Award and the Antonin Carême Medal, and was inducted into the National Restaurant Association College of Diplomats in 2006. In 1988, the Louisiana Legislature gave him the title of "Louisiana's Culinary Ambassador to the World." In 1995, Folse was one of fifty people recognized in *Nation's Restaurant News*' "Profiles of Power." In October 2008 Folse was awarded the Louisiana Governor's Cultural Economic Development Arts Award and Southern Foodways Alliance Lifetime Achievement Award. Folse is a past president of ACF and RCA, and a past chairman of Distinguished Restaurants of North America. Currently, he serves as the marketing specialist on the Louisiana Seafood Promotion & Marketing Board.

Thirty years of culinary excellence later, Folse is still adding ingredients to the corporate gumbo he calls Chef John Folse & Company, which is as diverse as the Louisiana landscape, and he would not want it any other way.

Stuffed Crown Roast of Wild Boar
Prep time: 3 hours
Yields: 10 servings

Comment:
The crown roast, or bone-in pork loin roast, has always been the choice table centerpiece in Creole kitchens during the holiday season. It may be stuffed with everything from meat and fruit to rice and vegetables.

Ingredients for roast:
1 (9–11 pound) crown roast of wild boar, rib ends frenched
Shoepeg corn bread dressing (see recipe below)
¼ cup minced garlic
¼ cup sliced green onions
1 tablespoon salt
1½tablespoons black pepper
Granulated garlic to taste
Salt and black pepper to taste
¼ cup chopped basil

¼ cup chopped thyme
¼ cup chopped sage
1 cup diced onions
1 cup diced carrots
1 cup diced celery
1 cup diced apples

Method:
Preheat oven to 350° F. NOTE: If you are not familiar with crown roast, have your butcher "french," or clean end of rib bones of any meat or sinew. Ask butcher for a quick demonstration in tying roast into crown shape. In a small mixing bowl, combine minced garlic, green onions, 1 tablespoon salt, and 1½ tablespoons black pepper. Using a paring knife, make 8–10 (¾-inch) slits in loin and season generously with mixture. Season roast inside and out with granulated garlic, salt, pepper, and remaining herbs. Tie roast into crown shape and place on a large sheet of foil. Place in a baking pan with a 2-inch lip. Fill center of roast with shoepeg corn bread dressing, placing any excess into corners of roasting pan. Surround outside of roast with onions, carrots, celery, and apples. Fold foil up side of crown roast and over rib ends to cover loosely during cooking process. Bone should be protected well to keep from burning or turning overly brown while baking. Place in middle of oven and cook 2 hours. Open foil and brown roast 45 minutes or until meat thermometer reaches 155–160° F. A sauce may be made from drippings by allowing pan to sit for 1 hour and skimming excess fat. Thicken drippings slightly with 1 tablespoon cornstarch.

Shoepeg Corn Bread Dressing
Prep time: 2–2½ hours
Yields: 6 servings

Comment:
Most often in south Louisiana, meat and oyster dressings are found as accompaniments on the holiday table. Bread or corn bread dressings are definitely a southern tradition but are seldom seen in bayou country. As an added flavor enhancer, add 1 pint of shucked oysters with liquid to the pot during the final cooking phase.

Ingredients:
1 cup yellow cornmeal
1 (11-ounce) can shoepeg corn
½ cup flour
2 teaspoons baking powder
1 teaspoon salt
2 tablespoons sugar
1 egg
5 tablespoons melted butter, divided
¾ cup milk
4 chicken leg quarters
½ cup diced onions
½ cup diced celery
½cup diced red bell peppers
1 tablespoon minced garlic
¼ teaspoon rubbed sage
⅛ teaspoon dried basil
⅛ teaspoon dried thyme
¼ cup minced pimientos

¼ cup sliced green onions
¼ cup chopped parsley
1 pint oysters with liquid (optional)
Salt and black pepper to taste

Method:
Preheat oven to 375° F. In a mixing bowl, combine cornmeal, flour, baking powder, 1 teaspoon salt, and sugar. In a separate bowl, whisk egg, 2 tablespoons melted butter, and milk. Add egg mixture to cornmeal and blend well. Pour corn bread batter into a well-greased 9-inch cake pan and bake 15–20 minutes. Remove and cool. Separate chicken legs from thighs. In a 2-quart stockpot, combine chicken, onions, celery, bell peppers, and garlic. Cover with 6 cups cold water, bring to a rolling boil, then reduce to simmer. Cook 30–40 minutes or until chicken is tender and falling from bones. Remove chicken and allow to cool. Retain stock and seasoning. Bone and finely chop cooled chicken. Return meat to pot with stock and seasonings. Stir in shoepeg corn, sage, basil, thyme, remaining butter, pimientos, green onions, and parsley. Bring to a rolling boil, reduce to simmer, and cook 15 minutes. If desired, add oysters and liquid and cook 2 minutes longer. Strain stock and reserve 3 cups. Crumble corn bread into a large mixing bowl. Season reserved stock with salt and pepper. Add chicken and seasonings to crumbled corn bread along with 2½ cups of seasoned stock. Stir until well blended. Dressing should be very moist, but not watery. If desired, use unbaked dressing to stuff roasts, such as crown roast of pork (see recipe above). To serve dressing as a side, pour mixture back into cake pan and drizzle with 3–4 tablespoons stock. Bake uncovered 20–30 minutes or until dressing begins to brown lightly around edges. Dressing may be made the evening before cooking, but should be baked immediately prior to serving. Do not overcook, as dressing will tend to dry out.

Fire-Roasted Tenderloin of Wild Boar
Prep time: 30 minutes
Yields: 6 servings

Comment:
Tenderloin is the most tender of all game meat. Although you may use it in any recipe, the quick grilling or sautéing of this cut will yield the best result. It also adapts well to most spices and flavors.

Ingredients:
3 pounds wild boar tenderloin, trimmed and cut into 1-inch thick slices
Kosher salt and ground black pepper to taste
Granulated garlic to taste
¼ cup olive oil
2 medium onions, sliced
1 teaspoon minced garlic
½ teaspoon dried thyme
½ teaspoon dried basil
½ teaspoon dried oregano
½ teaspoon mustard seed
¼ teaspoon cayenne pepper
2 green bell peppers, julienned
2 red bell peppers, julienned
1 cup diced tomatoes
1¼ cups demi-glace
4 tablespoons unsalted butter
3 tablespoons peanut oil

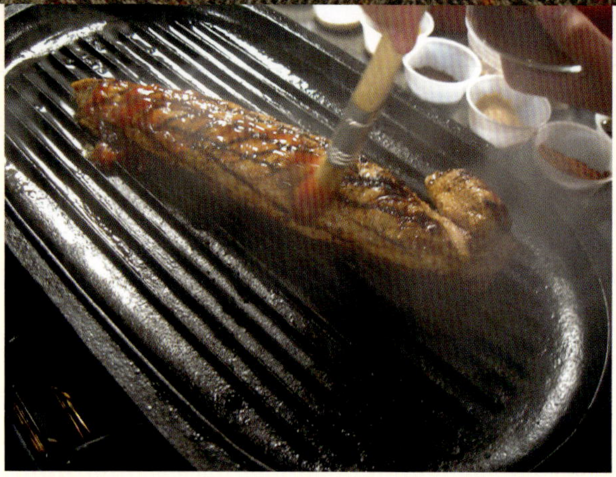

Method:
Place pieces of boar between 2 layers of plastic wrap and gently pound to ½ inch thick. Season the meat to taste using salt, pepper, and granulated garlic, then set aside. In a large skillet, heat olive oil over medium-high heat and sauté onions until lightly browned. Add minced garlic, thyme, basil, oregano, mustard, and cayenne and cook 5 minutes. Add bell peppers and cook until tender. Add tomatoes, demi-glace, and butter, stirring constantly until butter is incorporated. Season to taste with salt, pepper, and granulated garlic. Remove from heat and keep warm. In a cast-iron skillet, heat peanut oil over medium-high heat. Add seasoned tenderloins, 4–6 at a time, and sear meat for 3–4 minutes or until done, turning once. Remove from heat and keep warm. To serve, place 2 pieces of wild boar on each plate and cover with sauce.

Corn Bread–Stuffed Wild Boar Chops
Prep time: 2½ hours
Yields: 6 servings

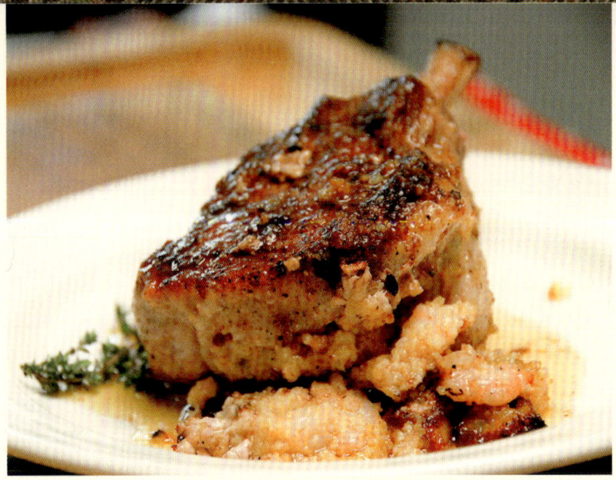

Comment:
In the 1800s pork was preferred to other meats because of its availability and versatility. Crown roast was the most sought-after dish but was too much for a small family. Instead, many people simply stuffed the center-cut chops with a variety of local fillings to create a delicious pork meal.

Ingredients:
6 (1-inch) center-cut wild boar chops
3 cups crumbled corn bread
Salt and cracked black pepper to taste
Granulated garlic to taste
Louisiana hot sauce to taste
½ cup butter
1 cup diced onions
½ cup diced celery
¼ cup diced red bell pepper
¼ cup diced garlic
1 cup (150–200 count) shrimp
¼ cup sliced green onions
¼ cup chopped parsley
1 tablespoon chopped sage
1 tablespoon chopped thyme
1 tablespoon chopped basil
1 quart hot chicken stock
4 cups sauce Acadian (see recipe below)

Method:
Preheat oven to 375° F. Split chops down center to form a pocket for stuffing. Season chops inside and out with salt, pepper, granulated garlic, and hot sauce. Set aside. In a cast-iron skillet, melt butter over medium-high heat. Sauté onions, celery, bell pepper, and diced garlic 3–5 minutes or until vegetables are wilted. Stir in shrimp, green onions, parsley, sage, thyme, and basil. Blend well and cook 10 minutes or until shrimp are pink. Sprinkle in corn bread. Add chicken stock, one ladle at a time, to keep stuffing moist and prevent it from sticking. Season with salt, pepper, and hot sauce. Once stuffing is full-flavored, remove from heat and allow to cool. Stuff equal amounts of corn bread mixture into centers of chops. Place chops in an ovenproof casserole dish and top with sauce Acadian, or substitute 2 cans of cream of shrimp soup. Bake, covered, 1½ hours or until chops are tender.

Sauce Acadian
Prep time: 1 hour
Yields: 2 quarts

Comment:
Technically speaking, this sauce is a Nantua, which is classically made by adding crawfish butter and paprika to a béchamel. In Louisiana, we have further enriched the recipe to create sauce Acadian.

Ingredients:
2 pounds crawfish shells, heads included
2 cups chopped onions
1 cup chopped celery
1 cup chopped carrots
6 cloves garlic
2 bay leaves
1 teaspoon dried thyme
15 whole peppercorns
1 gallon cold water
2 cups dry white wine
½ cup tomato sauce
½ cup white butter roux
1 pint heavy whipping cream
½ ounce brandy
Salt and white pepper to taste

Method:
In a 2-gallon stockpot over medium-high heat, combine crawfish shells and heads, onions, celery, carrots, garlic, bay leaves, thyme, peppercorns, water, and wine. Bring mixture to a low boil, reduce heat to simmer, and cook 30 minutes. During cooking process, skim off all impurities that rise to surface. When stock is cooked, strain it through cheesecloth or a fine sieve. Discard shells and vegetables. Return hot stock to pot, bring to a low boil, and reduce to approximately 2 quarts. Add tomato sauce and white roux, whisking constantly until roux is well blended and sauce is slightly thickened. Reduce heat to simmer and cook 15 minutes. Add heavy whipping cream and brandy. Season with salt and white pepper. Strain sauce for a second time, adjust seasonings, and allow to cool.